D1032452

ACADEMY
FIELD
69. OREGON

ACS SYMPOSIUM SERIES **283**

Polycyclic Hydrocarbons and Carcinogenesis

Ronald G. Harvey, EDITOR
University of Chicago

Developed from a symposium sponsored by
the Division of Organic Chemistry
at the 188th Meeting
of the American Chemical Society,
Philadelphia, Pennsylvania,
August 26–31, 1984

ACHS 1985

American Chemical Society, Washington, D.C. 1985

Library
L.U.P.
Indiana, Pa.
616.994 P768l
c.1

Library of Congress Cataloging in Publication Data

Polycyclic hydrocarbons and carcinogenesis.
  (ACS symposium series, ISSN 0097–6156; 283)

  "Developed from a symposium sponsored by the
Division of Organic Chemistry at the 188th Meeting of
the American Chemical Society, Philadelphia,
Pennsylvania, August 26–31, 1984."

  Includes bibliographies and index.

  1. Polycyclic aromatic hydrocarbons—Physiological
effect—Congresses. 2. Polycyclic aromatic
hydrocarbons—Congresses. 3. Carcinogenesis—
Congresses.

  I. Harvey, Ronald G., 1927–   . II. American
Chemical Society. Division of Organic Chemistry.
III. American Chemical Society. Meeting (188th: 1984:
Philadelphia, Pa.) IV. Series.

RC268.7.P64P65   1985   616.99'4071   85–13384
ISBN 0–8412–0924–3

Copyright © 1985

American Chemical Society

All Rights Reserved. The appearance of the code at the bottom of the first page of each chapter in this volume indicates the copyright owner's consent that reprographic copies of the chapter may be made for personal or internal use or for the personal or internal use of specific clients. This consent is given on the condition, however, that the copier pay the stated per copy fee through the Copyright Clearance Center, Inc., 27 Congress Street, Salem, MA 01970, for copying beyond that permitted by Sections 107 or 108 of the U.S. Copyright Law. This consent does not extend to copying or transmission by any means—graphic or electronic—for any other purpose, such as for general distribution, for advertising or promotional purposes, for creating a new collective work, for resale, or for information storage and retrieval systems. The copying fee for each chapter is indicated in the code at the bottom of the first page of the chapter.

The citation of trade names and/or names of manufacturers in this publication is not to be construed as an endorsement or as approval by ACS of the commercial products or services referenced herein; nor should the mere reference herein to any drawing, specification, chemical process, or other data be regarded as a license or as a conveyance of any right or permission, to the holder, reader, or any other person or corporation, to manufacture, reproduce, use, or sell any patented invention or copyrighted work that may in any way be related thereto. Registered names, trademarks, etc., used in this publication, even without specific indication thereof, are not to be considered unprotected by law.

PRINTED IN THE UNITED STATES OF AMERICA

ACS Symposium Series

M. Joan Comstock, *Series Editor*

Advisory Board

Robert Baker
U.S. Geological Survey

Martin L. Gorbaty
Exxon Research and Engineering Co.

Roland F. Hirsch
U.S. Department of Energy

Herbert D. Kaesz
University of California—Los Angeles

Rudolph J. Marcus
Office of Naval Research

Vincent D. McGinniss
Battelle Columbus Laboratories

Donald E. Moreland
USDA, Agricultural Research Service

W. H. Norton
J. T. Baker Chemical Company

Robert Ory
USDA, Southern Regional
  Research Center

Geoffrey D. Parfitt
Carnegie–Mellon University

James C. Randall
Phillips Petroleum Company

Charles N. Satterfield
Massachusetts Institute of Technology

W. D. Shults
Oak Ridge National Laboratory

Charles S. Tuesday
General Motors Research Laboratory

Douglas B. Walters
National Institute of
  Environmental Health

C. Grant Willson
IBM Research Department

FOREWORD

The ACS SYMPOSIUM SERIES was founded in 1974 to provide a medium for publishing symposia quickly in book form. The format of the Series parallels that of the continuing ADVANCES IN CHEMISTRY SERIES except that, in order to save time, the papers are not typeset but are reproduced as they are submitted by the authors in camera-ready form. Papers are reviewed under the supervision of the Editors with the assistance of the Series Advisory Board and are selected to maintain the integrity of the symposia; however, verbatim reproductions of previously published papers are not accepted. Both reviews and reports of research are acceptable, because symposia may embrace both types of presentation.

CONTENTS

Preface..vii

1. Polycyclic Aromatic Hydrocarbon Carcinogenesis: An Introduction.........1
   Anthony Dipple

2. Stereoselective Metabolism and Activations of Polycyclic Aromatic
   Hydrocarbons...19
   Shen K. Yang, Mohammad Mushtaq, and Pei-Lu Chiu

3. Synthesis of the Dihydrodiol and Diol Epoxide Metabolites of
   Carcinogenic Polycyclic Hydrocarbons...........................35
   Ronald G. Harvey

4. The Bay Region Theory of Polycyclic Aromatic Hydrocarbon
   Carcinogenesis...63
   Roland E. Lehr, Subodh Kumar, Wayne Levin, Alexander W. Wood,
   Richard L. Chang, Allan H. Conney, Haruhiko Yagi, Jane M. Sayer,
   and Donald M. Jerina

5. Effects of Methyl and Fluorine Substitution on the Metabolic Activation
   and Tumorigenicity of Polycyclic Aromatic Hydrocarbons.............85
   Stephen S. Hecht, Shantu Amin, Assieh A. Melikian, Edmond J. LaVoie,
   and Dietrich Hoffmann

6. Mechanisms of Interaction of Polycyclic Aromatic Diol Epoxides
   with DNA and Structures of the Adducts..........................107
   Nicholas E. Geacintov

7. X-ray Analyses of Polycyclic Hydrocarbon Metabolite Structures........125
   Jenny P. Glusker

8. Polycyclic Aromatic Hydrocarbon–DNA Adducts: Formation, Detection,
   and Characterization...187
   Alan M. Jeffrey

9. The Intercalation of Benzo[a]pyrene and 7,12-Dimethylbenz[a]anthracene
   Metabolites and Metabolite Model Compounds into DNA.............209
   P. R. LeBreton

10. A Mechanism for the Stereoselectivity and Binding of Benzo[a]pyrene
    Diol Epoxides to DNA..239
    Kenneth J. Miller, Eric R. Taylor, and Josef Dommen

11. One-Electron Oxidation in Aromatic Hydrocarbon Carcinogenesis.......289
    Ercole L. Cavalieri and Eleanor G. Rogan

12. Hydroperoxide-Dependent Oxygenation of Polycyclic Aromatic
    Hydrocarbons and Their Metabolites............................307
    Lawrence J. Marnett

13. The Mutational Consequences of DNA Damage Induced
    by Benzo[a]pyrene...327
    Eric Eisenstadt

14. **Chemical Properties of Ultimate Carcinogenic Metabolites of Arylamines and Arylamides**..341
    Fred F. Kadlubar and Frederick A. Beland

15. **The In Vitro Metabolic Activation of Nitro Polycyclic Aromatic Hydrocarbons**...371
    Frederick A. Beland, Robert H. Heflich, Paul C. Howard, and Peter P. Fu

Author Index..397

Subject Index...397

PREFACE

THE ENVIRONMENTAL ORIGIN OF MANY HUMAN CANCERS is gaining increasing acceptance. Significant levels of cancer-causing agents are commonly present in polluted air, automobile exhaust, tobacco smoke, and many common foods. Direct evidence concerning the role of chemical carcinogens in the etiology of human cancer is difficult to obtain due to necessary restrictions on human experimentation. However, there is a mounting volume of indirect evidence that supports the importance of chemical agents as causative factors.

Epidemiological studies show wide variation in the incidences of various types of cancers in different populations. These geographic differences are often dramatic. For example, the incidence of female breast cancer is particularly high in the United States and relatively low in Japan. On the other hand, the occurrence of stomach cancer in these two countries is approximately reversed, being exceptionally high in Japan and relatively low in the United States. These differences appear to be related more to diet and lifestyle factors than to inherent genetic differences. Thus, the cancer profile of the Japanese population in Hawaii shows a marked shift in the relative incidence of breast and stomach cancer away from the levels in Japan toward those in the United States. In general the cancer patterns of migrant populations tend to shift toward those patterns characteristic of the new environment.

Carcinogenesis research has demonstrated the tumorigenic activities of a large number of chemical substances in experimental animals. These include molecules of diverse chemical classes, organic and inorganic, and natural products as well as compounds synthesized in the laboratory or produced by industry.

Man has served as the unintentional guinea pig for the identification of some major classes of carcinogens. These include the polycyclic aromatic hydrocarbons (PAH), or polyarenes, which have been identified as the active components of soot, which was recognized by the London surgeon Percivall Pott two centuries ago as responsible for cancer of the scrotum in chimney sweeps. Subsequently, polycyclic hydrocarbons have been implicated as agents responsible for skin cancer in other occupations such as shale oil distillation and mule spinning in the cotton industry.

The carcinogenicity of aromatic amines, such as benzidine and 2-naphthylamine, was first recognized by Rehn in the 1890s as an occupational hazard in the German dyestuffs industry. Compounds in this class induce tumors in man predominantly in the urinary bladder. The nitrosamines, another major class of carcinogens, were first recognized as such by their induction of tumors in the livers of workmen in the chemical industry who were using N-nitrosodimethylamine as a solvent.

Polycyclic aromatic hydrocarbons hold a unique place in carcinogenesis research. The first pure compounds recognized as carcinogens more than 50 years ago were the PAH benzo[a]pyrene and dibenz[a,h]anthracene. Because only certain polyarenes exhibit tumorigenic activity and the level of activity is highly dependent upon molecular structure (e.g., number of fused rings, molecular shape, and presence of methyl or other groups in particular molecular regions), the PAH are ideally suited to studies of structure–activity relationships. The polycyclic hydrocarbons are also exceptional in their ability to induce various types of tumors selectively, dependent upon their mode of administration and other experimental conditions. Thus, oral administration of 10 mg of 7,12-dimethylbenz[a]anthracene to female Sprague–Dawley rats was shown by Huggins to elicit mammary tumors with 100% incidence. Intravenous injection of a lipid emulsion of the same hydrocarbon to male or female Long–Evans rats selectively induced leukemia with similarly high incidence. In contrast, intramuscular injection of this hydrocarbon into the legs of Long–Evans rats gave predominantly local sarcomas at the site of injection. Other malignancies may be induced by PAH under appropriate experimental conditions. PAH-induced tumors are widely employed as standards in experimental oncology.

Polycyclic aromatic hydrocarbons' potential importance in human cancer is strongly suggested by their environmental occurrence and their exceptional carcinogenic potency; PAH as a class rank second only to the potent hepatocarcinogenic aflatoxins. The environmental prevalence of PAH is largely a consequence of PAH formation as products of combustion of fossil fuels and other organic matter. Although human populations are chronically exposed to low levels of polyarenes, individual levels of exposure may vary widely, determined by lifestyle, particularly cigarette smoking, diet, and occupation.

Research in PAH carcinogenesis has made major advances in the past decade. Most notable has been identification of diol epoxide metabolites as the active forms of benzo[a]pyrene, 7,12-dimethylbenz[a]anthracene, and other carcinogenic PAH. This finding has stimulated enormous research activity and opened the way to determination of the detailed molecular mechanism of action of this important class of carcinogenic molecules.

The symposium upon which this book is based brought together leading investigators concerned with the mechanisms of carcinogenesis of PAH at the molecular level. The individual chapters in this book are not merely verbatim reports of the symposium proceedings but rather critical reviews of symposium topics with extensive references to investigations in other laboratories. Since the pertinent literature references are scattered in journals in diverse fields ranging from synthetic and theoretical chemistry to oncology and molecular biology, it has been difficult for nonspecialists to keep abreast of recent advances. This book provides a convenient summary of current developments that cuts across these diverse academic disciplines.

Several additional chapters on relevant topics that could not be included in the symposium due to time limitations are also included in this volume. Chapter 10 presents a unified theoretical treatment of the covalent binding to DNA of the reactive diol epoxide metabolite of benzo[a]pyrene implicated as the active form of this hydrocarbon. Chapter 15 on the in vitro metabolic activation of nitro polycyclic aromatic hydrocarbons is the first review of this very active area of investigation. Interest in this topic has been stimulated by the discovery that nitration of pyrene and other PAH by oxides of nitrogen occurs in the atmosphere to form nitro–PAH derivatives which are often highly mutagenic. Chapter 14 on the carcinogenic metabolites of arylamines and arylamides reviews the large body of literature on the mechanism of carcinogenesis of this important class of PAH compounds. Since nitro-PAH are reducible by bacterial enzymes to polycyclic arylamines and the fused aromatic ring systems of both these classes of PAH compounds may undergo activation to diol epoxide derivatives, multiple overlapping mechanistic pathways exist for the metabolic activation of unsubstituted PAH and their nitro- and amino-substituted derivatives, compounding the mechanistic complexity.

This book is expected to be of interest to investigators active in all aspects of carcinogenesis research as well as to graduate students, educators, and others seeking an introduction to this important field of research.

It is hoped that this volume will contribute toward the ultimate elucidation of the molecular mechanism of induction of cancer by PAH. This knowledge will provide a rational basis for the design of approaches for the prevention and cure of this dread disease.

Additional support for the symposium was provided by the U.S. Department of Energy. I particularly thank Walter Trahanovsky of the American Chemical Society for his personal contribution toward making this project a success and my wife Helene for her support and understanding throughout this project. Any opinions, findings, conclusions, or recommendations expressed herein are those of the authors and do not necessarily reflect the views of the U.S. Department of Energy.

RONALD G. HARVEY
University of Chicago
Chicago, Illinois

March 22, 1985

Polycyclic Aromatic Hydrocarbon Carcinogenesis
An Introduction

ANTHONY DIPPLE

LBI-Basic Research Program, Chemical and Physical Carcinogenesis Laboratory, National Cancer Institute, Frederick Cancer Research Facility, Frederick, MD 21701

Since polycyclic aromatic hydrocarbons are widely distributed throughout the atmosphere and water sources of the world, it is essentially impossible to avoid exposure to nanogram quantities of these substances on a daily basis. It is particularly important, therefore, that the mechanism of action of these carcinogens be understood. This introduction provides a general background on experimental carcinogenesis and structure-activity relationships for hydrocarbons. It also traces the key steps involved in the discovery that polycyclic aromatic hydrocarbons are the carcinogenic components of complex mixtures such as soots and tars and the more recent discovery that bay region dihydrodiol epoxides are probably the metabolites of hydrocarbons that initiate the carcinogenic process.

Although there are earlier reports of lifestyle-associated cancers in the scientific literature, the 1775 observation of Percival Pott, surgeon to St. Bartholomew's Hospital in London, that scrotal cancer in chimney sweepers originates from their occupational exposure to soot (1) represents the key historical development in the fields of chemical carcinogenesis in general and polycyclic aromatic hydrocarbon carcinogenesis in particular. This observation was followed a century later by von Volkmann's reports of occupational skin cancers in workers in the coal tar industry in Germany (2), and by the early 1900's it was widely recognized that soot [produced by the inefficient combustion of coal and containing up to 40% coal tar (3)], coal tar [produced by the destructive distillation of coal], and pitch [the residue after distilling coal tar] are all carcinogenic for man. At that time, it was conceivable that a single carcinogenic substance might be responsible for all the known occupational cancers (4), but attempts to characterize the carcinogens in these complex combustion and pyrolysis products had to await the development of an experimental system for determining carcinogenic activity. It was not until

0097-6156/85/0283-0001$06.00/0
© 1985 American Chemical Society

1915 that such a system was developed by Yamagiwa and Ichikawa (5) who succeeded in producing malignant skin tumors in rabbits by the persistent topical application of coal tar. Three years later Tsutsui (6) produced tumors in mice by repeated application of tars to the skin, and this particular assay, which was quickly adopted in other laboratories, has proved to be of lasting value and is still used frequently today.

The series of events which then led to the identification of certain polycyclic aromatic hydrocarbons as the first pure chemical carcinogens has been described in depth by Sir Ernest Kennaway (7), a key participant in the discovery, and only a brief outline of this exciting story is   included herein. In 1921, Bloch and Dreifuss (8) in Switzerland had established that the carcinogen in coal tar was of high boiling point, was free of nitrogen and sulphur, formed a stable picrate, and was probably a complex hydrocarbon. Kennaway pursued these leads after joining what is now the Institute of Cancer Research in London in 1922, and, amongst a number of notable studies on carcinogenic tars, he demonstrated that pyrolyses of isoprene or acetylene in an atmosphere of hydrogen gave rise to carcinogenic distillates, thereby proving that carcinogenic activity resided in some compound containing only carbon and hydrogen (9).

A second vital observation was made when Mayneord, a physicist, joined in the research effort and decided to examine the conspicuous fluorescence of the many carcinogenic distillates present in Kennaway's laboratory. He found that most of the carcinogenic tars exhibited a common fluorescence spectrum (λ_{max} 400, 418 and 440 nm) but, in subsequent studies with Hieger, none of the hydrocarbons available at that time exhibited these spectral characteristics (7). The spectrum of benz[a]anthracene was found to be similar to, but of longer wavelength than, that of the carcinogenic preparations but this similarity directed Kennaway's attention to Clar's report of the synthesis of dibenz[a,h]anthracene (10). Tumors were obtained when this hydrocarbon was repeatedly painted on to mice and thus it was established that the properties necessary to elicit tumors in animals were contained within the structure of a single pure chemical compound (11).

Since the spectrum of dibenz[a,h]anthracene was not identical to that of the carcinogenic tars, carcinogenic activity obviously was not confined to a unique chemical structure and indeed Cook, who had joined this research effort in 1929, soon synthesized a number of new hydrocarbons, and many of these were carcinogenic. The outstanding problem of the carcinogen in the tars was ultimately resolved by a massive experiment beginning with the distillation of 2 tons of gas-works pitch. Thereafter, by fractional distillation, differential extractions, fractional crystallization and by following the fluorescence spectrum and carcinogenic activity of the various fractions, Cook, Hewett and Hieger (12) were able to obtain gram quantities of a crystalline material which was carcinogenic and exhibited the fluorescence which had been associated with carcinogenic activity. This material was resolved into three separate substances, perylene and two unknown isomers of perylene which were identified by structure-determining syntheses as benzo-

[a]pyrene and benzo[e]pyrene. Benzo[a]pyrene was the source of
both carcinogenic activity and the characteristic fluorescence of
the carcinogenic tars. This finding completed a remarkable episode
of research during which the carcinogenic activity (or, at least,
some of the of the carcinogenic activity) associated with complex
materials like soots and coal tars came to be attributed to a
specific chemical component, benzo[a]pyrene.

Human Exposure

The carcinogenic activity of soots, tars and oils in man is beyond
dispute (13-16) and, in addition to the skin cancers which were
noted initially, there have also been several reports indicating
that higher incidences of respiratory tract and upper gastrointes-
tinal tract tumors are associated with occupational exposures to
these carcinogens (summarized in refs. 13-16). While present day
working conditions are drastically improved over those of the 19th
century, contemporary studies [for example on Danish chimney
sweeps (17)] continue to find increased cancer risks associated
with exposure to these polycyclic aromatic hydrocarbon-containing
materials. Nevertheless, because any given hydrocarbon is but
one component of these complex occupational carcinogens, the position taken by Working Groups of the International Agency for
Research on Cancer over the last ten years (13-16) is that individual hydrocarbons, such as benzo[a]pyrene, have not been proven to
be carcinogens for man. This lack of direct proof of their carcinogenicity for man should not be overinterpreted. It does not
imply that the human race is resistant to these potent experimental
carcinogens and most carcinogenesis researchers adopt the view-point
that, until proof to the contrary is obtained, chemicals found to
be carcinogens in animals should be regarded as probable carcinogens
for man. The polycyclic aromatic hydrocarbons have to be regarded
in this way since some of them are very potent experimental
carcinogens and the metabolism and DNA binding products of benzo[a]
pyrene in human cells and tissues are very similar to those seen
in susceptible experimental animals (reviewed in 16). Some investigators believe that the hydrocarbons are proven skin carcinogens
and probable respiratory tract carcinogens for man (18).
    The presence of polycyclic aromatic hydrocarbons in the
environment is of obvious concern and, apart from specific occupational environments, human exposure to these compounds derives
from combustion products released into the atmosphere. Estimates
of the total annual benzo[a]pyrene emissions in the United States
range from 900 tons (19) to about 1300 tons (20). These totals
are derived from heat and power generation (37-38%), open-refuse
burning (42-46%), coke production (15-19%) and motor vehicle
emissions (1-1.5%) (19,20). Since the vast majority of these
emissions are from stationary sources, local levels of air pol-
lution obviously vary. Benzo[a]pyrene levels of less than
1 μg/1,000 m^3 correspond to clean air (20). At this level,
it can be estimated that the average person would inhale about
0.02 μg of benzo[a]pyrene per day, and this could increase
to 1.5 μg/day in polluted air (21).

Polycyclic particulates released into the air can be washed out by rain or settle out under gravity, thereby contaminating water and soil and providing other possible routes for human exposure. In fact, hydrocarbons are widespread through the world's waters (22) and can enter our food chain by being taken up by plankton, by filter-feeding mollusks, and by fish (23). The average level of benzo[a]pyrene in drinking water is about 0.01 µg/l (19) so that daily intake from drinking water is of a similar order to that from breathing reasonably clean air. These intakes, together with the hydrocarbons present in our uncooked foods (24), are probably largely unavoidable and because of this, and the high carcinogenic potency exhibited by the hydrocarbons, it is important that the mechanism of action of these ubiquitous carcinogens be investigated in order to determine how to minimize their threat to man.

Experimental Carcinogenesis

The process of carcinogenesis remains poorly understood and this is perhaps not surprising given that it occurs over a period of many months in experimental animals and many years in man. While mouse skin remains a convenient and popular system for monitoring the carcinogenic activity of polycyclic aromatic hydrocarbons, it is by no means the only tissue sensitive to this class of carcinogens. To a large extent, the tissue affected is determined by the route of administration of the carcinogen and the animal species under investigation. For example, 7,12-dimethylbenz[a]anthracene is a particularly potent carcinogen for the mammary gland of young female Sprague-Dawley rats after oral or intravenous administration (25,26), dietary benzo[a]pyrene leads to leukemia, lung adenoma and stomach tumors in mice (27), and either of these hydrocarbons can induce hepatomas in male mice when injected on the first day of life (28). Nevertheless, the mouse skin system has proved to be particularly valuable because of the rapidity of tumor induction, the ease of detection of tumors and because the multi-stage nature of the carcinogenic process was experimentally established in this system.

Tumors are induced in mouse skin either by the repeated application of small doses of polycyclic hydrocarbons, by a single application of a large dose, or by the single application of a sub-carcinogenic dose of hydrocarbon (initiation) followed by repeated applications of a noncarcinogenic agent such as croton oil or its active constituent 12-0-tetradecanoyl-phorbol-13-acetate (promotion) (29). The characteristics of the latter initiation-promotion system indicate that initiation is essentially irreversible. This follows from the fact that even when treatment with promoters is begun several months after initiation with a hydrocarbon, a high yield of tumors is still ultimately obtained. In contrast, promotion is reversible to some extent, and it is only effective following, not preceding, the initiating events. These discoveries suggest that only the initiation stage of carcinogenesis has an absolute requirement for the chemical carcinogen.

More detailed reviews of the complex initiation-promotion literature
should be consulted for a full appreciation of this topic (30,31).
      While several general mechanisms through which chemicals might
initiate the carcinogenic process have been conceived, the weight
of evidence at present suggests that the polycyclic aromatic hydrocarbons initiate carcinogenesis through a mutagenic mechanism.
This involves an initial covalent interaction between a metabolite
of the hydrocarbon and the DNA of the target tissue. While these
interactions are becoming fairly clearly understood (32), it is
important to remember that this chemical interaction in itself
does not constitute the process of initiation. This follows because
initiation is operationally defined as an irreversible process
and several studies of the transformation of mammalian cells in
vitro show that until cells exposed to a carcinogen have been
permitted to undergo cell division, the transforming effects of
carcinogens can be reversed (33,34). Thus, even the initiation
stage of carcinogenesis is complex, and is dependent not only on
the properties of the hydrocarbon administered, but it also requires complex contributions from the animal tissue involved,
first in metabolizing the hydrocarbon to an appropriate carcinogenic metabolite and secondly in making permanent the potentially
initiating damage generated by this metabolite. In order that
tumors should eventually arise following the initiation stage,
appropriate promotional stimuli are also required and promotion
itself has been resolved into several stages (29,35). Given this
degree of complexity, it would be grossly optimistic to expect any
obvious relationship between the structure of polycyclic aromatic
hydrocarbons and their carcinogenic activities. However, at the
outset of investigations into the carcinogenic hydrocarbons, the
structure-activity approach was the only means available to attempt
to determine the mechanism of action of these compounds. While it
did not provide the key evidence leading to our present level of
understanding, it has contributed considerably to the latter, and,
with the benefit of hindsight, it is clear that the success of the
structure-activity approach was limited primarily by our inability
to interpret the information it yielded, rather than by the information itself.

Structure and Activity

Immediately after the first few carcinogenic hydrocarbons were
identified, scientists puzzled over the developing structure-activity relationships and tried to identify the structural features of
the hydrocarbons which are associated with their carcinogenic
activity.
      For unsubstituted polycyclic aromatic hydrocarbons (see examples
on page 6), it seems that a minimum of four benzene rings is required
for, but does not guarantee, carcinogenic activity. Thus, only two
of the six possible arrangements of four benzene rings represent
compounds with definite although weak carcinogenic activity.
These are benzo[c]phenanthrene and benz[a]anthracene which, as
Hewett (36) noted in 1940, are both phenanthrene derivatives. In
contrast, linear structures such as naphthacene are not associated

Structures of Carcinogenic and Noncarcinogenic Unsubstituted
Polycyclic Aromatic Hydrocarbons

Inactive			Carcinogenic	

Naphthacene   Triphenylene   Pyrene   Benz[a]anthracene   Benzo[c]phenanthrene

Picene[a]   Chrysene[a]   Benzo[a]pyrene   Dibenzo[b,def]chrysene

Benzo[e]pyrene[a]   Dibenzo[fg,op]naphthacene   Benzo[rst]pentaphene   Dibenzo[def,p]chrysene

Anthanthrene   Benzo[c]pentaphene   Naphtho[1,2,3,4-def]chrysene

Benzo[b]fluoranthene

Fluoranthene   Dibenzo[b,k]chrysene   Benzo[j]fluoranthene   Benzo[b]fluoranthene

[a] These compounds show tumor-initiating activity.

with carcinogenic activity. Structures such as triphenylene and
dibenzo[fg,op]naphthacene, which behave chemically like condensed
polyphenyls, are also inactive as carcinogens. While pyrene itself, does not exhibit any carcinogenic activity, the more potent
carcinogens amongst the unsubstituted hydrocarbons are benzo[a]-
pyrene and several compounds which can be considered to be benzo-
derivatives of benzo[a]pyrene. The presence of a benzo[a]pyrene
structure within a more complex molecule does not guarantee carcinogenic activity however, because anthanthrene is not a carcinogen.
In addition to the benzenoid polycyclics, structures containing
five-membered rings are also present in the environment and some
of these are also carcinogenic.

Structure-activity relationships become even more complex when
the effect of various substituents on carcinogenic activity is considered (37). While benz[a]anthracene is considered to be a
fairly weak carcinogen, 7,12-dimethylbenz[a]anthracene is one of
the most potent of the hydrocarbon carcinogens. A number of other
dimethylbenz[a]anthracenes, some trimethylbenz[a]anthracenes and
all of the twelve possible methylbenz[a]anthracenes have been
synthesized and their carcinogenic activities evaluated. These
findings have been reviewed many times (37-40) and though there
are slight differences from one study to another the overall conclusions are fairly clear (Table I). Thus, for the methylbenz[a]-
anthracenes, the most potent carcinogen is 7-methylbenz[a]anthra-
cene but substitution at positions 6, 8 or 12 is also associated
with substantial activity. None of the other methylbenz[a]anthra-
cenes have exhibited substantial carcinogenic activity, with the
possible exception of 5-methylbenz[a]anthracene (41), but in the
earlier tests on mouse skin (38) substitution in the angular ring
i.e., on the 1-, 2-, 3- or 4-positions, was found to yield totally
inactive compounds while the remaining isomers i.e. the 9-, 10-,
and 11-methylbenz[a]anthracenes all exhibited some slight activity.
The concept that substitution in the angular ring was inversely
associated with carcinogenic activity was strengthened by the
findings with dimethylbenz[a]anthracenes. Even though both 7-
and 12-methylbenz[a]anthracene are potent carcinogens, the 1,7-,
1,12-, 4,7-, and 4,12-dimethylbenz[a]anthracenes are all inactive
(37). Similarly, the potent activity of 7,12-dimethylbenz[a]-
anthracene can be destroyed by the presence of methyl groups on the
2-or 3- positions but the 4-methyl derivative remains active.
The wealth of structure-activity data for the hydrocarbons, together with the fact that they were the first pure chemicals recognized to exhibit carcinogenic potency, attracted a great deal of
attention over the years, with numerous attempts being made to
define the structural features associated with carcinogenic activity. Attention was directed largely towards the presence of a
phenanthrene structure within most of the carcinogenic hydrocarbons
(38) and then subsequently, to the presence of an aromatic bond
with a high degree of double bond character analogous to that of
the 9,10-bond in phenanthrene. While many investigators contributed to the development of so-called electronic theories of carcinogenesis, the most widely applicable description of a hydrocarbon
carcinogen in these terms was developed by the Pullmans (42).

They found that, with few exceptions, the carcinogens and noncar-
cinogens among the unsubstituted hydrocarbons could be distinguished by indices describing the reactivity of K-regions and
L-regions towards addition reactions (Figure 1). Carcinogens
were characterized by a high reactivity at the K-region together
with a low reactivity at the L-region (where such a structural
feature was present). The interpretation of this correlation
implied that some interaction between a K-region and some cellular
constituent was responsible for initiating the carcinogenic process,
while some alternative interaction at a reactive L-region could
destroy the carcinogenic properties of an otherwise potential
carcinogen. These ideas had a major influence on thinking about
the mechanisms of action of the hydrocarbon carcinogens from the
early 1940's to the early 1970's. While it is no longer thought
that the K-region plays a major role in the activation of carcinogens, it remains possible that inactivating reactions may occur
at L-regions. Specific exceptions to the K- and L-region hypothesis
were anthanthrene which was expected to be a carcinogen but is
not, and the overall effect of methyl groups in the angular 1,2,3,
4-ring of benz[a]anthracene derivatives in reducing carcinogenic
activity.

     Once it was appreciated that a vicinal dihydrodiol epoxide
might be the metabolite of benzo[a]pyrene responsible for carcinogenic activity (43), Jerina and Daly (44) were able to suggest
that a "bay region" was the structural feature required for carcinogenic activity and that the active metabolites for many hydrocarbons would be found to be bay region dihydrodiol epoxides i.e.
vicinal dihydrodiol epoxides wherein the epoxide ring is adjacent
to a bay region (Figure 2). (The term bay region is used to
describe a concave area of the periphery of aromatic hydrocarbons
and was initially introduced because protons in such a region
i.e. at 1 and 12 in 7-methylbenz[a]anthracene or 10 and 11 in
benzo[a]pyrene (Figure 2) exhibit distinctive nmr properties).
This suggestion neatly accounted for the exceptions to the K-region hypothesis, above, since anthanthrene does not contain a bay
region and substitutions in the angular ring of benz[a]anthracene
and its homologues might be expected to interfere with metabolic
activation to a bay region dihydrodiol epoxide. In addition, the
presence of a phenanthrene structure within a more complex hydrocarbon is necessary for a bay region to be present as well as for
a K-region to be present. It appears, then, that the exceptions
to the earlier structure-activity relationships were signalling
the current understanding of the structural features required for
carcinogenic activity but, as detailed in the following section,
experimental advances in understanding the mechanism of metabolic
activation of polycyclic hydrocarbons had to be made before these
signals could be interpreted.

Metabolic Activation of Hydrocarbon Carcinogens

In early studies of the metabolism of hydrocarbons, it was noted
that vicinal _trans_ dihydrodiols were frequently found as hydrocarbon metabolites and, because of their _trans_ configuration, Boyland

Table I. Effect of a Methyl Group at a Single Starred Position on
Carcinogenic Activity of Benz[a]anthracene Derivatives

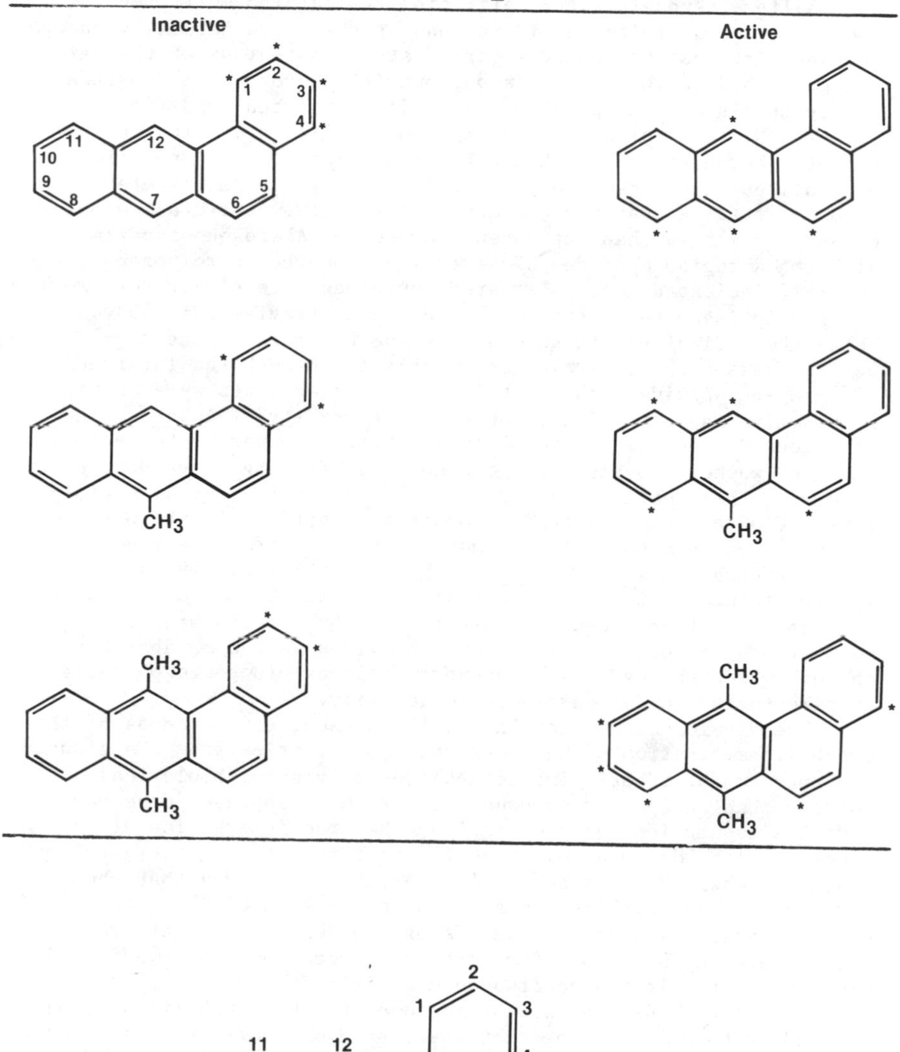

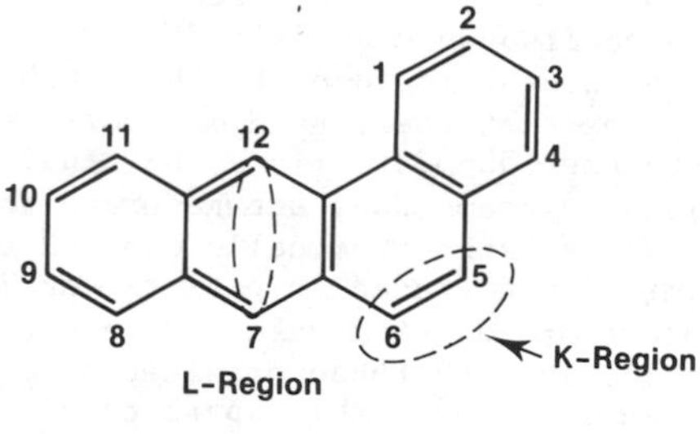

Figure 1. Benz[a]anthracene with regions of low bond localization
energy (K-region) and low para localization energy (L-region)
indicated.

(45) suggested that these diols probably arose from an intermediate
epoxide. Moreover, he argued that such epoxides might be the
metabolites responsible for initiating the carcinogenic process.
The structure-activity considerations at that time naturally enough
focussed interest on epoxides formed at the K-regions of the carcinogenic hydrocarbons (Figure 3), but it was not until 1964 that
the synthesis of such putative metabolites was achieved (46).
    The Millers' pioneering work on metabolic activation in the
aromatic amine field (47) had established the role of specific
metabolites in the carcinogenic action of N-2-fluorenylacetamide
by demonstrating that the N-hydroxy metabolite was overall a more
potent carcinogen than the parent compound. Analogous experiments
with the K-region epoxides of several hydrocarbon carcinogens,
however, indicated that these arene epoxides were either very weak
or totally inactive as chemical carcinogens (reviewed in 48).
While these findings did not support the hypothesis that hydrocarbons expressed their carcinogenic activity through the intermediacy
of K-region epoxides, the fact that such metabolites were chemically reactive and potentially subject to a variety of inactivating
reactions during the course of application to experimental animals
led many workers to feel that the negative findings were not con-
clusive. This feeling was strengthened by the biological activities exhibited by the K-region epoxides in various in vitro systems,
which showed them to be toxic, mutagenic, and effective inducers
of transformation in vitro (48). Thus, by the early 1970s, the
wealth of information on biological activities in various systems
and the lack of an acceptable alternative hypothesis was leading
to a growing acceptance of K-region epoxides as the metabolites
through which the hydrocarbons exert their carcinogenic potential,
despite their lack of carcinogenic activity.
    The developments which led to the present day concepts of the
metabolic activation of hydrocarbons did not arise from the classical approach of identifying metabolites of greater biological
potency than the parent compound, but from an approach dependent
upon the assumption (or presumption) that the interaction of carcinogens with DNA is a key event in the initiation of the carcinogenic process. Brookes and Lawley (49) found in 1964 that when
radioactive hydrocarbons are applied to the skin of mice, they
become covalently bound to the DNA of the skin. Moreover, the
extents of binding to DNA for various hydrocarbons followed fairly
closely their relative carcinogenic activities.
    Thereafter, Brookes sought to identify the metabolites involved
in binding to DNA assuming that these same metabolites were involved
in the carcinogenic process (50-53). Since the binding of hydrocarbon to DNA in cellular systems does not generate enough material
to examine directly, DNA isolated from cells exposed to radioactive
hydrocarbons was enzymically degraded to deoxyribonucleosides and
the radioactive hydrocarbon-deoxyribonucleoside adducts were
compared chromatographically with those obtained from some putative
reactive metabolite. In 1973 (53), this approach clearly showed
that in mouse skin or mouse embryo cells in culture, the carcinogen
7-methylbenz[a]anthracene did not bind to DNA through a K-region
epoxide intermediate and, unlike the correlative biological activi-

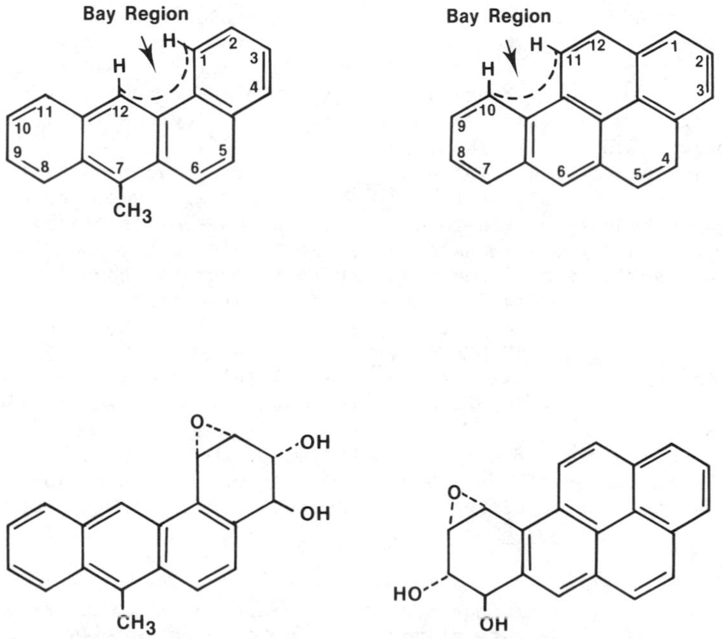

Figure 2.   7-Methylbenz[a]anthracene and benzo[a]pyrene indicating those regions defined as bay regions and the structures of the corresponding bay region dihydrodiol epoxides.

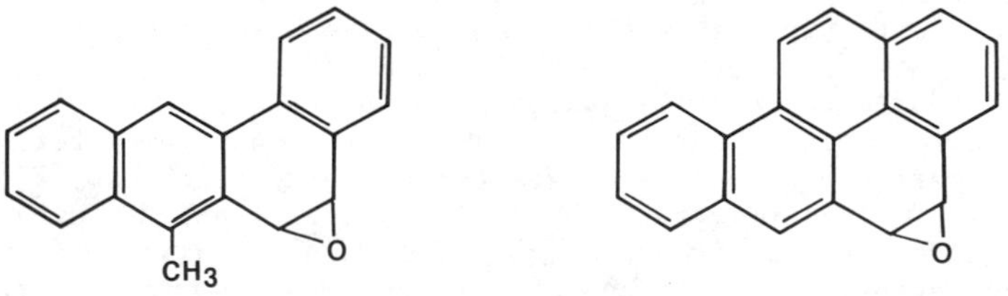

Figure 3.   The K-region epoxides of 7-methylbenz[a]anthracene and benzo[a]pyrene.

ty data, this direct measurement of events occurring within the
biological system could not be circumvented.

The search for an alternative reactive metabolite for the polycyclic hydrocarbon carcinogens was soon successful. Also in 1973,
Borgen et al. (54) reported that, in the presence of a microsomal
system from hamster liver, trans 7,8-dihydro-7,8-dihydroxybenzo[a]-
pyrene (a metabolite of benzo[a]pyrene) was bound to DNA in vitro
some ten times more extensively than was benzo[a]pyrene itself.
They concluded that this trans 7,8-dihydrodiol "is further metabolized to an active alkylating agent", though they made no specific
suggestion as to its structure. Sims and his colleagues, who had
long been proponents of the role of hydrocarbon epoxides in carcinogenesis, rapidly realized that the alkylating activity could
arise from epoxidation of the nonaromatic 9,10-double bond in
the trans 7,8-dihydrodiol (Figure 4). They synthesized a small
amount of this dihydrodiol epoxide, and were able to show that its
products of reaction with DNA in vitro were chromatographically in-
distinguishable from those obtained when benzo[a]pyrene itself
was bound to DNA in cellular systems through metabolic activation
(43).

This general sequence of metabolic steps through which hydrocarbons become bound to DNA has subsequently been found to apply
to several other polycyclic aromatic hydrocarbon carcinogens and
this supports the bay region dihydrodiol epoxide generalization of
Jerina and Daly, discussed earlier. Moreover, several studies in
which metabolites involved in the dihydrodiol epoxide pathway
have been tested for carcinogenic activity are largely supportive
of the idea that this route of metabolic activation is also involved
in the carcinogenic action of hydrocarbon carcinogens. The most
thorough studies have been done in the case of benzo[a]pyrene, so
it is convenient to summarize these as representative of the most
extensive developments in the area in general. For this carcinogen, all of the possible stereoisomers and enantiomers of the bay
region dihydrodiol epoxide metabolites have been synthesized and
their biological activities evaluated (Table II) (55-58 and references cited therein).

Overall, findings on the tumorigenicity of the compounds
listed in Table II indicates that benzo[a]pyrene expresses its
carcinogenic potential through metabolic conversion to a bay region
dihydrodiol epoxide. The (+)-enantiomer of benzo[a]pyrene-7,8-
epoxide is a more potent carcinogen than the (-)-enantiomer but
neither of these has demonstrated greater activity than benzo[a]-
pyrene itself. Similarly, the dihydrodiol and anti and syn dihydrodiol epoxides derived from the (+) 7,8-epoxide are all more
potent carcinogens than their enantiomers derived from the (-) 7,8-
epoxide. However, in comparison with the carcinogenic activity of
benzo[a]pyrene, the situation is less clear, with the (-) 7,8-dihy-
drodiol being the only metabolite which consistently exhibits an
activity equal to or greater than that of the parent hydrocarbon.
In the newborn mouse system (Table II), the (+) anti dihydrodiol
epoxide is clearly more effective than benzo[a]pyrene but this is
not the case in initiation-promotion studies or in complete carcinogenesis studies on mouse skin. Nevertheless, there is always

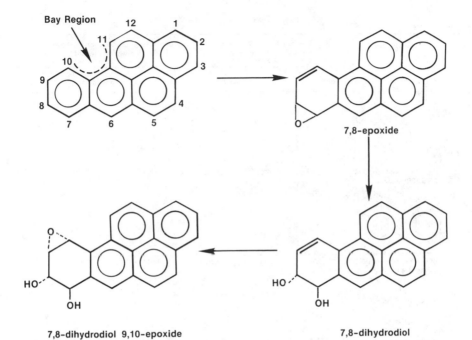

7,8-dihydrodiol 9,10-epoxide **7,8-dihydrodiol**

Figure 4. The bay region dihydrodiol epoxide route of metabolism of benzo[a]pyrene.

Table II. Carcinogenic Activities of Benzo[a]pyrene (BP) Metabolites

Compound	Tumor Initiation in Mouse Skin Metabolite Activity/Activity of BP at Same Dose			Lung Adenoma Induction in Newborn Mice Metabolite Activity/Activity of BP at Same Dose		
	Dose (μmol)	% Mice with Tumors	Av. Tumors per Mouse	Dose (μmol)	% Mice with Adenoma	Lung Adenoma per Mouse
(+) BP[7R,8S]-epoxide	0.1[a]	38/68	0.76/2.1	0.7[a]	71*	2.14*
(-) BP[7R,8R]-dihydrodiol	0.1[b]	77/77	3.8/2.6	0.14[d]	88*	9.18*

Continued on next page.

Table II. Continued

Compound	Tumor Initiation in Mouse Skin Metabolite Activity/Activity of BP at Same Dose			Lung Adenoma Induction in Newborn Mice Metabolite Activity/Activity of BP at Same Dose		
	Dose (μmol)	% Mice with tumors	Av. tumors per mouse	Dose (μmol)	% Mice with lung adenoma	Adenoma per mouse
(+) BP[7R,8S]-dihydrodiol [9S,10R]-epoxide	0.1[c]	47/68	1.1/2.1	0.014[e]	89/4	7.55/0.03
(-) BP[7R,8S]-dihydrodiol [9R,10S]-epoxide	0.1[c]	10/68	0.1/2.1	0.014[e]	11/4	0.13/0.03
(-) BP[7S,8R]-epoxide	0.1[a]	11/68	0.11/2.1	0.7[a]	7*	0.11*
(+) BP[7S,8S]-dihydrodiol	0.1[b]	31/77	0.44/2.6	0.14[d]	6*	0.06*
(-) BP[7S,8R]-dihydrodiol [9R,10S]-epoxide	0.1[c]	6/68	0.06/2.1	0.014[e]	1/4	0.01/0.03
(+) BP[7S,8R]-dihydrodiol [9S,10R]-epoxide	0.1[c]	17/68	0.17/2.1	0.014[e]	4/4	0.22/0.03

*Activities for benzo[a]pyrene under the same conditions were not reported.

[a] Epoxide data from Ref. 59; benzo[a]pyrene data from Ref. 56.

[b] From Ref. 55.

[c] From Ref. 56.

[d] From Ref. 57.

[e] From Ref. 58.

some difficulty in the interpretation of carcinogenesis tests with
highly reactive compounds and thus, the high carcinogenic activity
of the 7,8-dihydrodiol (as well as the high activity of analogous
diols from other hydrocarbons) strongly indicates that the dihydrodiol epoxide route of activation is involved in polycyclic aromatic
hydrocarbon carcinogenesis.

Acknowledgments

Research sponsored by the National Cancer Institute, DHHS, under
Contract No. N01-CO-23909 with Litton Bionetics, Inc.

Literature Cited

1.   Pott, P. "Chirurgical Observations" (1775), Reprinted in
      National Cancer Inst. Monogr, 1963, 10, 7-13.
2.   von Volkmann, R. Beiträge zur Chirurgie Leipzig, 1875.
3.   Ross, P. Br. Med. J. 1948, 2, 369-74.
4.   Haddow, A. Persp. in Biol. Med. 1974, 17, 543-88.
5.   Yamagiwa, K.; Ichikawa, K. Mitt. Med. Fak. Tokyo 1915, 15,
      295-344.
6.   Tsutsui, H. Gann 1918, 12, 17-21.
7.   Kennaway, E.L. Br. Med. J. 1955, 2, 749-52.
8.   Bloch, B.; Dreifuss, W. Schweiz. Med. Wochenschr. 1921, 51,
      1035-7.
9.   Kennaway, E.L. Br. Med. J. 1925, 2, 1-4.
10.   Clar, E. Ber. Dtsch. Chem. Ges. 1929, 62, 350-9.
11.   Kennaway, E.L.; Hieger, I. Br. Med. J. 1930, 1, 1044-6.
12.   Cook, J.W.; Hewett, C.L.; Hieger, I. J. Chem. Soc. 1933, 395-
      405.
13.   "Evaluation of Carcinogenic Risk" International Agency for
      Research on on Cancer Monographs, Vol. 3, 1972.
14.   "Chemicals and Human Cancer" International Agency for Research
      on Cancer Monographs, Suppl. 1, 1979.
15.   "Chemicals and Industries Associated with Human Cancer" Inter-
      national Agency for Research on Cancer Monographs, Suppl. 4,
      1982.
16.   "Evaluation of Carcinogenic Risk" International Agency for
      Research on Cancer Monographs, Vol. 32, 1983.
17.   Hansen, E.S. Am. J. Epidemiol. 1983, 117, 160-4.
18.   Schmähl, D.; Habs, M. In "Environmental Carcinogens: Poly-
      cyclic Aromatic Hydrocarbons", Grimmer, G., Ed., CRC: Boca
      Raton, 1983; p. 237.
19.   Baum, E.J. In "Polycyclic Hydrocarbons and Cancer"; Gelboin,
      H.V.; Ts'O, P.O.P., Eds.; Academic: New York, 1978; Vol.
      1, p 45.
20.   "Particulate Polycyclic Organic Matter," National Academy of
      Sciences 1972.
21.   Woo, Y.T.; Arcos, J.C. In "Carcinogens in Industry and the
      Environment"; Sontag, J.M., Ed.; Dekker: New York, 1981;
      p. 167.
22.   Brown, R.A.; Huffman, H.L. Science 1976, 191, 847-9.

23. Shabad, L.M. J. Natl. Cancer Inst. 1980, 64, 405–10.
24. Grasso, P.; O'Hare, C. In "Chemical Carcinogens" Searle,
    C.E., Ed., American Chemical Society, Washington, D.C., 1976;
    p. 701.
25. Huggins,C.; Briziarelli,G.; Sutton,H. J. Exp. Med. 1959,
    109, 25–42.
26 Huggins,C.; Grand,L.C.; Brillantes, F.P. Nature 1961, 189,
    204–7.
27. Rigdon, R.H.; Neal,J. Proc. Soc. Exp. Biol. Med. 1969, 130,
    146–8.
28. Roe, F.J.C.; Waters, M.A. Nature 1967, 214, 299–330.
29. Boutwell, R.K. Progr. Exp. Tumor Res. 1964, 4, 207–50.
30. Slaga, T.J. "Mechanisms of Tumor Promotion, Vol. II"; CRC
    Press: Boca Raton, Florida, 1984.
31. Iversen, O.H.; Astrup, E.G., Cancer Investign. 1984, 2,
    51–60.
32. Grunberger, D.; Weinstein, I.B. In "Chemical Carcinogens
    and DNA" Grover, P.L., Ed.; CRC Press, Boca Raton, 1979,
    Vol. II p. 59.
33. Borek, C.; Sachs, L. Proc. Natl. Acad. Sci., U.S.A. 1967,
    57, 1522–7.
34. Kakunaga, T. Cancer Res. 1975, 35, 1637–42.
35. Slaga, T.J.; Fischer, S.M.; Nelson, K.; Gleason, G.L. Proc.
    Natl. Acad. Sc., U.S.A. 1980, 77, 2251–54.
36. Hewett, C.L. J. Chem. Soc. 1940, 293–303.
37. Dipple, A. In "Chemical Carcinogens" Searle, C.E., Ed.; ACS
    Monogr., 173 American Chemical Society, Washington, D.C., 1976,
    pp. 245–314.
38. Badger, G.M. Br. J. Cancer 1948, 2, 309–50.
39. Arcos, J.C.; Argus, M.F. "Chemical Induction of Cancer, Vol.
    IIA", Academic, New York, 1974.
40. Dipple, A.; Lawley, P.D.; Brookes, P. Eur. J. Cancer 1968,
    4, 493–506.
41. Lacassagne, A.; Zajdela, F.; Buu-Hoi, N.P.; Chalvet, O. Bull.
    Cancer 1962, 49, 312–7.
42. Pullman, A.; Pullman, B. Adv. Cancer Res. 1955, 3,117–69.
43. Sims, P.; Grover, P.L.; Swaisland, A.; Pal, K.; Hewer, A.
    Nature 1974, 252, 326–8.
44. Jerina, D.M.; Daly, J.W. In "Drug Metabolism: Parke, D.V.,
    Smith, R.L., Eds., Taylor and Francis, London, 1976, pp. 13–32.
45. Boyland, E., Biochem. Soc. Symp. 1950, 5, 40–54.
46. Newman, M.S.; Blum, S. J. Am. Chem. Soc. 1964, 86, 5598–600.
47. Miller, E.C.; Miller, J.A. Pharmacol. Rev. 1966, 18, 805–38.
48. Sims, P.; Grover, P.L. Adv. Cancer Res. 1974, 20, 165–274.
49. Brookes, P.; Lawley, P.D. Nature 1964, 202, 781–4.
50. Brookes, P.; Heidelberger, C. Cancer Res. 1969, 29, 157–65.
51. Dipple, A.; Brookes, P.; Mackintosh, D.S.; Rayman, M.P. Bio-
    chemistry 1971, 10, 4323–30.
52. Baird, W.M.; Brookes, P. Cancer Res. 1973, 33, 2378–23.
53. Baird, W.M.; Dipple, A.; Grover, P.L.; Sims, P.; Brookes, P.
    Cancer Res. 1973, 33, 2386–92.

54. Borgen, A.; Darvey, H.; Castagnoli, N.; Crocker, T.C.;
     Rasmussen, R.E.; Yang, I.Y. J. Med. Chem. 1973, 16, 502-6.
55. Levin, W.; Wood, A.W.; Chang, R.L.; Slaga, T.J.; Yagi, H.;
     Jerina, D.M.; Conney, A.H. Cancer Res. 1977, 37, 2721-25.
56. Slaga, T.J.; Bracken, W.J.; Gleason, G.; Levin, W.; Yagi, H.;
     Jerina, D.M.; Conney, A.H. Cancer Res. 1979, 39, 67-71.
57. Kapitulnik, J.; Wislocki, P.G.; Levin, W.; Yagi, H.; Thakker,
     D.R.; Akagi, H.; Koreeda, M.; Jerina, D.M.; Conney, A.H.
     Cancer Res. 1978, 38, 2661-65.
58. Buening, M.K.; Wislocki, P.G.; Levin, W.; Yagi, H.; Thakker,
     D.R.; Akagi, H.; Koreeda, M.; Jerina D.M.; Conney, A.H.
     Proc. Natl. Acad. Sci., U.S.A. 1978, 75,5358-61.
59. Levin, W.; Buening, A.; Wood, A.; Chang, R.L.; Kedzierski,
     B.; Thakker, D.H.; Boyd, D.R.; Gadaginamath, G.S.; Armstrong,
     R.N.; Yagi, H.; Karle, J.M.; Slaga, T.J.; Jerina, D.M.;
     Conney, A.H. J. Biol. Chem. 1980, 255, 9067-74.

RECEIVED March 5, 1985

Stereoselective Metabolism and Activations of Polycyclic Aromatic Hydrocarbons

SHEN K. YANG, MOHAMMAD MUSHTAQ, and PEI-LU CHIU

Department of Pharmacology, F. Edward Hébert School of Medicine, Uniformed Services University of the Health Sciences, Bethesda, MD 20814-4799

Current understandings on the stereoselective metabolism and activation pathways of the weak carcinogen benz[a]anthracene and two of the potent carcinogens, benzo[a]pyrene and 7,12-dimethylbenz[a]anthracene, are reviewed. Different stereoselective pathways of metabolism occur in the formations of the procarcinogenic dihydrodiols, the bay-region dihydrodiolepoxides, and the K-region dihydrodiols by rat liver microsomal enzymes. Recent evidence suggests that a methyl substituent at the C-12 position of benz[a]anthracene enhances the carcinogenicity of the methylated hydrocarbon and also changes the stereoselective metabolism in the formation and hydration of the K-region 5,6-epoxide as well as the procarcinogenic 3,4-epoxide.

Polycyclic aromatic hydrocarbons (PAHs) are common particulate environmental pollutants and may be responsible for some cancer induction in man. The biological properties of PAHs, such as mutagenicity, carcinogenicity, and covalent binding to cellular macromolecules, require metabolic activation by the cytochrome P-450 containing drug-metabolizing enzyme systems. The metabolism of PAHs has been studied intensively in the past thirty years and the recent rapid progress in the understanding of their activation pathways is largely due to the recognition of benzo[a]pyrene 7,8-dihydrodiol-9,10-epoxide as the major carcinogenic and mutagenic metabolite of benzo[a]pyrene (BaP) (Figure 1; for reviews, see 1-4 and references therein).

BaP is metabolically activated predominantly to the 7R,8S-dihydrodiol-9S,10R-epoxide (anti form) and to a minor extent to 7R,8S-dihydrodiol-9R,10S-epoxide (syn form) via 7R,8S-epoxide and 7R,8R-dihydrodiol (Figure 1). Evidence for the formation of the 7S,8S-dihydrodiol and the 7S,8R-dihydrodiol-9R,10S-epoxide from the metabolism of BaP in vivo on mouse skin has been reported (5).

BaP is also stereoselectively metabolized to the 4R,5R-dihydrodiol via the 4S,5R-epoxide and to the 9R,10R-dihydrodiol via the 9S,10R-epoxide (Figure 1).

This chapter not subject to U.S. copyright.
Published 1985, American Chemical Society

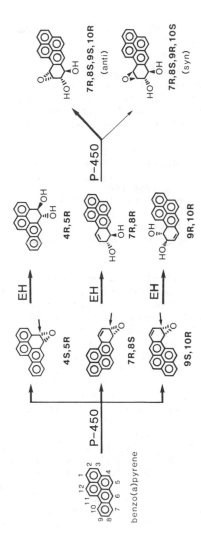

Figure 1. The major pathways in the metabolism of BaP to BaP epoxides, dihydrodiol, and 7,8-dihydrodiol-9,10-epoxides. The absolute configurations are as shown. The position of trans-addition of water is shown by an arrow. The optical purity of the 4,5-epoxide formed in BaP metabolism is dependent on the cytochrome P-450 isozymes present in the microsomal enzyme system. EH = epoxide hydrolase.

Absolute Configurations of the Dihydrodiol Metabolites of Benzo[a]pyrene

The configuration of the 4R,5R-dihydrodiol was established by application of the exciton chirality method (6). To minimize undesired interactions between the electric transition dipoles of the two p-N,N-dimethylaminobenzoate chromophores and the dihydrodiol chromophore, a 4,5-dihydrodiol enantiomer was first reduced to 1,2,3,3a,4,5,7,8,9,10-decahydro and 4,5,7,8,9,10,11,12-octahydro derivatives (6). We found that it is not necessary to reduce the chrysene chromophore of a BaP 4,5-dihydrodiol enantiomer (Figure 2). Similarly, the absolute configurations of the K-region dihydrodiol enantiomers of BA (7), 7-bromo-BA (8), 7-fluoro-BA (9), 7-methyl-BA (10), and 7,12-dimethyl-BA (DMBA) (7) can also be determined by the exciton chirality method without further reduction.

The absolute configuration of the 7,8-dihydrodiol metabolite was also established to be 7R,8R by the exciton chirality method (11,12). Our result (Figure 2) is in agreement with those reported earlier (11,12).

The absolute configuration of the 9,10-dihydrodiol metabolite was established to be 9R,10R both by nuclear magnetic resonance spectroscopy and by the structures of the hydrolysis products formed from the syn and anti 9,10-dihydrodiol-7,8-epoxides which were synthesized from the same 9,10-dihydrodiol enantiomer (13). The absolute configuration of a BaP trans-9,10-dihydrodiol enantiomer, after conversion to a tetrahydro product, can also be determined by the exciton chirality method (Figure 2) (19,20).

Optical Purity of the Dihydrodiol Metabolites of Benzo[a]pyrene

The optical purities of the dihydrodiol metabolites of BaP have been determined by three methods; (i) derivatization of the dihydrodiol metabolites with either (-)menthoxyacetyl chloride (15-16) or (-)-α-methoxy-α-trifluoromethylphenylacetyl chloride (17), (ii) circular dichroism spectra (14,16,18), and (iii) direct separation of enantiomers by chiral stationary phase HPLC (19,20). Method i requires the availability of relatively large amounts of racemic dihydrodiol standards and is very time-consuming. Method ii requires microgram quantities of dihydrodiol metabolites and the availability of a spectropolarimeter. Method iii can analyze sub-microgram quantity of unlabeled dihydrodiol metabolites and sub-nanogram quantity of radiolabeled dihydrodiol metabolites. If the enantiomers of a dihydrodiol can be resolved by the chiral stationary phase HPLC, method iii has the advantages of speed and sensitivity. The enantiomers of the non-K-region dihydrodiols of BaP can either be analyzed as the dihydrodiol or as the tetrahydrodiol (Figure 3).

The optical purities of the dihydrodiol metabolites formed in BaP metabolism by liver microsomes from Sprague-Dawley rats (1,14,15) are higher than those from liver microsomes from rats of Long-Evans strain (17). Repeated experiments in our laboratory using both rat strains indicate that small differences indeed exist (Table I). However, the percentages of R,R enantiomers are consistently higher than those reported by another laboratory (17).

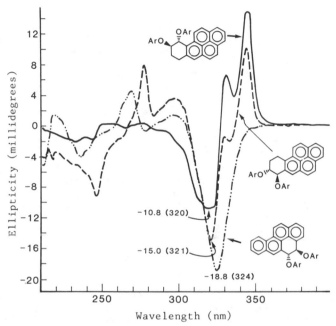

Figure 2. The exciton chirality CD spectra of <u>bis</u>-p-N,N-dimethyl-
aminobenzoyl derivatives of BaP 7,8,9,10-tetrahydro-<u>trans</u>-7,8-diol
(1.0 A_{246}/ml, derived from BaP <u>trans</u>-7,8-dihydrodiol metabolite),
BaP 7,8,9,10-tetrahydro-<u>trans</u>-9,10-diol (1.0 A_{244}/ml, derived from
BaP <u>trans</u>-9,10-dihydrodiol metabolite), and BaP <u>trans</u>-4,5-dihydro-
diol metabolite (1.0 A_{270}/ml). CD spectra are expressed by ellipti-
city at the indicated concentration as described (<u>9</u>,<u>50</u>). Ar = p-N,N-
dimethylaminobenzoyl.

Studies using either molecular oxygen-18 or oxygen-18 water indicated that the 4R,5R-dihydrodiol is derived by water attack at the C-4 position of the metabolically formed 4,5-epoxide intermediate (15,21). These results established that 4S,5R-epoxide is formed as the metabolic precursor of the 4R,5R-dihydrodiol. Hydration studies of the optically pure BaP 4,5-epoxide enantiomers indicated that the 4S,5R-epoxide is hydrated exclusively at the S-center (C-4 position) whereas 85% of the 4R,5S-epoxide is hydrated at the S-center (C-5 position) (22 and Figure 4).

Table I. Optical Purity of the Dihydrodiol Metabolites Formed in the Metabolism of Benzo[a]pyrene by Liver Microsomes from Untreated, Phenobarbital (PB)-, 3-Methylcholanthrene (3MC)-, and Polychlorinated Biphenyls (PCBs, Aroclor 1254)-Treated Rats

Pretreatment and Strain of Rats	% R,R-Dihydrodiol Enantiomer[*]		
	4,5-	7,8-	9,10-
Untreated			
Sprague-Dawley	96.3	96.9	99.6
Long-Evans	95.2	97.4 (93)[**]	98.6
PB-treated			
Sprague-Dawley	95.1	98.2	99.3
Long-Evans	94.4	94.9 (92)	96.6
3MC-treated			
Sprague-Dawley	99.6	99.6	99.6
Long-Evans	98.9 (96)	99.2 (96)	99.7 (96)
PCBs-treated			
Sprague-Dawley	99.2	99.2	99.6
Long-Evans	98.4	99.2	98.7

[*]Each entry is an average of data obtained from two separate experiments using different microsomal preparations. Enantiomeric composition was determined by CD spectral data (18) and by CSP-HPLC (19,20).
[**]Data in parentheses are from refs. 17 and 23.

It was recently reported that ≥97% of BaP 4,5-epoxide metabolically formed from the metabolism of BaP in a reconstituted enzyme system containing purified cytochrome P-450c (P-448) is the 4S,5R enantiomer (24). The epoxide was determined by formation, separation and quantification of the diastereomeric trans-addition products of glutathione. Recently a BaP 4,5-epoxide was isolated from a metabolite mixture obtained from the metabolism of BaP by liver microsomes from 3-methylcholanthrene-treated Sprague-Dawley rats in the presence of the epoxide hydrolase inhibitor 3,3,3-trichloropropylene oxide, and was found to contain a 4S,5R/4R,5S enantiomer ratio of 94:6 (Chiu et al., unpublished results). However, the content of the 4S,5R enantiomer was ≤60% when liver microsomes from untreated and phenobarbital-treated rats were used as the enzyme sources. Because BaP 4R,5S-epoxide is also hydrated predominantly to 4R,5R-dihydro-

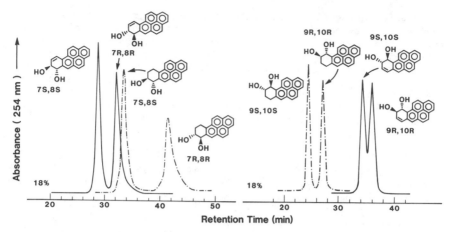

Figure 3. Mechanism of microsomal EH-catalyzed hydration of the K-region epoxide enantiomers of BA, BaP, and DMBA. The percentages of the <u>trans</u>-addition product by water for each enantiomeric epoxide are indicated. The enantiomeric composition of the dihydrodiol enantiomers formed from the hydration of DMBA 5S,6R-epoxide was determined using 1 mg protein equivalent of liver microsomes from phenobarbital-treated rats per ml of incubation mixture and this hydration reaction is highly dependent on the concentration of the microsomal EH (<u>49</u>). The epoxide enantiomer formed predominantly from the respective parent hydrocarbon by liver microsomes from 3-methylcholanthrene-treated rats is shown in the box.

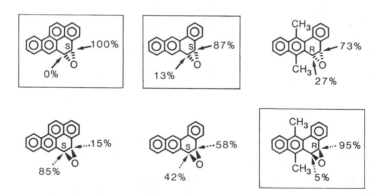

Figure 4. CSP-HPLC separation of the enantiomers of BaP dihydro- and tetrahydro- <u>trans</u>-7,8-diol and <u>trans</u>-9,10-diol. A column (4.6 mm ID x 25 cm) of γ-aminopropylsilanized silica with ionically bonded (R)-N-(3,5-dinitrobenzoyl)phenylglycine was eluted with 18% (v/v) of ethanol/acetonitrile (2:1, v/v) in hexane at a solvent flow rate of 2 ml/min.(Reproduced with permission from Ref. 20. Copyright 1984 Elsevier.)

Library
I.U.P.
Indiana, Pa.

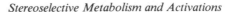

diol (22), the dihydrodiols formed in the metabolism of BaP by liver
microsomes from untreated and phenobarbital-treated rats are also
highly enriched in 4R,5R enantiomers (Table I).

It was shown that microsomal epoxide hydrolase-catalyzed trans-
addition of water to BaP 9,10-epoxide occurs stereospecifically at
the C-9 position (15). Since BaP is metabolized essentially to an
optically pure 9R,10R-dihydrodiol (13 and 15; Table I), the 9,10-
epoxide formed in BaP metabolism must have 9S,10R absolute stereochemistry (Figure 1). Similarly, the 7,8-epoxide formed in BaP
metabolism is hydrated specifically at the C-8 position to form the
7R,8R-dihydrodiol (14,21). Hence the enzymatically formed 7,8-
epoxide intermediate has 7R,8S absolute stereochemistry (Figure 1).
Although the 7R,8R-dihydrodiol is formed almost exclusively from BaP
metabolism in rat liver microsomes (Table I) and in bovine bronchial
explants (25), the 7S,8S-dihydrodiol is also formed from BaP metabolism in mouse skin epidermis in vivo (5).

Stereoselective Metabolism at the 1,2,3,4-ring of Benz[a]anthracene

In BA metabolism, the procarcinogenic BA trans-3,4-dihydrodiol (26)
constitutes 1.5-4% of all the metabolites formed by rat liver microsomes (27) and a major component of the free dihydrodiols formed by
mouse skin maintained in short-term organ culture (28). In this
system (28), the noncarcinogenic dihydrodiols may be preferentially
removed by conjugation reactions to yield water soluble products.

BA trans-3,4-dihydrodiol cannot be separated from BA trans-8,9-
dihydrodiol in several HPLC conditions (27-29). Quantification of BA
trans-3,4-dihydrodiol by HPLC can only be accomplished after conver-
ting the 3,4-dihydrodiol to its diacetate (25,26). The BA trans-3,4-
dihydrodiol formed in BA metabolism by liver microsomes from phenobarbital-treated rats was determined to have a 3R,4R/3S,4S enantiomer ratio of 69:31 (30). Recently we have determined the optical
purity of the BA trans-3,4-dihydrodiol formed in the metabolism of
BA by three liver microsomes prepared from untreated rats and rats
that had been pretreated with an enzyme inducer. As shown in Table
II, cytochrome P-450 isozymes contained in liver microsomes from 3-
methylcholanthrene- or phenobarbital-treated rats had similar
stereoselectivity toward the 3,4-double bond of BA. BA trans-3,4-
dihydrodiol is formed via the 3,4-epoxide intermediate (31).

In contrast to the findings that both enantiomers of BaP trans-
7,8-dihydrodiol can be metabolized to the bay-region 7,8-dihydro-
diol-9,10-epoxides, the metabolic pathways of the enantiomeric BA
trans-3,4-dihydrodiols are quite different. The more tumorigenic
3R,4R-dihydrodiol (26) is poorly converted (<16% of total metabolites) by rat liver microsomes to its bay-region 1,2-epoxide,
whereas the less tumorigenic 3S,4S-dihydrodiol is not converted to
the 3,4-dihydrodiol-1,2-epoxides (32; Figure 5). The major metabolites formed in the metabolism of BA trans-3,4-dihydrodiol are bis-
dihydrodiols (32). In the absence of hydroxyl groups, 3,4-dihydro-BA
is a more potent tumorigen than both of the BA trans-3,4-dihydrodiol
enantiomers (33). These results indicate that the absolute stereochemistry of the hydroxyl groups of BA trans-3,4-dihydrodiol play
important roles in the interaction with the cytochrome P-450
isozymes. It would be of interest to ascertain if similar pathways
of metabolism exist in the metabolism of the enantiomeric BA trans-

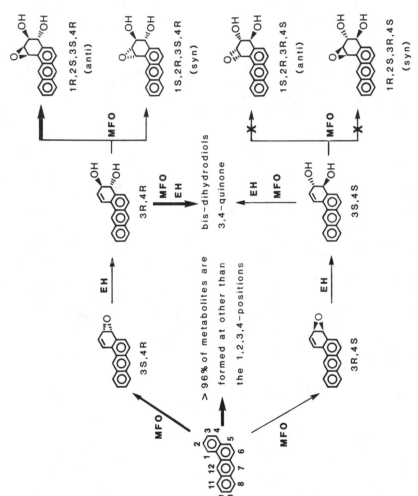

Figure 5. Metabolic activation pathways of BA. MFO abbreviates for the cytochrome P-450-containing mixed-function oxidases. The absolute configurations of the metabolites are as shown.

3,4-dihydrodiols in intact cells and under in vivo conditions.

Metabolism at the 1,2,3,4-ring of 7,12-Dimethylbenz[a]anthracene

DMBA trans-3,4-dihydrodiol is a metabolite of DMBA (34) and a potent mutagen to cultured bacterial (35,36) and mammalian cells (37,38) as well as a potent tumorigen (39,40).

As found in BA metabolism, DMBA trans-3,4-dihydrodiol is a minor metabolite (0.2-4.9% of all the metabolites) formed by rat liver microsomes and nuclei (41) and both enantiomeric 3,4-dihydrodiols are formed in DMBA metabolism by rat liver microsomes (Table II). The 3,4-dihydrodiol is a major component of the free dihydrodiols formed in mouse skin maintained in short-term culture (28). The optical purities of these dihydrodiols were determined by a CSP-HPLC method (43). The metabolic fates of the enantiomeric DMBA 3,4-dihydrodiols are not yet known. Studies in our laboratory indicate that the products formed in liver microsomal metabolism of DMBA 3,4-dihydrodiol bind extensively to the components of liver microsomes and the expected 1,2,3,4-tetrols of DMBA were not detected in the acetone/ethyl acetate extract of the incubation mixture (unpublished results). It is known that these products bind extensively to DNA (44).

The metabolites that bind to DNA in cultured mammalian cells treated with DMBA are derived from both syn and anti forms of DMBA 3,4-dihydrodiol-1,2-epoxides (45,46). Since both enantiomers of the DMBA trans-3,4-dihydrodiols are formed from DMBA metabolism, the metabolites that bind to DNA (44-46) may be derived from the metabolism of either or both of the enantiomeric DMBA trans-3,4-dihydrodiols (Figure 6). The exact structures of the diastereomeric 3,4-dihydrodiol-1,2-epoxide formed in vitro or in vivo are yet to be established. It would be of interest to see whether both of the DMBA trans-3,4-dihydrodiol enantiomers can be metabolized to form the bay-region 3,4-dihydrodiol-1,2-epoxide. In comparison, BA 3S,4S-dihydrodiol is not activated to bay-region 3,4-dihydrodiol-1,2-epoxide (32). The determination of the relative carcinogenic activities of the four possible 3,4-dihydrodiol-1,2-epoxides of DMBA (Figure 6) are not yet possible due to the lack of procedures for their syntheses. When available, a better understanding of the activation pathways can be achieved.

Stereoselective Metabolism at the K-Region of Benz[a]anthracene

The major enantiomer of BA trans-5,6-dihydrodiol formed in BA metabolism by rat liver microsomes is the 5R,6R enantiomer (Table II). Both enantiomeric BA trans-5,6-dihydrodiols may be derived from the epoxide hydrolase-catalyzed hydration of either one of the two enantiomeric BA 5,6-epoxides (22) (Figure 4). The 5,6-epoxide, analyzed as glutathione conjugates, formed in the metabolism of BA in a reconstituted system containing rat liver cytochrome P-450c (P-448) was found to contain >97% of the 5S,6R enantiomer (47). These findings indicate that the 5S,6R-epoxide is enzymatically hydrated by trans-addition of water preferentially at the S-center (C-5 position, Figure 4) to yield the BA 5R,6R-dihydrodiol. The formations of BA 5S,6R-epoxide and BaP 4S,5R-epoxide by rat liver cyto-

Table II.   Enantiomeric Composition of the 3,4- and 5,6-Dihydrodiols
Formed in the Metabolism of BA and DMBA by Liver Microsomes from Rats
of the Sprague-Dawley Strain

| Enzyme Inducer | % of Dihydrodiol Enantiomer[*] | | | |
| | BA | | DMBA | |
	3R,4R	5R,6R	3R,4R	5R,6R
None	83	77	57	89
Phenobarbital	91 (69)[**]	81 (68)	62	95
3-Methylcholanthrene	90	84 (81)	64	94

[*]BA 3,4-dihydrodiol metabolites were isolated by a reversed-phase
HPLC using a Vydac C18 column (Chiu et al., unpublished results).
DMBA dihydrodiol metabolites were isolated as described (42). The
enantiomeric composition was determined either by CD spectral data
or by CSP-HPLC (7,19,20).
[**]Data in parentheses are from ref. 30 using rats of Long-Evans
strain.

chrome P-450c as the initial K-region epoxidation products at the K-
region of the respective parent hydrocarbon are predicted by a
substrate binding model proposed by Jerina et al. (48).

Stereoselective Metabolism at the K-region of 7,12-Dimethylbenz[a]-anthracene

In contrast to the metabolism of BA and BaP, the 5,6-dihydrodiols
formed in the metabolism of DMBA by liver microsomes from untreated,
phenobarbital-treated, and 3-methylcholanthrene-treated rats are
found to have 5R,6R/5S,6S enantiomer ratios of 11:89, 6:94, and
5:95, respectively (7,49 and Table II). The enantiomeric contents of
the dihydrodiols were determined by a CSP-HPLC method (7,43). The
5,6-epoxide formed in the metabolism of DMBA by liver microsomes
from 3MC-treated rats was found to contain predominantly (>97%) the
5R,6S-enantiomer which is converted by microsomal epoxide hydrolase-
catalyzed hydration predominantly (>95%) at the R-center (C-5 position, see Figure 3) to yield the 5S,6S-dihydrodiol (49). In the
metabolism of 12-methyl-BA, the 5S,6S-dihydrodiol was also found to
be the major enantiomer formed (50) and this stereoselective reaction is similar to the reactions catalyzed by rat liver microsomes
prepared with different enzyme inducers (unpublished results).
Labeling studies using molecular oxygen-18 indicate that 5R,6S-
epoxide is the precursor of the 5S,6S-dihydrodiol formed in the
metabolism of 12-methyl-BA (51).
    The presence of a methyl group at the C-12 position of BA is
apparently the major contributing factor determining the formation
of a 5R,6S-epoxide in the metabolism of 12-methyl-BA and DMBA.

Stereochemically this is opposite to the 5S,6R-epoxide formed predominantly from the metabolism at the K-region of BA (47).

Stereoselective Metabolism at the 8,9,10,11-rings of 7,12-Dimethyl-benz[a]anthracene and Benz[a]anthracene

The 8,9- and 10,11-dihydrodiols formed in the metabolism of BA and
DMBA respectively are all highly enriched (>90%) in R,R enantiomers
(Table III). Labeling experiments using molecular oxygen-18 in the
in vitro metabolism of the respective parent compounds and subsequent mass spectral analyses of dihydrodiol metabolites and their
acid-catalyzed dehydration products indicated that microsomal
epoxide hydrolase-catalyzed hydration reactions occurred exclusively
at the nonbenzylic carbons of the metabolically formed epoxide
intermediates (unpublished results). These findings indicate that
the 8,9- and 10,11-epoxide intermediates, formed in the metabolism
of BA and DMBA respectively, contain predominantly the 8R,9S and
10S,11R enantiomer, respectively. These stereoselective epoxidation
reactions are relatively insensitive to the cytochrome P-450 isozyme
contents of different rat liver microsomal preparations (Table III).

Table III.   Enantiomeric Compositions of the 8,9- and 10,11-
Dihydrodiols Formed in the Metabolism of BA and DMBA by
Liver Microsomes from Rats of the Sprague-Dawley Strain

Enzyme Inducer	% of Dihydrodiol Enantiomer[*]			
	BA		DMBA	
	8R,9R	10R,11R	8R,9R	10R,11R
None	94	>99	91	91
Phenobarbital	90 (89)[**]	95 (83)	96	86
3-Methylcholanthrene	>99 (98)[**]	>99 (98)[**]	99	98

[*]BA 8,9- and 10,11-dihydrodiol metabolites were isolated as
described (29). DMBA dihydrodiol metabolites were isolated as
described (42). The enantiomeric composition was determined by
CSP-HPLC (7,19,20,43).
[**]Data in parentheses are from ref. 30 using rats of Long-Evans
strain.

On the Model of the Substrate Binding Site of Cytochrome P-450c

In the livers of rats pretreated with 3-methylcholanthrene, >70%
of the total cytochromes P-450 is cytochrome P-450c (P-448) (52).
The finding that BaP is highly stereoselectively metabolized to
dihydrodiols and bay-region 7,8-dihydrodiol-9,10-epoxides (summarized in Figure 1) by liver microsomes from 3-methylcholanthrene-
treated rats led Jerina et al. (48) to propose a model of the
substrate binding site for cytochrome P-450c (Figure 7) which

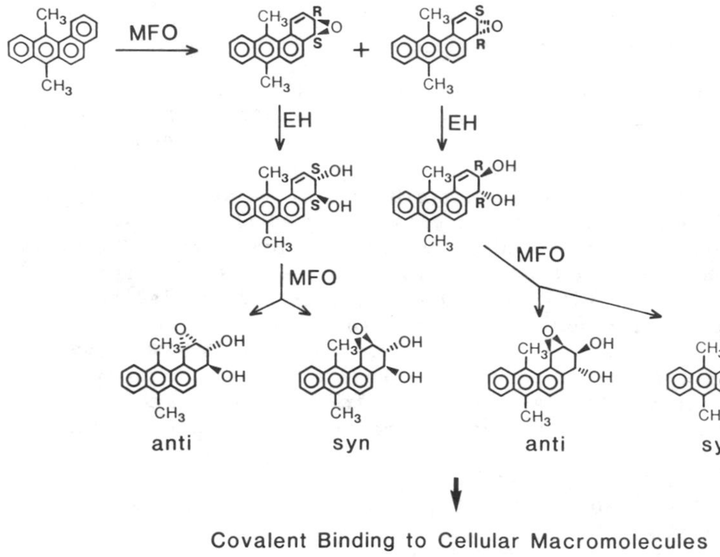

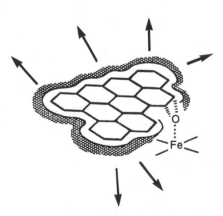

Figure 6. Metabolic activation pathways of DMBA. The absolute configurations of the metabolites are as shown.

Figure 7. Steric model proposed by Jerina, et al. for the catalytic binding site of cytochrome P-450c (P-448) to account for the stereoselective metabolism of polycyclic aromatic hydrocarbons (48). The boundary should be enlarged in the directions shown to accommodate substrates whose mechanism of stereoselective oxygenation does not fit the steric model originally proposed.

correctly predicts the stereochemistries of some non-K-region dihydrodiols formed in the metabolism of phenanthrene, chrysene, BA, and
BaP. The model also correctly predicted the major K-region epoxide
enantiomers formed from the metabolism of BA and BaP by cytochrome
P-450c in a reconstituted rat liver enzyme system (22,47). In view
of the findings that the 5R,6S-epoxide is the predominant stereoiso-
mer formed in the K-region metabolism of 12-methyl-BA and DMBA, the
proposed model (48) therefore cannot be generalized to include
hydrocarbons such as 12-methyl-BA and DMBA. The model (48) also
predicted that a 3S,4R-epoxide should be formed preferentially.
However, the formation of both trans-3,4-dihydrodiol enantiomers in
the metabolism of DMBA (Table II) was not predicted. The 1,2- and
3,4- positions of BA, monomethyl-BA, and DMBA do not fit the catalytic binding site of the proposed model. Unless the proposed boundary
of the substrate binding site is expanded, the 1,2-double bond of BA
trans-3R,4R-dihydrodiol, the 3,4-double bond of BA trans-1,2-dihy-
drodiol, the 7,8- and 9,10- double bonds of cholanthrene, the 1,2-
and 2,3- double bonds of BaP all do not fit the catalytic binding
site in the proposed model although metabolism are known to occur at
these double bonds (1,2,4,53-56). BA trans-1,2-dihydrodiol enantiomers can be stereoselectively metabolized to the 1,2-dihydrodiol-
3,4-epoxide (53). BA 3R,4R-dihydrodiol is metabolized to the bay-
region anti-3,4-dihydrodiol-1,2-epoxide (32). BaP 2,3-epoxide is a
major intermediate in the metabolism of BaP (55). The formation of
1-hydroxy-BaP as a metabolite of BaP suggests that BaP 1,2-epoxide
is also a metabolic intermediate (56). Trans-1,2- and trans-3,4-
dihydrodiols are among the major metabolites formed from the metabolism of 8-methyl-BA (57,58) and of 8-hydroxymethyl-BA (59). Thus
these and other compounds mentioned above must fit into the catalytic binding site of cytochrome P-450c in some way, but these were
not considered by the model originally proposed (48). In order to
accommodate substrates such as those mentioned above, the minimal
size of the catalytic binding site of cytochrome P-450c must be
enlarged considerably (Figure 7). If enlarged, the proposed model
would have lost its predictive value. A model for the substrate
binding site for cytochrome P-450c (or any one of the cytochrome P-
450 isozymes) must be able to account for the formations of known
metabolites, regardless whether they are chiral or achiral.

Acknowledgments

This work was supported by National Cancer Institute grant no.
CA29133 and Uniformed Services University of the Health Sciences
Protocol no. R07502. The opinions or assertions contained herein are
the private ones of the authors and are not to be construed as
official or reflecting the views of the Department of Defense or the
Uniformed Services University of the Health Sciences. The experi-
ments reported herein were conducted according to the principles set
forth in the "Guide for the Care and Use of Laboratory Animals",
Institute of Animal Resources, National Research Council, DHEW Pub.
No. (NIH) 78-23.

Literature Cited

1.   Yang, S. K.; Roller, P. P.; Gelboin, H. V. "Carcinogenesis -A

Comprehensive Survey Polynuclear Aromatic Hydrocarbons";
Freudenthal, R.; Jones, P. W., Eds.; Raven Press, New York,
1978, pp. 285-301

2. Gelboin, H. V. Physiol Rev. 1980, 60, 1107-66
3. Conney, A. H. Cancer Res. 1982, 42, 4875-917
4. Cooper, C. S.; Grover, P. L.; Sims, P. "Progress in Drug Meta-
    bolism", Vol. 7, Bridges, J. W.; Chasseaud, L. F., Eds.; John
    Wiley & Sons Ltd: London, 1983, pp. 295-396
5. Koreeda, M.; Moore, P. D.; Wislocki, P. G.; Levin, W.; Conney,
    A. H.; Yagi, H.; Jerina, D. M. Science 1977, 199, 778-81
6. Kedzierski, B.; Thakker, D. R.; Armstrong, R. N.; Jerina, D. M.
    Tetrahedron Lett. 1981, 22, 405-8
7. Yang, S. K.; Fu, P. P. Biochem. J. 1984, 223, 775-82
8. Fu, P. P.; Yang, S. K. Carcinogenesis 1983, 4, 979-88
9. Chiu, P.-L.; Fu, P. P.; Yang, S. K. Biochem. Biophys. Res.
    Commun. 1982, 106, 1405-12
10. Yang, S. K.; Fu, P. P. Chem.-Biol. Interac. 1984, 49, 71-88
11. Nakanishi, K.; Kasai, H.; Cho, H.; Harvey, R. G.; Jeffrey, A.
    M.; Jennette, K. W.; Weinstein, I. B. J. Am. Chem. Soc. 1977,
    99, 258-60
12. Yagi, H.; Akagi, H.; Thakker, D. R.; Mah, H. D.; Koreeda, M.;
    Jerina, D. M. J. Am. Chem. Soc. 1977, 99, 2358-9
13. Yagi, H.; Jerina, D. M. J. Am. Chem. Soc. 1982, 104, 4026-7
14. Yang, S. K.; McCourt, D. W.; Leutz, J. C.; Gelboin, H. V.
    Science 1977, 196, 1199-201
15. Yang, S. K.; Roller, P. P.; Gelboin, H. V. Biochemistry 1977,
    16, 3680-6
16. Harvey, R. G.; Cho, H., Anal. Biochem. 1977, 80, 540-6
17. Thakker, D. R.; Yagi, H.; Akagi, H.; Koreeda, M.; Lu, A. Y. H.;
    Levin, W.; Wood, A.W.; Conney, A. H.; Jerina, D. M. Chem.-Biol.
    Interac. 1977, 16, 281-300
18. Chiu, P.-L.; Weems, H. B.; Wong, T. K.; Fu, P. P.; Yang, S. K.
    Chem.-Biol. Interac. 1983, 44, 155-68
19. Weems, H. B.; Yang, S. K. Anal. Biochem. 1982, 125, 156-61
20. Yang, S. K.; Weems, H. B.; Mushtaq, M.; Fu, P. P. J.
    Chromatogr. 1984, 316, 569-84
21. Thakker, D. R.; Yagi, H.; Levin, W.; Lu, A. Y. H.; Conney, A.
    H.; Jerina, D. M. J. Biol. Chem. 1977, 252, 6328-34
22. Armstrong, R. N.; Kedzierski, B.; Levin, W.; Jerina, D. M. J.
    Biol. Chem. 1981, 256, 4726-33
23. Jerina, D. M.; Yagi, H.; Thakker, D. R.; Karle, J. M., Mah, H.
    D.; Boyd, D. R.; Gadaginamath, G.; Wood, A. W.; Buening, M.;
    Chang, R. L.; Levin, W.; Conney, A. H. In "Advances in Pharma-
    cology and Therapeutics", Vol. 9, Cohen, Y., Ed.; Pergamon
    Press: New York, 1979, pp. 53-62
24. Armstrong, R. N.; Levin, W.; Ryan, D. E.; Thomas, P. E.; Mah,
    H. D.; Jerina, D. M. Biochem. Biophys. Res. Commun. 1981, 100,
    1077-84
25. Weinstein, I. B.; Jeffrey, A. M.; Jennette, K. W.; Blobstein,
    S. H.; Harvey, R. G.; Harris, C.; Autrup, H.; Kasai, H.;
    Nakanishi, K. Science 1976, 193, 592
26. Wislocki, P. G.; Buening, M. K.; Levin, W.; Lehr, R. E.;
    Thakker, D. R.; Jerina, D. M.; Conney, A. H. J. Natl. Cancer
    Inst. 1979, 63, 201-4
27. Thakker, D.R.; Levin, W.; Yagi, H.; Ryan, D.; Thomas, P. E.;

Karle, J. M.; Lehr, R. E.; Jerina, D. M.; Conney, A. H. <u>Mol.</u> <u>Pharmacol</u>. 1979, 15, 138-153

28. MacNicoll, A. D.; Grover, P. L.; Sims, P. <u>Chem.</u>-<u>Biol</u>. <u>Interac</u>., 1981, 29, 169-88

29. Yang, S. K. <u>Drug</u> <u>Metab</u>. <u>Disp</u>. 1982, 10, 205-11

30. Thakker, D. R.; Levin, W.; Yagi, H.; Turujman, S.; Kapadia, D.; Conney, A. H.; Jerina, D. M. <u>Chem.</u>-<u>Biol</u>. <u>Interac</u>. 1979, 27, 145-61

31. Yang, S. K.; Fu, P. P.; Roller, P. P.; Harvey, R. G.; Gelboin, H. V. <u>Fed</u>. <u>Proc</u>. 1978, 37, 597

32. Thakker, D. R.; Levin, W.; Yagi, H.; Tada, M.; Ryan, D. E.; Thomas, P. E.; Conney, A. H.; Jerina, D. M. <u>J</u>. <u>Biol</u>. <u>Chem</u>. 1982, 257, 5103-5110.

33. Levin, W.; Thakker, D. R.; Wood, A. W.; Chang, R. L., Lehr, R. E.; Jerina, D. M.; Conney, A. H. <u>Cancer</u> <u>Res</u>., 1978, 38, 1705-1710

34. Yang, S. K.; Chou, M. W.; Roller, P. P. <u>J</u>. <u>Am</u>. <u>Chem</u>. <u>Soc</u>., 1979, 101, 237-9

35. Wislocki, P. G.; Gadek, K. M.; Fu, P. P.; Chou, M. W.; Yang, S. K.; Lu, A. Y. H. <u>Cancer</u> <u>Res</u>. 1980, 40, 3661-4

36. Malaveille, C.; Bartsch, H.; Tierney,B.; Grover, P. L.; Sims, P. <u>Biochem</u>. <u>Biophys</u>. <u>Res</u>. <u>Commun</u>., 1978, 83, 1468-73

37. Huberman, E.; Chou, M. W.; Yang, S. K. <u>Proc</u>. <u>Natl</u>. <u>Acad</u>. <u>Sci</u>., <u>USA</u>, 1979, 76, 862-6

38. Marquardt, H.; Baker, S.; Tierney, B.; Grover, P. L.; Sims, P. <u>Biochem</u>. <u>Biophys</u>. <u>Res</u>. <u>Commun</u>., 1978, 85, 357-362

39. Wislocki, P. G.; Juliana, M. M.; MacDonald, J. S.; Chou, M. W.; Yang, S. K.; Lu, A. Y. H. <u>Carcinogenesis</u> 1981, 2, 511-4

40. Slaga, T. J.; Gleason, G.; DiGiovanni, J.; Sukumaran, K. B.; Harvey, R. G. <u>Cancer</u> <u>Res</u>. 1979, 39, 1934-6

41. Chou, M. W.; Yang, S. K.; Sydor, W.; Yang, C. S. <u>Cancer</u> <u>Res</u>. 1981, 41, 1559-64

42. Chou, M.W. and Yang, S.K. (1979) <u>J</u>. <u>Chromatogr</u>. 185, 635-54

43. Yang, S. K.; Weems, H. B. <u>Anal</u>. <u>Chem</u>. 1984, 56, 2658-62

44. Chou, M. W.; Yang, S. K. <u>Proc</u>. <u>Natl</u>. <u>Acad</u>. <u>Sci</u>., <u>USA</u>, 1978, 75, 5466-70

45. Bigger, C. A.; Sawicki, J. T.; Blake, D. M.; Raymond, L. G.; Dipple, A. <u>Cancer</u> <u>Res</u>. 1983, 43, 5647-51

46. Dipple, A.; Pigott, M.; Moschel, R. C.; Constantino, N. <u>Cancer</u> <u>Res</u>. 1983, 43, 4132-5

47. Van Bladeren, P. J.; Armstrong, R. N., Cobb, D.; Thakker, D. R.; Ryan, D. E.; Thomas, P. E.; Sharma, N. D.; Boyd, D. R.; Levin, W.; Jerina, D. M. <u>Biochem</u>. <u>Biophys</u>. <u>Res</u>. <u>Commun</u>., 1982, 106, 602-9

48. Jerina, D. M.; Michaud, D. P.; Feldman, R. J.; Armstrong, R. N.; Vyas, K. P.; Thakker, D. R.; Yagi, H.; Thomas, P. E.; Ryan, D. E.; Levin, W. "Microsomes, Drug Oxidation, and Drug Toxicity", Sato, R.; Kato, R., Eds.; Japan Scientific Societies Press: Tokyo, 1982, pp. 195-201

49. Mushtaq, M.; Weems, H. B.; Yang, S. K. <u>Biochem</u>. <u>Biophys</u>. <u>Res</u>. <u>Commun</u>. 1984, 125, 539-45

50. Fu, P. P.; Chou, M. W.; Yang, S. K. <u>Biochem</u>. <u>Biophys</u>. <u>Res</u>. <u>Commun</u>. 1982, 106, 940-6

51. Yang, S. K.; Weems, H. B.; Fu, P. P. Ninth International Pharmacology Congress, 1984, abstract no. 510, London

52. Thomas, P. E.; Reik, L.M.; Ryan, D. E.; Levin, W. J. Biol.
     Chem. 1981, 256, 1044-52
53. Chou, M. W.; Chiu, P.-L., Fu, P. P.; Yang, S. K.
     Carcinogenesis 1983, 4, 629-38
54. Li, X. C.; Fu, P. P.; Chou, M. W.; Yang, S. K. "Polynuclear
     Aromatic Hydrocarbons: Formation, Metabolism, and Measurement",
     Cooke, M.; Anthony, J.D., Eds.; Battelle Press: Columbus, 1983,
     pp. 583-98
55. Yang, S. K.; Roller, P. P.; Fu, P. P.; Harvey, R. G.; Gelboin,
     H. V. Biochem. Biophys. Res. Commun. 1977, 77, 1176-82
56. Selkirk, J. K.; Croy, R. G.; Gelboin, H. V. Cancer Res. 1976,
     36, 922-6
57. Yang, S. K.; Chou, M. W.; Weems, H. B.; Fu, P. P. Biochem.
     Biophys. Res. Commun. 1979, 90, 1136-41
58. Yang, S. K.; Chou, M. W.; Fu, P. P.; Wislocki, P. G.; Lu, A. Y.
     H. Proc. Natl. Acad. Sci., USA 1982, 79, 6802-6
59. Yang, S. K.; Chou, M. W.; Evans, F. E.; Fu, P. P. Drug Metab.
     Disp. 1984, 12, 403-13

RECEIVED March 5, 1985

Synthesis of the Dihydrodiol and Diol Epoxide Metabolites of Carcinogenic Polycyclic Hydrocarbons

RONALD G. HARVEY

Ben May Laboratory, University of Chicago, Chicago, IL 60637

Methods for the synthesis of the biologically active dihydrodiol and diol epoxide metabolites of both carcinogenic and noncarcinogenic polycyclic aromatic hydrocarbons are reviewed. Four general synthetic routes to the trans-dihydrodiol precursors of the bay region anti and syn diol epoxide derivatives have been developed. Syntheses of the oxidized metabolites of the following hydrocarbons via these methods are described: benzo(a)pyrene, benz(a)anthracene, benzo-(e)pyrene, dibenz(a,h)anthracene, triphenylene, phenanthrene, anthracene, chrysene, benzo(c)phenanthrene, dibenzo(a,i)pyrene, dibenzo(a,h)pyrene, 7-methyl-benz(a)anthracene, 7,12-dimethylbenz(a)anthracene, 3-methylcholanthrene, 5-methylchrysene, fluoranthene, benzo(b)fluoranthene, benzo(j)fluoranthene, benzo(k)-fluoranthene, and dibenzo(a,e)fluoranthene.

Polycyclic aromatic hydrocarbons (PAH) are widespread environmental contaminants and one of the most potent classes of carcinogenic chemicals. They are byproducts of combustion, and significant levels are produced in automobile exhaust, refuse burning, smoke stack effluents, and tobacco smoke. It is strongly suspected that PAH may play an important role in human cancer.

Recent advances in PAH carcinogenesis research over the past decade have led to identification of diol epoxide metabolites as the principal active forms of the PAH investigated to date (1,2). Benzo-(a)pyrene (BP) has been most intensively investigated, and it has been demonstrated that a diol epoxide metabolite anti-BPDE is the active intermediate which binds covalently to DNA in human and other mammalian tissues (3,4). Anti-BPDE was also demonstrated to be a powerful mutagen in both bacterial and mammalian cells (5). These findings stimulated an outpouring of research directed towards elucidation of the molecular mechanism of PAH carcinogenesis.

The scope and direction of these biological investigations have been largely determined by the development of methods for the synthesis of the PAH metabolites. The diol epoxides are not isolable as products of metabolism due to their exceptional chemical reactivity.

0097-6156/85/0283-0035$07.75/0
© 1985 American Chemical Society

Their dihydrodiol precursors are only isolable in tiny quantities, insufficient for complete investigations of their biological and nucleic acid binding properties.

This chapter reviews methods for the synthesis of the dihydrodiol and diol epoxide derivatives of both carcinogenic and noncarcinogenic PAH (Figure 1). Many of these syntheses were developed in our laboratories under con-tract to the National Cancer Institute, and the synthetic compounds were furnished to the Chemical Repository at the Illinois Institute of Technology Research Institute for distribution to qualified investigators for chemical and biological studies. This program has contributed importantly to the remarkable advances made in this field over the past decade. Since metabolism of PAH in mammalian cells affords trans-isomeric diols and diol epoxides exclusively (6), this article will be limited to only these stereoisomers. Relatively few cis-diols and diol epoxides have been investigated (7), and their biological properties are largely unknown.

Benzo(a)pyrene (Method I)

The diol epoxide derivative of benzo(a)pyrene, trans-7,8-dihydroxy-anti-9,10-epoxy-7,8,9,10-tetrahydrobenzo(a)pyrene also known as (±) -7β,8α-dihydroxy-9α,10α-epoxy-7,8,9,10-tetrahydrobenzo(a)pyrene,was the first diol epoxide to be synthesized. Interest in this compound was stimulated by the report by Borgen et al. (8) that a metabolite of benzo(a)pyrene, tentatively identified as the trans-7,8-diol (1) became covalently bound to DNA in the presence of rat liver microsomes. Sims et al. (9) suggested that the active metabolite was a diol epoxide derivative of unspecified stereo chemistry.

Synthesis of the trans-7,8-diol from the available 7-keto derivative of tetrahydro-BP was accomplished via the procedure outlined in Figure 2 (7,10,11). Prévost reaction of 9,10-dihydro-BP with silver benzoate and I₂ gave the trans-dibenzoate ester which underwent bromination with NBS followed by base-catalyzed dehydrobromination and methanolysis to yield 1. In subsequent studies it was demonstrated that introduction of the olefinic bond by bromination-dehydrobromination gives erratic results, and this transformation may be achieved more reliably in one step with 2,3-dichloro-5,6-dicyano-1,4-benzoquinone (DDQ) (7,12). Epoxidation of 1 with m-chloroperbenzoic acid took place stereospecifically to afford the anti isomeric diol epoxide (anti-BPDE) in which the epoxide oxygen atom is on the opposite face to the benzylic hydroxyl group. The syn diastereomer, which has the oxygen atom on the same face as the benzylic hydroxyl, was synthesized stereospecifically from 1 by conversion to the bromohydrin with NBS in moist DMSO followed by base-catalyzed cyclization. Closely related syntheses were also reported by Yagi et al. (13) without proof of the steric assignments. This general synthetic approach will be referred to in this chapter as Method I.

The anti stereospecificity of epoxidation by the peracid is interpreted as due to association of the reagent with the allylic hydroxyl group which directs the entering oxygen atom to the same face of the molecule. The stereospecificity of bromohydrin formation is explicable in terms of steric approach control involving initial attack of the bulky bromine atom on the face opposite to the benzylic hydroxyl group (7).

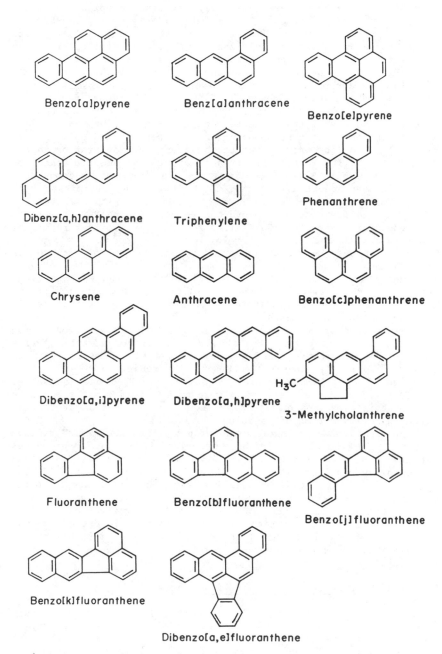

Figure 1. Structural formulae of polycyclic aromatic hydrocarbons.

Both diastereomers are relatively reactive, complicating their isolation and purification. The half-lives of anti-and syn-BPDE in water at pH 7 are 2 hr and 30 min., respectively. Although they tend to decompose on chromatographic absorbants, these diol epoxides can be purified by rapid chromatography on low activity alumina columns or by HPLC in the presence of triethylamine as a stabilizer.

The stereochemical assignments were made on the basis of chemical and NMR spectral evidence (7,10) (Figure 3). Thus, reaction of both diastereomers with sodium tert-butylthiolate afforded the respective products of trans-stereospecific ring-opening (2a and 3a). Acetylation gave the corresponding triacetates (2b and 3b). Analysis of the coupling patterns of the nonaromatic ring protons in the NMR spectra of the latter permitted clearcut assignment of the isomeric structures. Further confirmation was provided by demonstration that the adduct from anti-BPDE (2a) gave a precipitate with potassium triacetylosmate and readily formed an acetonide, consistent with the presence of cis-hydroxyl groups in the 8,9-positions. The adduct from syn-BPDE (3a) failed to enter into either reaction, consistent with the all-trans relation of the hydroxyl groups. These assignments were also supported by aqueous solvolysis studies (14-17) which showed that the primary products are the tetraols arising from cis and trans addition of water to the epoxide ring. Analysis of the chemical shift and coupling constant patterns of the NMR spectra of the tetraacetate derivatives confirmed the stereochemical isomer assignments.

Resolution of the enantiomers of anti-BPDE was achieved by reaction of the racemic dihydrodiol with (−)-menthoxyacetyl chloride followed by preparative HPLC separation and basic methanolysis to give the optically pure (+) and (−) dihydrodiols (18-20). Epoxidation of the (−)-dihydrodiol gave (+)anti-BPDE and vice versa. The CD spectra of the enantiomers were mirror images. The absolute stereochemistries were assigned as shown by application of the exciton chirality method of Nakanishi (3,19,21).

An alternative synthesis of 1 through reduction of BP 7,8-dione with NaBH$_4$ in ethanol has also been described (22). BP 7,8-dione was prepared from 9,10-dihydro-BP by conversion to the cis-7,8-tetrahydrodiol by reaction with OsO$_4$ followed by oxidation with DDQ (23) (Figure 4). This method entails the same number of steps as the original procedure and would appear to offer no advantage.

Syntheses of the trans-7,8-dihydrodiol derivatives of 7-and 8-methylbenzo(a)pyrene via Method I have also been reported (24-26). Conversion of the latter to trans-7,8-dihydroxy-anti-9,10-epoxy-8-methyl-7,8,9,10-tetrahydrobenzo(a)pyrene, has also been described (26).

Benz(a)anthracene (Methods II and III)

Although benz(a)anthracene (BA) is generally considered noncarcinogenic (27), it is a weak tumor initiator when administered with a phorbol ester promoter (28). More importantly, BA is a convenient model for the highly potent carcinogenic PAH 7,12-dimethylbenz(a)-anthracene and 3-methylcholanthrene (27), both of which are BA derivatives but which offer more serious synthetic problems.

Five vicinal trans-dihydrodiols of BA are known. One of these,

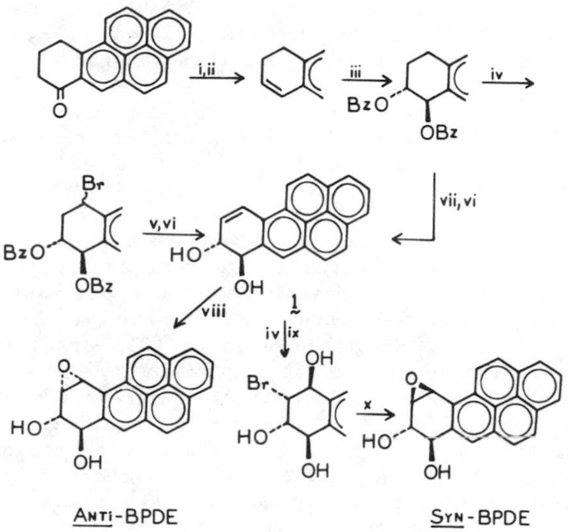

Figure 2.   Synthesis of trans-7,8-dihydroxy-anti(and syn)-9,10-epoxy-7,8,9,10-tetrahydrobenzo(a)pyrene (anti- and syn-BPDE) via Method I (9,10). Reagents: (i) NaBH$_4$; (ii) H$^+$; (iii) AgOBz, I$_2$; (iv) NBS; (v) DBN; (vi) NaOMe; (vii) DDQ; (viii) m-CPBA; (ix) DMSO, H$_2$O; (x) t-BuOK.

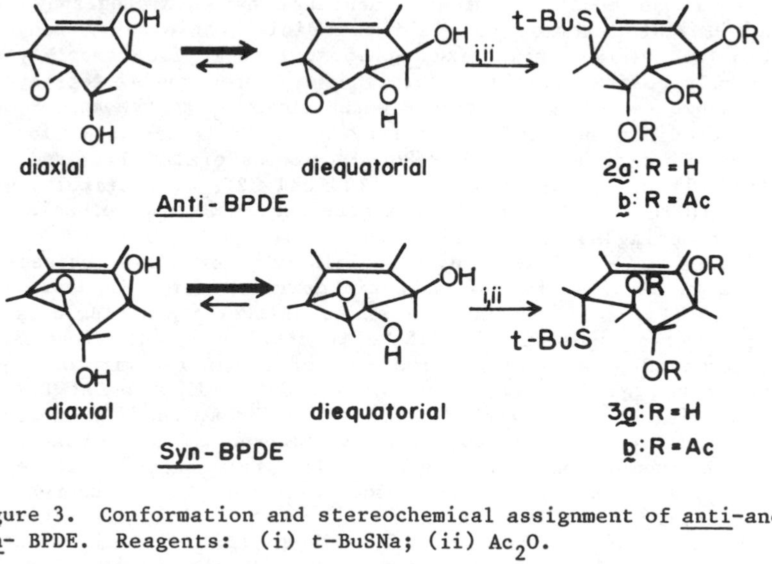

Figure 3.   Conformation and stereochemical assignment of anti-and syn- BPDE.   Reagents: (i) t-BuSNa; (ii) Ac$_2$O.

the 5,6-dihydrodiol, is a K-region dihydrodiol, incapable of forming a diol epoxide. Syntheses of the 1,2-, 3,4-, 8,9-, and 10,11-dihydrodiols of BA were reported by our research group (7,12,29) and by Lehr et al. (30).

These dihydrodiols were synthesized by Lehr and coworkers by a common approach based on Method I employed for the preparation of the BP 7,8-dihydrodiol. The only modification required was the use of diacetate diesters in place of tetrahydrobenzoate derivatives in the preparation of the 1,2- and 3,4-dihydrodiols. While this synthetic approach is relatively straightforward and provides the desired compounds, it suffers from the disadvantage that the ketones required as starting compounds are not readily available and must themselves be synthesized by multistep procedures. Commonly, the method employed is the classical Haworth synthesis (31) entailing Friedel Crafts succinoylation of an aromatic precursor having one less ring, Wolff-Kishner or Clemmensen reduction of the resulting keto acid, and acid-catalyzed cyclization of the arylbutyric acid to a ketone. Limitations of this approach include the inherent problems of controlling the site of succinoylation and the direction of ring closure. The desired isomers are not always favored and mixtures of isomers difficult to separate are frequently obtained.

A more efficient synthetic approach to the BA 1,2-and 3,4-dihydrodiols involves reduction of BA with lithium in ammonia followed by base-catalyzed isomerization of the olefinic product into conjugation (29) (Figure 5). This synthetic approach is designated Method II. Techniques for stepwise reduction of PAH with Li/NH$_3$ were developed previously in our laboratories (32-34). The structures of the products of reactions of this type are predictable by molecular orbital theory (35), and the yields are generally high. Base-catalyzed isomerization of 1,4,7,12-te trahydro-BA gave 1,2,7,12-tetrahydro-BA and 3,4,7,12-tetrahydro-BA in approximately equal ratio. Separation was most conveniently achieved by chromatography following Prevost reaction to yield the diol-dibenzoates. Dehydrogenation of the meso ring with o-chloranil took place smoothly and essentially quantitatively to furnish the corresponding tetrahydrodiol dibenzoates. These diesters were converted to the corresponding dihydrodiols and diol epoxides by the methods described for the analogous BP derivatives (7). Overall yields of the 1,2- and 3,4-dihydrodiols of BA via this route (14% and 32%, respectively) were superior to those obtained by the longer conventional synthesis from anthracene using Haworth cyclization.

The BA 8,9- and 10,11-dihydrodiols were also synthesized directly from BA (7,12) via a synthetic approach which is designated Method III (Figure 6). Hydrogenation of BA over a platinum catalyst took place regiospecifically in the terminal ring to afford 8,9,-10,11-tetrahydro-BA (36,37). Treatment of the latter with one equivalent of DDQ gave a mixture of 8,9-and 10,11-dihydro-BA; Prevost reaction yielded the tetrahydrodiol dibenzoates, readily separable by fractional crystallization or chromatography. It is convenient that both isomers are obtainable in the same reaction sequence, since both are generally required for conversion to the corresponding dihydrodiols and diol epoxides required for biological studies.

Syntheses of the BA 3,4- and 8,9-dihydrodiols through reduction of the corresponding quinones with NaBH$_4$ (Figure 4) has been reported by Oesch (22,23). Attempted preparation of BA 1,2-dihydro-

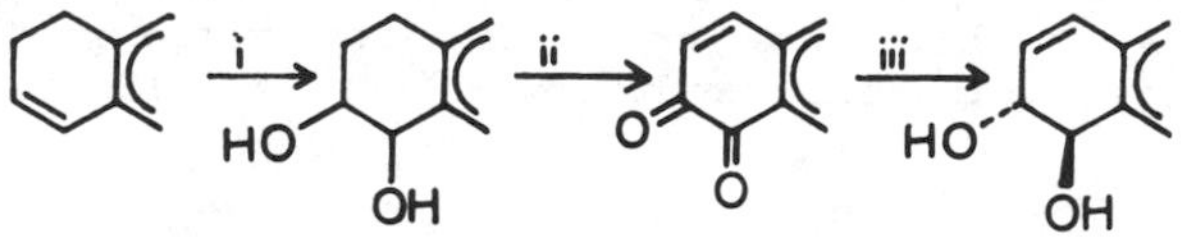

Figure 4. Alternative synthesis of <u>1</u> (<u>18</u>). Reagents: (i) OsO$_4$;
(ii) DDQ; (iii) NaBH$_4$.

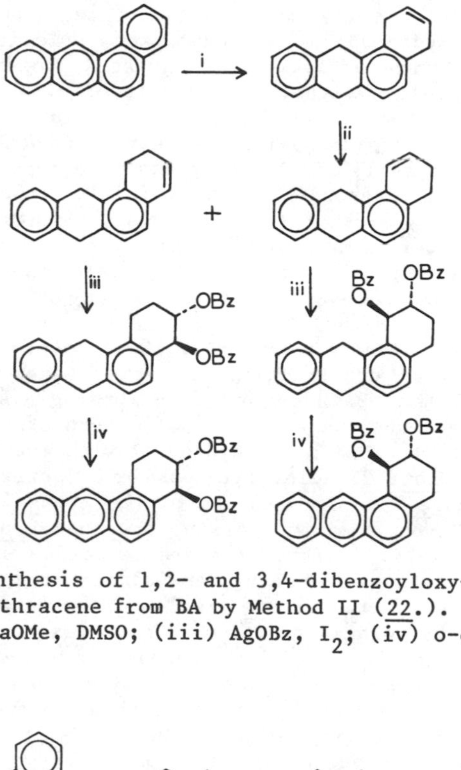

Figure 5.   Synthesis of 1,2- and 3,4-dibenzoyloxy-1,2,3,4-tetra
hydrobenz(a)anthracene from BA by Method II (<u>22</u>.). Reagents: (i)
Li,NH$_3$; (ii) NaOMe, DMSO; (iii) AgOBz, I$_2$; (iv) o-chloranil.

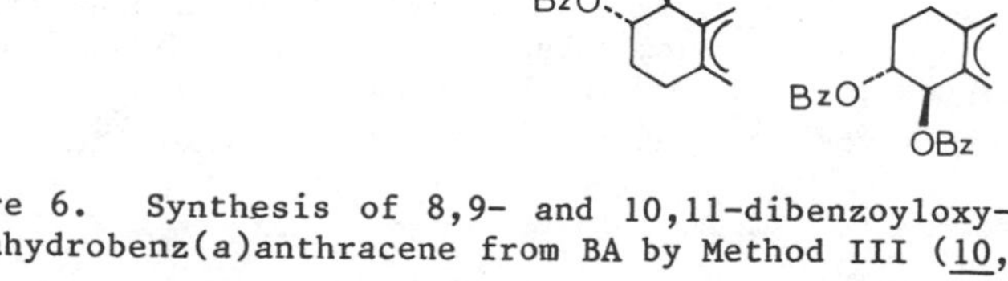

Figure 6.   Synthesis of 8,9- and 10,11-dibenzoyloxy-8,9,10,11-
tetrahydrobenz(a)anthracene from BA by Method III (<u>10,12</u>).

diol by this method gave instead the catechol, 1,2-dihydroxy-BA
(22). Since the quinones required as starting compounds were prepared by multistep syntheses from the related ketones, 4- and 8-oxo-
tetrahydro-BA, which in turn were synthesized by the multistep Haworth procedure, this route would appear to offer little practical
advantage.

Dibenz(a,h)anthracene

Dibenz(a,h)anthracene (DBA) is a weak carcinogen (38) and the first
PAH to be demonstrated to be carcinogenic (39). Syntheses of the
1,2- and 3,4-dihydrodiols of DBA and the corresponding diol epoxides
were accomplished from DBA by Method II (40). Reduction of DBA with
Li/NH$_3$ could not be stopped at the tetrahydro stage, but proceeded to
1,4,7,8,11,14-hexahydro-DBA. Base-catalyzed isomerization of this
diolefin gave a mixture of isomeric conjugated diolefins. Prévost
reaction of this mixture with 2 equivalents of silver benzoate and
iodine followed by dehydrogenation with DDQ furnished the dibenzoate
esters of 1,2-and 3,4-dihydroxy-1,2,3,4-tetrahydro-DBA in 1:9 ratio.
These isomers were readily separable by chromatography and underwent
conversion to the corresponding dihydrodiols by the usual procedures. The overall yields of the 1,2- and 3,4-dihydrodiols of DBA
via this sequence were 11% and 22% respectively, in eight steps.
These dihydrodiols were also individually synthesized from the related ketones, 1-oxo-and 4-oxo-1,2,3,4-tetrahydro-DBA, by appropriate modification of Method I (41). These ketones were themselves
prepared from phenanthrene via 8-oxo-8,9,10,11-tetrahydro-BA by standard methods of ring construction in eight steps each. The net
overall yields of the DBA 1,2- and 3,4-dihydrodiols from phenanthrene by these sequences were <0.5% in fourteen steps each.
    Epoxidation of the DBA 3,4-dihydrodiol with m-chloroperbenzoic
acid provided the anti 3,4-diol-1,2-epoxide (4) stereospecifically
in 85% yield (Figure 7). Epoxidation of the trans 1,2-dihydrodiol
with this peracid gave a mixture of the anti and syn diol epoxides
(5) in 1:3 ratio (40,41). Similar stereoselective syn epoxidation
of the 1,2-dihydrodiol of 7-methylbenz(a)anthracene was also observed (42). These are the first examples of preferential syn diol
epoxide formation by peracid reaction. This unexpected steric preference is associated with the unusual trans-diaxial conformation of
these sterically hindered bay region 1,2-dihydrodiols (43). This ef
fect is interpreted as due to the favorable geometry of an intermediate complex involving association of the peracid with the
benzylic hydroxyl group which directs epoxidation to the same face
of the molecule. In cases where the hydroxyl groups are diequatorial
the allylic hydroxyl group exerts the dominant influence.
    Synthesis of DBA 3,4-dihydrodiol from DBA 3,4-dione via bro-
mination and reduction of the 1,2-dibromo compound with sodium bo-
rohydride has also been described (44). The DBA 3,4-dione was synthesized by Kundu from 5-bromo-2-nitro-1-naphthol in nine steps in
4.2% overall yield (45). More recently, reduction of DBA 3,4-dione
with NaBH$_4$ in the presence of air has been reported to afford directly the 3,4-dihydrodiol (22).

Benzo(e)pyrene

In contrast to BP, benzo(e)pyrene (BeP) is generally considered to
be inactive as a complete carcinogen (38). The only vicinal di-

hydrodiol of BeP capable of forming a diol epoxide is the 9,10-
dihydrodiol (9), and it is not detected as a metabolite of BeP (46).
Nevertheless, it was of interest to synthesize this diol epoxide,
since it was predicted theoretically to be a moderately active carcinogen (47).

Synthetic routes to BeP 9,10-dihydrodiol are based on hexa-
hydropyrene (Figure 8). Conversion of this hydrocarbon to the 9-
keto derivative of octahydro-BeP (7) was readily accomplished by the
Haworth method (48,49). Reduction of 7 with NaBH$_4$ followed by acidic
dehydration and Prévost reaction gave the trans dibenzoate which
underwent dehydrogenation with DDQ to yield the tetrahydrodiol dibenzoate (8). A modification of this synthetic approach involving
aromatization of intermediate 6 prior to cyclization afforded 8 in
somewhat higher overall yield. Attempted introduction of the olefinic bond into 8 with excess DDQ failed, despite the fact that
similar reaction of the analogous tetrahydrodiol dibenzoate of BP
proceeded smoothly. This difference in reactivity results from the
fact that the dibenzoate groups of 8 are forced to adopt the diaxial
conformation due to steric interaction with the adjacent aromatic
ring. Consequently, hydride abstraction from the unsubstituted benzylic position of 8 is effectively blocked. The NMR spectrum of 8
confirms the diaxial orientation of the benzylic groups ($J_{9,10}$=
3.5Hz). Conversion of 8 to 9 was readily achieved by the bro-
mination-dihydrobromination method. Attempted synthesis of 9 by
reduction of BeP 9,10-dione with NaBH$_4$ in ethanol in the presence of
air gave only the corresponding catechol, 1,2-dihydroxy-BeP (22).
Similar reaction conducted under oxygen for a week furnished 9 in
moderate yield (35%).

Epoxidation of 9 with m-chloroperbenzoic acid afforded a mixture of the anti and syn diol epoxides (anti-and syn-BePDE). In our
initial studies only the anti isomer was isolated (48). Subsequently, it was found by Yagi et al. (50) that both diastereomers
are formed. In our experience, the relative ratio of isomers is
dependent upon experimental conditions. This is another example of
lack of stereospecificity of epoxidation of a diaxial dihydrodiol.

Triphenylene

This PAH is a common environmental contaminant. However, it is
inactive as a carcinogen in animal tests (51). The trans-1,2-di-
hydrodiol of triphenylene has been synthesized from phenanthrene by
a route analogous to that employed for the preparation of BeP 9,10-
dihydrodiol (48). Like the latter compound, epoxidation with peracid affords a mixture of the anti and syn diol epoxides (Figure 9)
(48,50).

Phenanthrene (Method IV)

Although phenanthrene is noncarcinogenic, some of its methylated
derivatives exhibit significant activity as mutagens (52,53). The
1,2- and 3,4-dihydrodiols of phenanthrene were first synthesized by
Jerina et al. (54) by a method involving reduction of the corresponding quinones with LiAlH$_4$. However, the yields in the re-

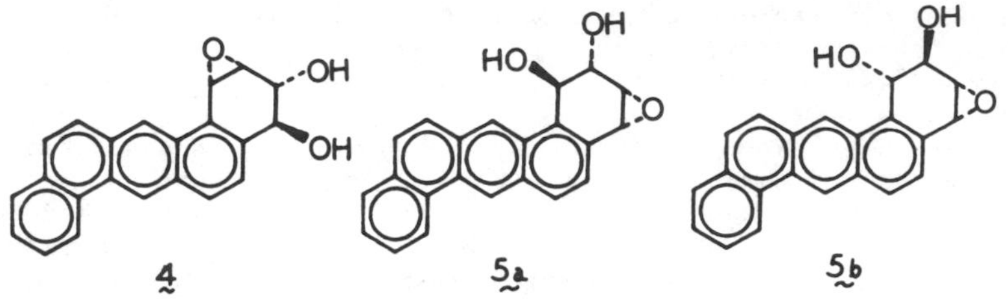

Figure 7. Structures of the DBA anti 3,4-diol-1,2-epoxide (4), anti-1,2-diol-3,4-epoxide (5a), and syn-1,2-diol-3,4-epoxide (5b).

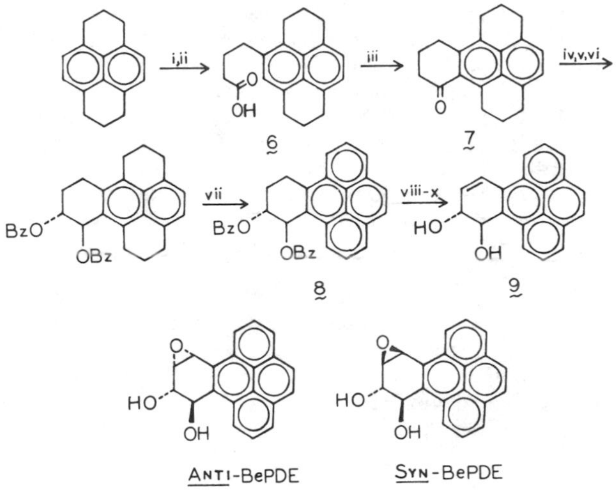

Figure 8. Synthesis of the 9,10-dihydrodiol of BeP (41,42). Reagents: (i) succinic anhydride, AlCl$_3$; (ii)H$_2$NNH$_2$, KOH; (iii) HF; (iv) NaBH$_4$; (v) H$^+$; (vi) AgOBz, I$_2$; (vii) DDQ; (viii) NBS; (ix) DBN; (x) NaOMe.

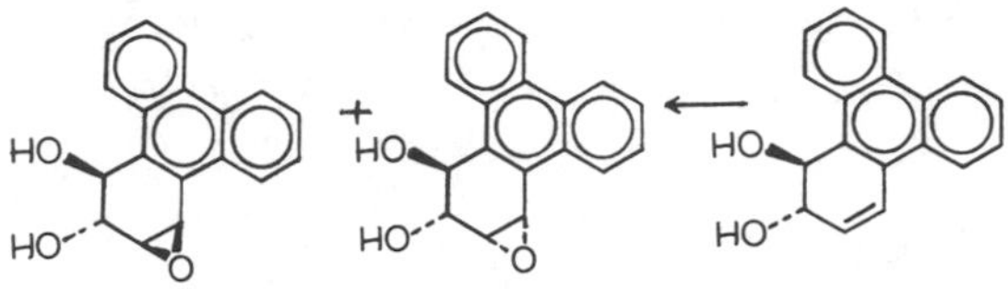

Figure 9. Epoxidation of the 1,2-dihydrodiol of triphenylene yields both anti and syn diol epoxide isomers.

duction step were only 4% and 1%, respectively. Subsequently, these dihydrodiols were synthesized by Lehr et al. (30) by Method I from the corresponding ketones, 1-oxo-and 4-oxo-1,2,3,4-tetrahydrophenanthrene. Subsequent reinvestigation of the quinone reduction route in our laboratories led to development of an improved procedure which afforded substantially improved yields (55). Reduction of the phenanthrene 1,2-and 3,4-diones with LiAlH$_4$ by this method gave the 1,2- and 3,4-dihydrodiols in yields of 46% and 27%, respectively (Figure 10). Further enhancement of these yields was achieved by carrying out these reductions with NaBH$_4$ in ethanol under air (22). It is believed that oxygen serves to reoxidize catechol byproducts to quinones.

The phenanthrene 1,2- and 3,4-diones are synthetically accessible from the related β-phenols. Oxidation of 2-phenanthrol with either Fremy's salt ((KSO$_3$)$_2$NO) or phenylseleninic anhydride gave phenanthrene 1,2-dione directly (55). Unexpectedly, oxidation of 3-phenanthrol with (KSO$_3$)$_2$NO yielded 2,2-dihydroxybenz(e)indan-1,3-dione (Figure 10). However, phenanthrene 3,4-dione was readily obtained from 3-phenanthrol by Fieser's method entailing diazonium coupling, reduction, and oxidation of the resulting 4-amino-3-phenanthrol with chromic acid (56).

The development of satisfactory methods for the stereoselective reduction of terminal ring PAH quinones to trans-dihydrodiols represented a significant advance. Combined with methods for the oxidation of β-phenols to ortho-quinones it provided the basis of what is probably the most general synthetic route to non-K-region dihydrodiols. This synthetic approach is designated Method IV. While reductions with LiAlH$_4$ appear to be generally stereospecific, the reactions with NaBH$_4$ afford somewhat better yields, but show lower stereoselectivity, yielding variable amounts of cis-as well as trans-dihydrodiols.

Synthesis of the anti and syn isomers of the 1,2-diol-3,4-epoxide of phenanthrene by epoxidation of the 1,2-dihydrodiol has been reported by Whalen et al. (57).

Anthracene

Anthracene is noncarcinogenic and is structurally incapable of forming a bay region diol epoxide. Anthracene 1,2-dihydrodiol is most conveniently synthesized from 2-anthranol by oxidation with phenylseleninic anhydride to anthracene 1,2-dione (55) followed by reduction with NaBH$_4$ in ethanol (22) or LiAlH$_4$ (55). Anthracene 1,2-dihydrodiol has also been synthesized via the Prévost reaction route (30).

Chrysene

Chrysene is a weak tumor initiator and is inactive as a complete carcinogen (38). The 1,2-dihydrodiol is more active as a mutagen than the 3,4- or the 5,6-dihydrodiols. The biological data support the hypothesis that the principal active metabolite of chrysene is the bay region anti-1,2-diol-3,4-epoxide (58).

Two synthetic approaches to the 1,2- and 3,4-dihydrodiols of chrysene (10 and 11) have been reported (41,59), and an additional novel method will be described herein (60). Syntheses of 10 and 11

from naphthalene (61) and phenanthrene (62) via 1-oxo- and 4-oxo-
1,2,3,4-tetrahydrochrysene by Method I was described by Karle et al.
(41). A more convenient synthetic route to these dihydrodiols directly from chrysene by Method III has been reported by Fu and Harvey
(59) (Figure 11). Hydrogenation of chrysene over a palladium catalyst afforded regiospecifically 5,6-dihydrochrysene, while similar
reaction over PtO$_2$ gave 1,2,3,4-tetrahydrochrysene, and hydrogenation over a mixed Pd-Pt catalyst furnished 1,2,3,4,5,6-hexahydro-
chrysene (37,59,63). Dehydrogenation of the hexahydrochrysene derivative with DDQ took place regioselectively to yield 3,4,5,6-te-
trahydrochrysene. This olefin underwent smooth transformation to
the 1,2-dihydrodiol (10) via the Prevost reaction, dehydrogenation,
and methanolysis. Although dehydrogenation of 1,2,3,4-tetrahydro-
chrysene could not be stopped at the dihydro stage, partial dehydrogenation was readily achieved by the bromination-dehydrobromination method to yield a mixture of 1,2-and 3,4-dihydrochrysene
from which the 3,4-dihydrodiol (11) was synthesized via the usual
Prévost reaction route.
     Epoxidation of 10 with m-chloroperbenzoic acid yielded the chr-
ysene anti-1,2-diol-3,4-epoxide, whereas similar reaction of 11 gave
a mixture of the corresponding anti and syn diol epoxides in a 5:3
ratio (57,59). These findings are in accord with previous observations that dihydrodiols free to adopt the diequatorial conformation
undergo anti stereospecific epoxidation, whereas bay region diaxial
dihydrodiols yield mixtures of anti and syn diastereomers. The syn-
1,2-diol-3,4-epoxide diastereomer of chrysene was synthesized from
10 via base-catalyzed cyclization of the bromotriol intermediate
(57,60). The optically pure (+) and (-) enantiomers of both the
anti and syn chrysene 1,2-diol-3,4-epoxides have also been prepared
(64).
     An alternative new synthetic approach to chrysene 1,2-dihydro-
diol based on Method IV has recently been developed (60). This
method (Figure 12) entails synthesis of 2-chrysenol via alkylation
of 1-lithio-2,5-dimethoxy-1,4-cyclohexadiene with 2-(1-naphthyl) e-
thyl bromide followed by mild acid treatment to ge nerate the diketone 12. Acid-catalyzed cyclization of 12 gave the unsaturated
tetracyclic ketone 13 which was transformed to 2-chrysenol via dehydrogenation of its enol acetate with o-chloranil followed by hydrolysis. Oxidation of 2-chrysenol with Fremy's salt gave chrysene
1,2-dione which underwent reduction with NaBH$_4$ in the presence of
oxygen to yield 11. This method is readily adaptable to synthesis on
any scale.

Benzo(c)phenanthrene

Benzo(c)phenanthrene (BcP) is exceptionally weak or inactive as a
carcinogen in experimental animals (51). On the other hand, the bay
region anti diol epoxide of BcP (14) exhibits high tumor initiating
activity on mouse skin (65).
     The 3,4-dihydrodiol of BcP was synthesized from 4-oxo-1,2,3,4-
tetrahydro-BcP (15) by Method I (66). The ketone 15 was itself
prepared from 4-oxo-1,2,3,4-tetrahydrophenanthrene via a multistep
sequence entailing Reformatsky reaction with methyl bromocrotonate,
dehydration of the resulting alcohol, isomerization to the arylbutyric acid, and cyclization of its acid chloride with SnCl$_4$. Full

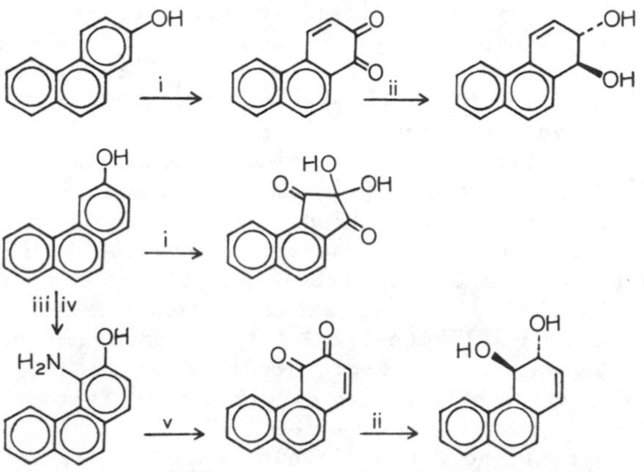

Figure 10. Synthesis of the 1,2- and 3,4-dihydrodiols of phenanthrene from the related phenols by Method IV (47). Reagents: (i) $(KSO_3)_2NO$ or $(PhSeO)_2O$; (ii) $LiAlH_4$ or $NaBH_4$,O_2; (iii) diazonium salt of sulfanilic acid; (iv) $Na_2O_3S_2$; (v) CrO_3.

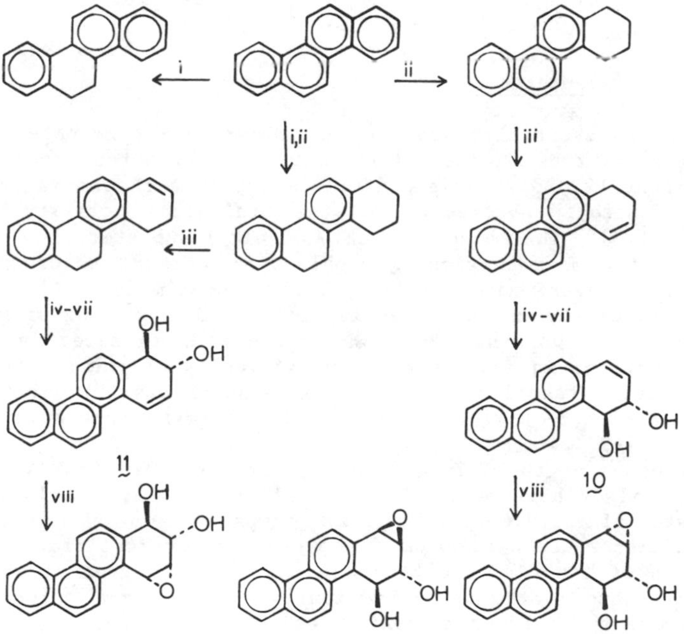

Figure 11. Synthesis of the chrysene 1,2- and 3,4-dihydrodiols and the corresponding diol epoxide derivatives from chrysene by Method III (51). Reagents: (i) H_2,Pd; (ii) H_2,Pt; (iii) DDQ; (iv) $AgOBz,I_2$; (v) NBS; (vi) DBN; (vii) NaOMe; (viii) m-CPBA.

experimental details have not been published. 4-Oxo-1,2,3,4-tetra-
hydrophenanthrene is synthetically accessible from naphthalene via
the Haworth synthesis (67).

    More convenient synthetic access to 15 is provided by the se-
quence in Figure 13 (68). Alkylation of the potassium salt of 2,6-
dimethoxy-1,4-cyclohexadiene with 2-(2-naphthyl)ethyl bromide in
liquid ammonia followed by mild acidic hydrolysis generated the diketone (16). Cyclization of 16 in polyphosphoric acid took place
smoothly in the desired direction to afford the partially saturated
ketone which underwent dehydrogenation with DDQ to 15.

    Synthesis of the 1,2-dihydrodiol of BcP by conventional methods
was blocked by the failure of attempts to synthesize its potential
synthetic precursors 1-keto-1,2,3,4-tetrahydro-BcP and 1,2-dihydro-
BcP (66). However, BcP 1,2-dihydrodiol was obtained in low yield
(~1%) by oxidation of BcP with ascorbic acid-ferrous sulfate (66).

Dibenzo(a,i)pyrene and Dibenzo(a,h)pyrene

These hexacyclic hydrocarbons are generally recognized as two of the
most potent unsubstituted carcinogenic PAH (38). The 3,4-dihydro-
diol of dibenzo(a,i)pyrene (17) and the 1,2-dihydrodiol of dibenzo-
(a,h) pyrene (18) have been synthesized from 4-oxo-1,2,3,4-tetra-
hydrodibenzo(a,i)pyrene and 1-oxo-1,2,3,4-tetrahydrodibenzo(a,h)py-
rene, respectively, by Method I. (69). Treatment of these dihydrodiols with m-chloroperbenzoic acid gave the corresponding anti diol
epoxides (66).

7-Methylbenz(a)anthracene

While BA is essentially inactive as a complete carcinogen, 7-me-
thylbenz(a)anthracene (MBA) exhibits relatively potent activity in
this respect (27,38). This difference typifies the often dramatic
effect of methyl substitution on the biological activity of PAH
compounds (70). The molecular basis of alkyl sub stitution effects
is one of the most intriguing problems in current carcinogenesis
research. However, much less progress has been made in elucidating
the details of the metabolic activation and DNA binding of methyl-
substituted than unsubstituted PAH because of the greater complexity
of their metabolism and the greater difficulty of the synthesis of
their active metabolites. Biological studies have implicated the
anti 3,4-diol-1,2-epoxide of MBA as its ultimate carcinogenic metabolite (71-73).

    Syntheses of the 1,2- and the 3,4-dihydrodiols of MBA via Me-
thods II and IV have been described (74). The 1,2- and 3,4-diol
dibenzoates of 1,2,3,4-tetrahydro-MBA prepared from MBA via the Li/-
NH$_3$ reduction route were readily separable by crystallization. Introduction of the olefinic bond into the 1,2-position of the 3,4-
diol dibenzoate by the usual bromination-dehydrobromination procedure was complicated by the greater facility of bromination by NBS on
the methyl group than the 1-position. This problem was solved (Figure 14) by allowing bromination to proceed to the dibromo stage,
followed by selective reduction of the bromomethyl group with NaBH$_4$
in diglyme. The monobromo derivative underwent dehydrobromination
with an amine base to furnish the 3,4-dihydrodiol dibenzoate ester
(19b), which on treatment with NaOMe in methanol yielded the free

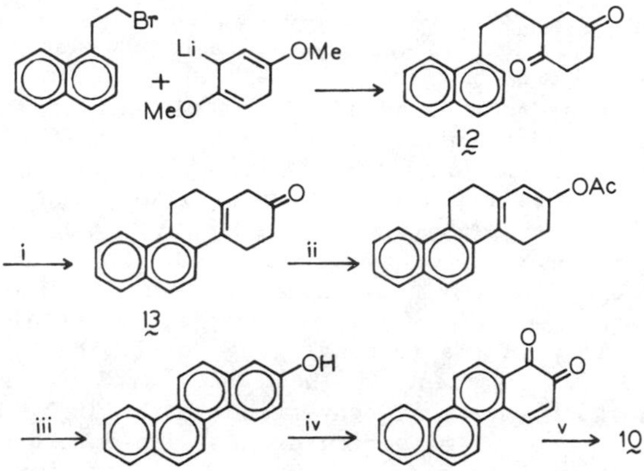

Figure 12. Synthesis of chrysene 1,2-dihydrodiol by Method IV (52). Reagents: (i) H^+; (ii) isopropenyl acetate; (iii) DDQ; (iv) $(KSO_3)_2NO$; (v) $NaBH_4,O_2$.

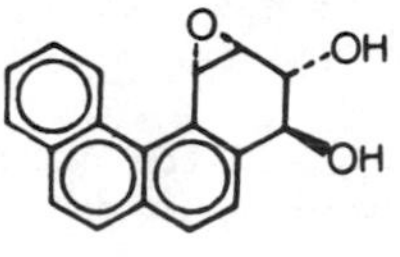

14

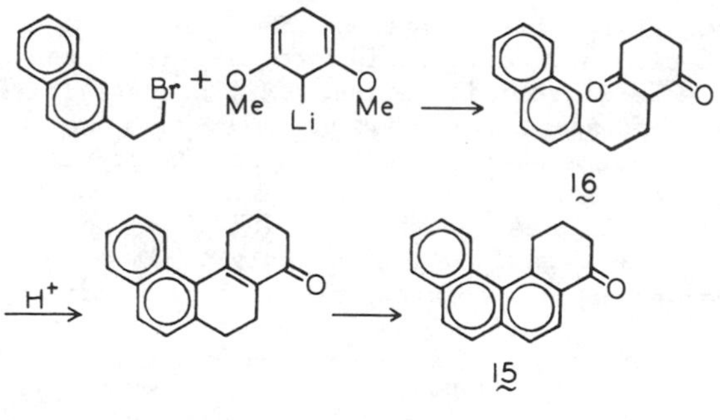

Figure 13. Synthesis of 4-oxo-1,2,3,4-tetrahydrobenzo(c)phenanthrene (15) (59).

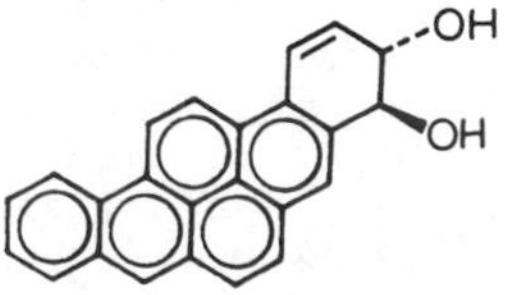

17

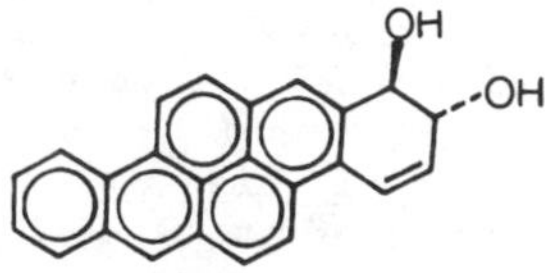

18

3,4-dihydrodiol (19a). A similar sequence of operations on the
1,2,3,4-tetrahydro-MBA 1,2-diol dibenzoate furnished the isomeric
1,2-dihydrodiol of MBA.

The 3,4-dihydrodiol was also synthesized via Method IV (74).
Oxidation of 3-hydroxy-MBA with Fremy's salt gave the 3,4-quinone
which underwent reduction with $LiAlH_4$ to give 19a. The yield in the
reduction step was only 15%, but it is likely that this could be
substantially improved by the use of the $NaBH_4/O_2$ system (18) developed after these studies were completed.

The 10,11-dihydrodiol of MBA was synthesized from MBA by Method
III (12). Hydrogenation of MBA over a platinum catalyst took place
regiospecifically in the terminal ring to provide 8,9,10,11-tetra-
hydro-MBA (75). Treatment of the latter with DDQ furnished 8,9-
dihydro-MBA which underwent conversion to the 10,11-dihydrodiol by
the usual procedures. Oxidation of MBA with ascorbic acid-ferrous
sulfate to afford low yields (< 0.2%) of the five possible dihydrodiols has also been described (76).

Epoxidation of the 3,4-dihydrodiol with m-chloroperbenzoic acid
afforded stereospecifically the corresponding anti diol epoxide
(74). Peracid oxidation of the bay region 1,2-dihydrodiol gave a
mixture of the anti and syn diol epoxide diastereomers. Assignment
of the major isomer as syn was made through analysis of the NMR
spectra of the acetates of the tetraols formed on hydrolysis of the
individual diol epoxides (42). Peracid oxidation of the 10,11-
dihydrodiol is reported to yield the corresponding anti diol epoxide
(12). However, it is likely for steric reasons that the syn isomer
is also formed.

7,12-Dimethylbenz(a)anthracene

7,12-Dimethylbenz(a)anthracene (DMBA) is the most potent carcinogenic PAH commonly employed in carcinogenesis research. Depending
upon its mode of administration, DMBA can selectively induce a diverse range of neoplasmas. For example, oral administration of DMBA
to female Sprague-Dawley rats elicits mammary cancer (77). Intramuscular administration of DMBA to male Long-Evans rats induces local sarcomas (27), and intravenous injection of a liquid emulsion of
DMBA to male or female Long-Evans rats leads to leukemia (78). When
solutions of DMBA are painted on the skins of mice, local skin tumors
are induced (79). Biological studies implicate the 3,4-diol-1,2-
epoxide as the probable active metabolite of DMBA (80-83).

The 3,4- and the 8,9-dihydrodiols of DMBA have been synthesized
by unambiguous routes (55,84). Synthesis of the 3,4-dihydrodiol of
DMBA (20) was accomplished by Method IV (55) (Figure 15). Oxidation
of 3-hydroxy-DMBA with either phenylseleninic anhydride or Fremy's
salt afforded DMBA-3,4-dione. Analogous oxidation of 4-hydroxy-DMBA
gave mainly DMBA 1,4-dione. Reduction of the 3,4-quinone with Li-
AlH_4 in ether yielded 20. However, this reaction is exceptionally
sensitive to traces of moisture, and yields are erratic ranging from
0-45%. More recently, we have discovered that analogous reductions
with $NaBH_4$ in ethanol in the presence of O_2 are more reproducible and
give consistently higher yields of 20 (~80%) (60).

Synthesis of the 8,9-dihydrodiol of DMBA (23) was accomplished
from 10,11-dihydro-DMBA (22) by Method I (84). The olefin 22 was
itself prepared through a synthetic sequence involving Diels-Alder

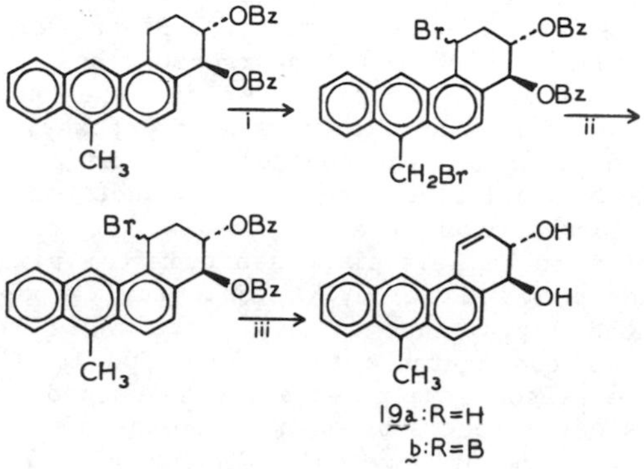

Figure 14.   Synthesis of MBA 3,4-dihydrodiol (19a) (64).   Reagents: (i) NBS; (ii) NaBH$_4$, diglyme; (iii) DBN.

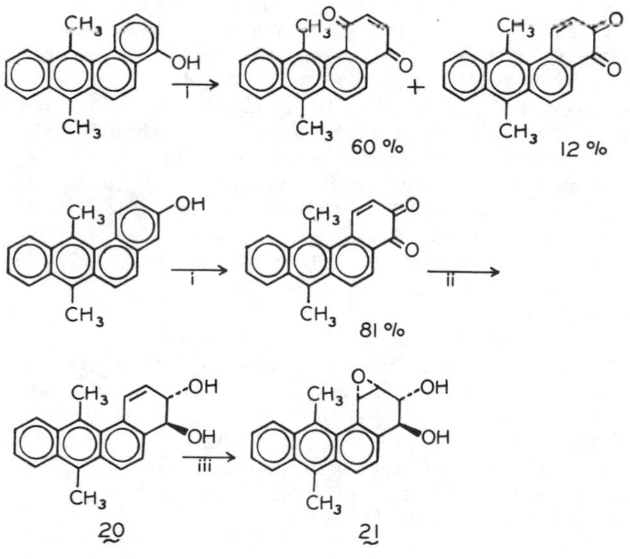

Figure 15.   Synthesis of DMBA 3,4-dihydrodiol (21) by Method IV
(47,75).   Reagents: (i) (PhSeO)$_2$O or (KSO$_3$)$_2$NO; (ii) LiAlH$_4$ or
NaBH$_4$,O$_2$; (iii) m-CPBA.

reaction of 1-methoxybuta-1,3-diene with phenanthrene-1,4-dione, followed by addition of methylmagnesium bromide, hydrogenation, and acid-catalyzed dehydration (Figure 16).

Initial attempts to prepare the anti 3,4-diol-1,2-epoxide of DMBA (21) by direct epoxidation of the 3,4-dihydrodiol of DMBA afforded mixtures of ill-defined products. Since DMBA itself undergoes facile oxidation with m-chloroperbenzoic acid to afford mixtures of meso region oxidation products, it appeared likely that analogous oxidation of 20 was the predominant pathway. On the other hand, when peracid oxidation was conducted in the presence of DNA, there was isolated a small percentage of an adduct tentatively identified as being formed from covalent binding of 21 to DNA (85). Since DMBA is known to be severely distorted from planarity due to steric interaction between the bay region methyl group and the hydrogen atom in the 1-position, it was conceivable that the diol epoxide 21 might be too unstable to isolate due to similar steric strain. For these reasons, we temporarily abandoned our efforts to synthesize 21. However, our subsequent success in synthesizing the bay region diol epoxide of 3-methylcholanthrene (86) which has a reactive meso region and the analogous diol epoxide of 5-methylchrysene (87) which has a bay region methyl group stimulated renewed efforts to synthesize 21. This was achieved by conducting the epoxidation of 20 under mild conditions, monitoring the course of reaction by HPLC, and isolating the diol epoxide from cold solution. Successful synthesis of the syn 3,4-diol-1,2-epoxide of DMBA has subsequently also been accomplished via the bromohydrin intermediate (60). However, both these isomeric diol epoxides have proven relatively unstable, decomposing on standing. The exceptional chemical reactivity of these diol epoxides may be an important determinant of the high tumorigenic potency of DMBA.

In contrast to 21, the diol epoxide derivative of the 8,9-dihydrodiol of DMBA was relatively stable. Although only the anti isomer was isolated and identified from epoxidation of the 8,9-dihydrodiol with m-chloroperbenzoic acid (84), it is likely that the syn isomer may also be formed in this reaction. The 8,9-dihydrodiol exists predominantly in the diaxial conformation as a consequence of steric interaction between the 8-hydroxyl and 7-methyl groups (88).

3-Methylcholanthrene

3-Methylcholanthrene (3-MC) is a potent carcinogen, intermediate in activity between DMBA and BP (27,77). It was first prepared in 1925 by Wieland from desoxycholic acid (89). Biological studies have tentatively identified the 9,10-dihydrodiol (24a) and/or its 1- or 2-hydroxy derivatives (24b and 24c) and the corresponding diol and triol epoxides (25a-c) as the proximate and ultimate carcinogenic forms, respectively, of 3-MC (90-93).

The 9,10-dihydrodiol of 3-MC (24a) was synthesized from 9-hydroxy-3-MC by Method IV (86). Oxidation of this phenol with Fremy's salt in the presence of Adogen 464, a quaternary ammonium phase transfer catalyst, furnished 3-MC 9,10-dione. Reduction of the quinone with $NaBH_4$-O_2 gave pure 24a in good yield. Treatment of 24a with m-chloroperbenzoic acid was monitored by HPLC in order to optimize the yield of the anti diol epoxide (25a) and minimize its decomposition.

9-Hydroxy-3-MC was not conveniently accessible by established methods, necessitating the development of a new synthesis of 3-MC and its derivatives (86,94). This method (Figure 17) is based upon the availability of ortho-lithiated arylamides through directed metalation of arylamides with alkyllithium reagents (95). Condensation of 2,2-dideuterio-4-methylindanone with the 2-lithio salt of N,N-diethyl-1-naphthamide or its 6-methoxy derivative at -60°C afforded smoothly the corresponding lactone. The deuterated ketone was employed to take advantage of the previously demonstrated partial suppression of the competing enolate anion pathway by the deuterium isotope effect (94,96). Reduction of the lactone with zinc and alkali gave the free carboxylic acid which on treatment with ZnCl$_2$ in acetic acid-acetic anhydride cyclized to the 6-acetoxy-3-MC derivative. Complete exchange of deuterium took place during this step. Selective reduction of the 6-acetoxy group was effected by treatment with HI in acetic acid. Reduction was essentially complete in 90 sec.; longer time gave more extensively reduced products. Treatment of 9-methoxy-3-MC with HBr in acetic acid gave 9-hydroxy-3-MC.

Recently resolution of the (+) and (−) enantiomers of 24a has been accomplished by reaction of 24a with (−)-menthoxyacetyl chloride and HPLC separation of the bis menthoxyacetate esters (97). Epoxidation of (+)-and (−)-24a provided optically pure (−)- and (+)-25a, respectively (97).

5-Methylchrysene

In contrast to chrysene and other monomethylchrysene isomers, which exhibit only minimal tumorigenic activity, 5-methylchrysene (5-MC) exhibits carcinogenic potency equal to that of BP (98). 5-MC, which is a component of tobacco smoke, is also more mutagenic toward Salmonella typhimurium than chrysene and the other methyl-chrysene isomers (99,100). Evidence from biological studies has implicated the anti-1,2-diol-3,4-epoxide metabolite (26), in which the epoxide ring in in the same bay region as the methyl group, as the principal active form in vivo (99,101,102). Although the anti-7,8-diol-9,10-epoxide (27), which is also a bay region diol epoxide, is formed metabolically, it appears to be less important biologically.

Syntheses of the 1,2- and 7,8-dihydrodiols of 5-MC have been described (60,87,103). The 1,2-dihydrodiol (30a) is most efficiently prepared from 1-hydroxy-5-MC (29a) by Method IV (87), with the difference that an α-phenol is employed rather than a β-phenol as in previous examples. Oxidation of 29a with (KSO$_3$)$_2$NO furnished a single isomeric quinone identified as 5-MC 1,2-dione (Figure 18). Formation of an ortho rather than a para quinone in the oxidation of an α-phenol with Fremy's reagent is unusual. Apparently the bay region methyl group serves to sterically block oxidative attack in the adjacent bay region site. Reduction of 5-MC 1,2-dione with NaBH$_4$ in ethanol gave 5-MC 1,2-dihydrodiol.

1-Hydroxy-5-MC (29a) was itself prepared (87) by a modification of the synthetic sequence employed in the preparation of 2-chrysenol (Figure 13). Reaction of 2-(1-(3-methylnaphthyl) ethyl bromide with the 1-lithio salt of 1,5-dimethoxycyclohexa-1,4-diene furnished the diketone 28a (Figure 18). Cyclization of the latter in polyphosphoric acid afforded the 1-keto derivative which underwent dehydrogenation over a palladium catalyst to yield 29.

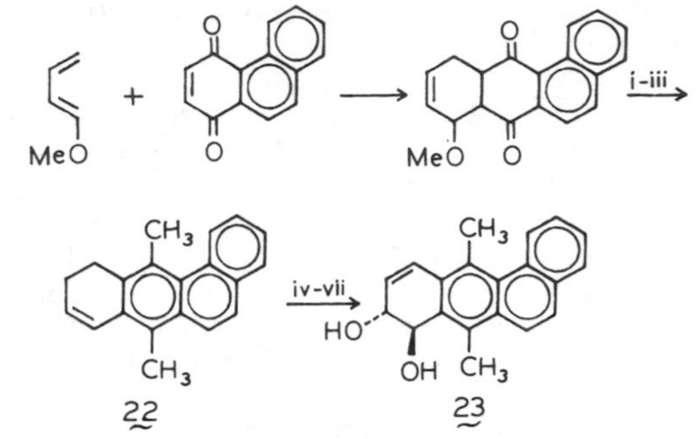

Figure 16.   Synthesis of DMBA 8,9-dihydrodiol (23) (74).   Reagents: (i) H$_2$,Pd; (ii) CH$_3$MgBr; (iii) H$^+$; (iv) AgOBz,I$_2$; (v)
NBS; (vi) DBN; (vii) NaOMe.

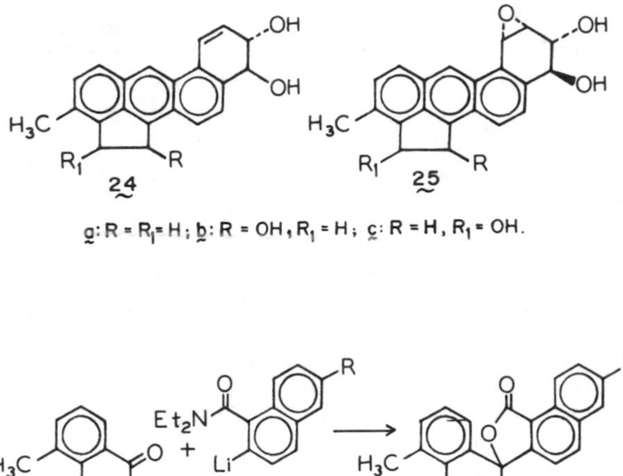

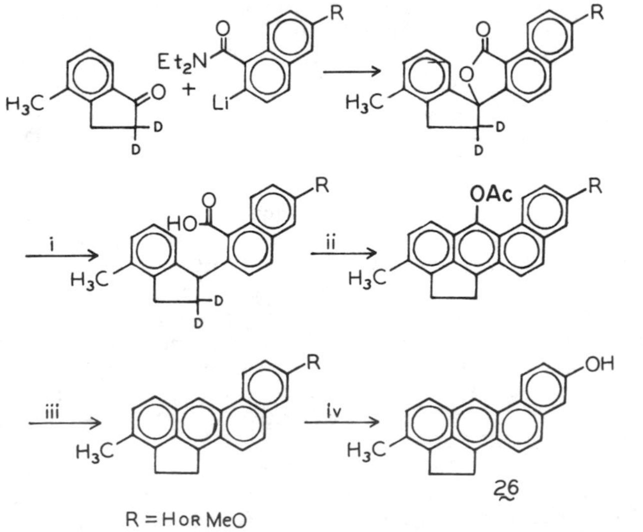

a: R = R$_1$= H; b: R = OH, R$_1$ = H; c: R = H, R$_1$ = OH.

R = H or MeO

Figure 17.   Synthesis of 9-hydroxy-3-MC (85).   Reagents: (i)
Zn,NaOH; (ii) ZnCl$_2$, HOAc, Ac$_2$O; (iii) HI, HOAc; (iv) HBr, HOAc.

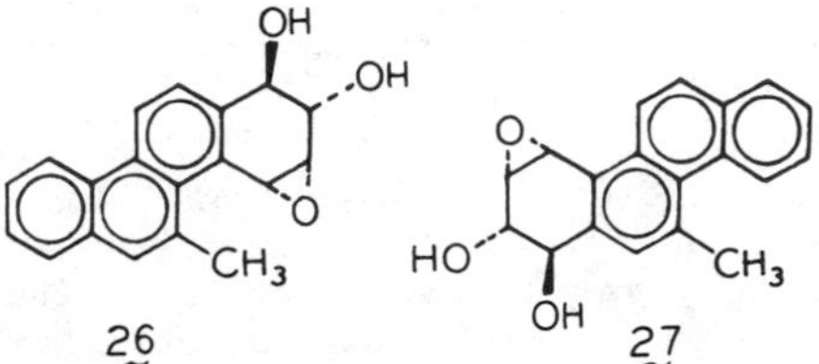

26 27

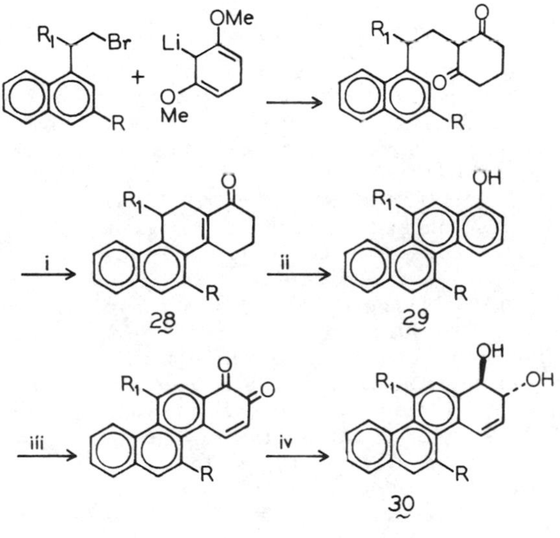

a: $R = CH_3$, $R_1 = H$. b: $R = H$; $R_1 = CH_3$.

Figure 18. Synthesis of 5-MC 1,2- and 7,8-dihydrodiols by Method IV (52,93). Reagents: (i) polyphosphoric acid; (ii) palladium; (iii) $(KSO_3)_2NO$; (iv) $NaBH_4$, O_2.

The 7,8-dihydrodiol of 5-MC (30b) was prepared from 8-hydroxy-
5-MC by Method IV (60,103). It was necessary to utilize the β-rather
than the α-phenol (i.e. 8-HO-5-MC rather than 7-HO-5-MC), since oxidation of the latter with Fremy's salt was anticipated to take place
predominantly in the para position. The 8-hydroxy-5-MC was prepared
in our laboratories by a modification of the procedure in Figure 18.
An alternative preparation of 8-hydroxy-5-MC via photocyclization of
methyl 3-phenyl-2-(1-(6-methoxynaphthyl) propenoate has also recently been described (103).

Epoxidation of the 1,2- and 7,8-dihydrodiols of 5-MC with m-
chloroperbenzoic acid furnished the corresponding anti diol epoxides
26 and 27. Compound 26 was the first diol epoxide bearing a methyl
group in the same bay region as the epoxide function to be synthesized. While the diol epoxide 26 is relatively reactive (104), it
is more stable than the structurally analogous DMBA 1,2-diol-3,4-
epoxide (21); it was obtained as a white crystalline solid.

The related syn diol epoxide isomers were synthesized from the
1,2- and 7,8-dihydrodiols of 5-MC by reaction with N-bromoacetamide
in moist DMSO followed by base-catalyzed cyclization by the usual
Method I procedures (60).

Fluoranthene

Fluoranthene is one of the more prevalent PAH in the human environment. Although fluoranthene is not active as a carcinogen, its
2- and 3-methyl derivatives have been shown to be active as tumor
initiators (105). The major mutagenic metabolite of fluoranthene in
the Ames assay has been identified as the 2,3-dihydrodiol (31)
(106). Synthesis of 31 by Method I (107,108) and its conversion to
the related anti and syn diol epoxide derivatives (32,33) has been
reported (108). The isomeric trans-1,10b-dihydrodiol (37) and the
corresponding anti and syn diol epoxide isomers (38,39) have also
been prepared (108) (Figure 19). Synthesis of 37 from 2,3-dihydro-
fluoranthene (109) could not be accomplished by Prévost oxidation.
An alternative route involving conversion of 2,3-dihydrofluoranthene
to the cis-tetrahydrodiol (34) with OsO_4 followed by dehydration,
silylation, and oxidation with peracid gave the α-hydroxyketone 35.
The trimethylsilyl ether derivative of the latter underwent stereoselective phenylselenylation to yield 36. Reduction of 36 with
$LiAlH_4$ followed by oxidative elimination of the selenide function
afforded 37. Epoxidation of 37 with t-BuOOH/VO(acac)$_2$ and de-
silylation gave 38, while epoxidation of the acetate of 37 and deacetylation furnished 39.

Benzofluoranthenes

Benzo(b)-, benzo(k)-, and benzo(j)fluoranthene are common environmental contaminants (38). The tumor-initiating activity of benzo-
(b)fluoranthene on mouse skin is about equal to that of DBA (38).
All three isomers are mutagenic in the Ames assay (110). Syntheses
of the 1,2-, 9,10-, and 11,12-dihydrodiols of benzo(b)fluoranthene
by Method I have been reported (111,112).

Dibenzo(a,e)fluoranthene

This PAH is a moderately potent carcinogen which produces sarcomas
at the site of injection (51). Metabolism studies suggest that the

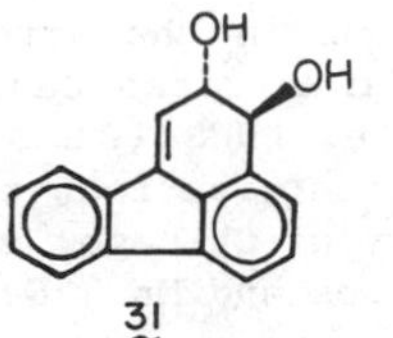

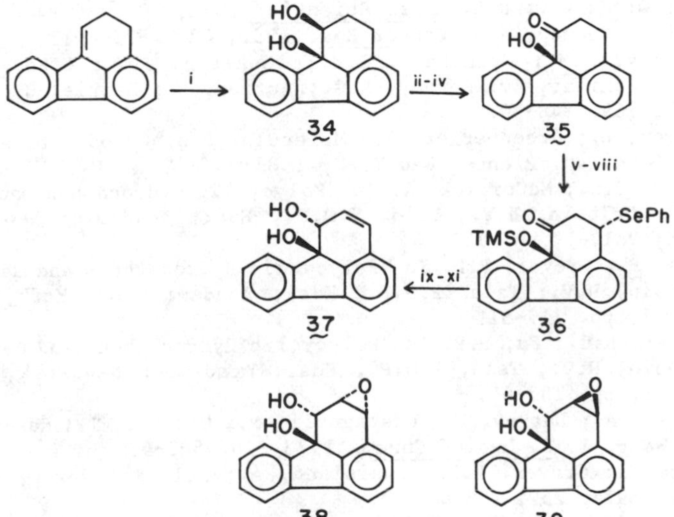

Figure 19. Synthesis of the fluoranthene 1,10b-dihydrodiol (37) and the corresponding anti and syn diol epoxide isomers (38 and 39). Reagents: (i) OsO$_4$; (ii) H$^+$, (iii) Me$_3$SiI; (iv) m-CPBA; (v) Me$_3$SiCl; (vi) LDA; (vii) Me$_3$SiCl; (viii) PhSeCl; (ix) LiAlH$_4$; (x) H$_2$O$_2$; (xi) K$_2$CO$_3$.

principal metabolites of dibenzo(a,e)fluoranthene which bind covalently to DNA are the diol epoxides formed from the 3,4-and 12,13-dihydrodiols (113). While chemical synthesis of these dihydrodiols have not been reported, their diol epoxide derivatives have been prepared by treatment of the dihydrodiols obtained by metabolic oxidation with m-chloroperbenzoic acid (114).

Acknowledgments

Investigations carried out in the author's laboratories were supported by grants from the American Cancer Society (BC-132) and the National Cancer Institute, DHHS (CA 36097). I also wish to acknowledge the important contributions to our recent work by my colleagues at the University of Chicago, especially Drs. Hongmee Lee, John Pataki, Stephen Jacobs and Ms. Cecilia Cortez.

Literature Cited

1. Review: Harvey, R.G. Amer. Scientist 1982, 70, 386-93.
2. Review: Conney, A.H. Cancer Res. 1982, 42, 4875-4917.
3. Jeffrey, A.M.; Weinstein, I.B.; Jennette, J.W.; Grzeskowiak, K.; Nakanishi, K.; Harvey, R.G.; Autrup, H.; Harris, C. Nature 1977, 269, 348-50.
4. Singer, B.; Greenberger, D. "Molecular Biology of Mutagens and Carcinogens"; Plenum: New York, 1983; p. 143.
5. Maher, V.M.; McCormick, J. In "Polycyclic Hydrocarbons and Cancer"; Gelboin, H.V.; Ts'o, P.O.P., Eds.; Academic: New York, 1978; Vol. 2, pp. 137-74.
6. Sims, P.; Grover, P.L. In "Polycyclic Hydrocarbons and Cancer"; Gelboin, H.V.; Ts'o, P.O.P., Eds.; Academic: New York, 1981; Vol. 3, pp. 117-81.
7. Harvey, R.G.; Fu, P.P. In "Polycyclic Hydrocarbons and Cancer"; Gelboin, H.V.; Ts'o, P.O.P., Eds.; Academic: New York, 1978; Vol. 1, pp. 133-65.
8. Borgen, A.; Darvey, H.; Castagnoli, N.; Crocker, T.; Rasmussen, R.; Wang, I. J. Medic. Chem. 1973, 16, 502-6.
9. Sims, P.; Grover, P.L.; Swaisland, A.; Pal, K.; Hewer, A. Nature, 1974, 252, 326-8.
10. Beland, F.A.; Harvey, R.G. J.C.S. Chem. Commun. 1976, 84-5.
11. McCaustland D.J.; Engel, J.F. Tetrahedron Lett. 1975, 2549-52.
12. Fu, P.P.; Harvey, R.G. Tetrahedron Lett. 1977, 2057-62.
13. Yagi, H.; Hernandez, O.; Jerina, D.M. J. Amer. Chem. Soc. 1975, 97, 6881-3.
14. Keller, J.W.; Heidelberger, C.; Beland, F.A.; Harvey, R.G. J. Amer. Chem. Soc. 1976, 98, 8276-7.
15. Yagi, H.; Thakker, D.R.; Hernandez, O.; Koreeda, M.; Jerina, D.M. J. Amer. Chem. Soc. 1977, 99, 1604-11.
16. Yang, S.K.; McCourt, D.M.; Gelboin, H.V. J. Amer. Chem. Soc. 1977, 99, 5130-4.
17. Yang, S.K.; McCourt, D.M.; Gelboin, H.V.; Miller, J.R.; Roller, P.P. J. Amer. Chem. Soc. 1977, 99, 5124-30.
18. Harvey, R.G.; Cho, H. Anal. Biochem. 1977, 80, 540-6.

19. Yagi, H.; Akagi, H.; Thakker, D.R.; Mah, H.; Koreeda, M.; Jerina, D.M. J. Am. Chem. Soc. 1976, 99, 2358-9.
20. Yang, S.K.; Gelboin, H.V.; Weber, J.D.; Fischer, D.L.; Sankaran, V.; Engle, J. F. Anal. Biochem. 1977, 78, 520-6.
21. Nakanishi, K.; Kasai, H.; Cho, H.; Harvey, R.G.; Jeffrey, A.M.; Jennette, K.W.; Weinstein, I.B. J. Am. Chem. Soc. 1977, 99, 258-60.
22. Platt, K.L.; Oesch, F. J. Org. Chem. 1983, 48, 265-8.
23. Platt, K.L.; Oesch, F. Tetrahedron Lett. 1982, 23, 163-6.
24. Fu, P.P.; Lai, C.; and Yang, S.K. J. Org. Chem. 1981, 46, 220-2.
25. Konieczny, M. and Harvey, R.G. Carcinogenesis 1982, 3, 573-5.
26. Lee, H.; Sheth, J.; and Harvey, R.G. Carcinogenesis 1983, 4, 1297-9.
27. Huggins, C.B.; Pataki, J.; Harvey, R.G. Proc. Natl. Acad. Sci. USA 1967, 58, 2253-60.
28. Scribner, J.D. J. Natl. Cancer Inst. 1973, 50, 1717-9.
29. Harvey, R.G.; Sukumaran, K.B. Tetrahedron Lett., 1977, 2387-90.
30. Lehr, R.E.; Schaeffer-Ridder, M.; Jerina, D.M. J. Org. Chem. 1977, 42, 736-44.
31. Berliner, E. Organic Reactions 1949, 5, 229-89.
32. Harvey, R.G. Synthesis 1970, 161-72.
33. Harvey, R.G.; Urberg, K. J. Org. Chem. 1968, 33, 2206-11.
34. Harvey, R.G.; Fu, P.P.; Rabideau, P.W. J. Org. Chem. 1976, 41, 2706-10.
35. Harvey, R.G. Synthesis 1970, 161-72.
36. Fu, P.P.; Harvey, R.G. Tetrahedron Lett. 1977, 415-8.
37. Fu, P.P.; Lee, H.M.; Harvey, R.G. J. Org. Chem. 1980, 45, 2797-2803.
38. "Monograph on the Evaluation of Carcinogenic Risks of the Chemical to Man: Certain Polycyclic Aromatic Hydrocarbons and Heterocyclic Compounds", Internat. Agency Res. Cancer, World Health Organization, 1973; Vol. 3.
39. Kenneway, E.L. Biochem. J. 1930, 24, 497-504.
40. Lee, H.M.; Harvey, R.G. J. Org. Chem. 1980, 45, 588-92.
41. Karle, J.M.; Mah, H.D.; Jerina, D.M.; Yagi, H. Tetrahedron Lett. 1977, 4021-4024.
42. Lee, H.; Harvey, R.G. Tetrahedron Lett. 1981, 1657-60.
43. Zacharias, D.; Glusker, J.P.; Fu, P.P.; Harvey, R.G. J. Amer. Chem. Soc. 1979, 101, 4043-51.
44. Kundu, N.G. J.C.S. Chem. Commun. 1979, 564-5.
45. Kundu, N.G. J. Org. Chem. 1979, 44, 3086-8.
46. MacLeod, M.C.; Cohen, G.M.; Selkirk, J.K. Cancer Res. 1979, 39, 3463-70.
47. Jerina, D.M.; Lehr, R.E.; Yagi, H.; Hernandez, O.; Dansette, P.M.; Wislocki, P.G.; Wood, A.W.; Chang, R.L.; Levin, W.; Conney, A.H. In "In Vitro Activation in Mutagenesis Testing"; De Serres, F.J.; Bend, J.R.; Philpot, R.M.; Eds.; Elsevier: Amsterdam, 1976.
48. Harvey, R.G.; Lee, H.M.; Shyamasundar, N. J. Org. Chem. 1979, 44, 78-83; correction 1005.
49. Lehr, R.E.; Taylor, C.W.; Kumar, S.; Mah, H.D.; Jerina, D.M. J. Org. Chem. 1978, 43, 3462-6.
50. Yagi, H.; Thakker, D.R.; Lehr, R.E.; Jerina, D.M. J. Org. Chem. 1979, 44, 3439-42.

51. "IARC Monographs on the Evaluation of the Carcinogenic Risk of
     Chemicals to Humans", Internat. Agency Res. Center, World
     Health Organization, 1983; Vol. 32.
52. Bucker, M.; Glatt, H.R.; Platt, K.L.; Avnir, D.; Ittak, Y.;
     Blum, J.; Oesch, F. Mutat. Res. 1979, 66, 337-48.
53. LaVoie, E.; Tulley, L.; Bedenko, V.; Hoffmann, D. In "Poly-
     nuclear Aromatic Hydrocarbons: Chemistry and Biological Ef-
     fects"; Bjorseth, A.; Dennis, A.J., Eds.; Battelle: Columbus,
     Ohio, 1979; p. 1041.
54. Jerina, D.M.; Selander, H.; Yagi, H.; Wells, M.C.; Davey, J.F.;
     Mahadevan, V.; Gibson, D.T. J. Amer. Chem. Soc. 1976, 98, 5988-
     96.
55. Sukumaran, K.B.; Harvey, R.G. J. Org. Chem. 1980, 45, 4407-13.
56. Fieser, L.F. J. Amer. Chem. Soc. 1929, 51, 940-52.
57. Whalen, D.L.; Ross, A.M.; Yagi, H.; Karle, J.M.; Jerina, D.M.
     J. Amer. Chem. Soc. 1978, 100, 5218-21.
58. Sims, P.; Grover, P.L. In "Polycyclic Hydrocarbons and Cancer";
     Gelboin, H.V.; Ts'o, P.O.P., Eds.; Academic: New York, 1983;
     Vol. 3, pp. 117-81.
59. Fu, P.P.; Harvey, R.G. J. Org. Chem. 1979, 44, 3778-84.
60. Harvey, R.G.; Pataki, J.P.; Lee, H. In "Polynuclear Aromatic
     Hydrocarbons," IXth International Symposium, Cooke, M.; Dennis,
     A.J., Eds.; Battelle: Columus, Ohio, in press.
61. Cook, J.W.; Schoental, R. J. Chem. Soc. 1945, 288-43.
62. Bachmann, W.E.; Struve, W.S. J. Org. Chem. 1939, 4, 456-63.
63. Fu, P.P.; Harvey, R.G. J.C.S. Chem. Commun. 1978, 585-6.
64. Yagi, H.; Vyas, K.P.; Tada, M.; Thakker, D.R.; Jerina, D.M. J.
     Org. Chem. 1982, 47, 1110-17.
65. Wood, A.W.; Chang, R.L.; Levin, W.; Ryan, D.E.; Thomas, P.E.;
     Croisy-Delcey, M.; Ittah, Y.; Yagi, H.; Jerina, D.M.; Conney,
     A.H. Cancer Res. 1980, 40, 2876-83.
66. Croisy-Delcey, M.; Ittah, Y.; Jerina, D.M. Tetrahedron Lett.
     1979, 2849-52.
67. Haworth, R.D. J. Chem. Soc. 1932, 1125-1133.
68. Pataki, J.; Harvey, R.G. J. Org. Chem. 1982, 47, 20-2.
69. Lehr, R.E.; Kumar, S.; Cohenour, P.T.; Jerina, D.M. Tetra-
     hedron Lett. 1979, 3819-22.
70. Dipple, A. In "Chemical Carcinogens"; Searle, C.E., Ed.; ACS
     Monograph 173, American Chemical Society: Washington, D.C.,
     1976; pp. 245-314.
71. Slaga, T.J.; Gleason, G.L.; Mills, G.; Ewald, L.; Fu, P.P.;
     Lee, H.M.; Harvey, R.G. Cancer Res. 1980, 40, 1981-4.
72. Vigny, P.; Duquesne, M.; Coulomb, H.; LaCombe, C.; Tierney, B.;
     Grover, P.L.; Sims, P. FEBS Lett. 1977, 75, 9-12.
73. Malaveille, C.; Tierney, B.; Grover, P.L.; Sims, P.; Bartsch,
     H. Biochem. Biophys. Res. Commun. 1977, 75, 427-33.
74. Lee, H.M.; Harvey, R.G. J. Org. Chem. 1979, 44, 4948-53.
75. Fu, P.P.; Lee, H.M.; Harvey, R.G. Tetrahedron Lett. 1978, 551-
     4.
76. Tierney, B.; Abercrombie, B.; Walsh, C.; Hewer, A.; Grover,
     P.L.; Sims, P. Chem.-Biol. Interact. 1978, 21, 289-98.
77. Pataki, J.; Huggins, C.B. Cancer Res. 1969, 29, 506-9.
78. Huggins, C.B.; Sugiyama, T. Proc. Natl. Acad. Sci. USA 1966,
     55, 74-81.

79. DiGiovanni, J.; Diamond, L.; Harvey, R.G.; Slaga, T.J. Carcinogenesis 1983, 4, 403-7.

80. Slaga, T.J.; Gleason, G.; DiGiovanni, J.; Sukumaran, K.B.; Harvey, R.G. Cancer Res. 1979, 39, 1934-6.

81. Ivanovic, V.; Geacintov, N.E.; Jeffrey, A.M.; Fu, P.P.; Harvey, R.G.; Weinstein, I.B. Cancer Lett. 1978, 4, 131-40.

82. Bigger, C.A.; Sawicki, J.T.; Blake, D.M.; Raymond, L.G.; Dipple, A. Cancer Res. 1983, 43, 5647-51.

83. Malaveille, C.; Bartsch, H.; Tierney, B.; Grover, P.L.; Sims, P. Biochem. Biophys. Res. Commun. 1978, 83, 1468-73.

84. Harvey, R.G.; Fu, P.P.; Cortez, C.; Pataki, J. Tetrahedron Lett. 1977, 3533-6.

85. Jeffrey, A.M.; Weinstein, I.B.; Harvey, R.G. Proc. Am. Assoc. Cancer Res. 1979, 20, 131.

86. Jacobs, S.; Cortez, C.; Harvey, R.G. Carcinogenesis 1983, 4, 519-22.

87. Pataki, J.; Lee, H.; Harvey, R.G. Carcinogenesis 1983, 4, 399-402.

88. Tierney, B.; Hewer, A.; MacNicoll, A.D.; Gervasi, P.G.; Rattle, H.; Walsh, C.; Grover, P.L.; Sims, P. Chem.-Biol. Interact. 1978, 23, 243-   .

89. Wieland,   ; Schlichting,   Z. Physiol. Chem. 1925, 150, 267-
        .

90. King, H.W.S.; Osborne, M.R.; Brookes, P. Int. J. Cancer 1977, 20, 564-71.

91. Thakker, D.R.; Levin, W.; Wood, A.W.; Conney, A. H.; Stoming, T.A.; Jerina, D.M. J. Am. Chem. Soc. 1978, 100, 645-7.

92. Malaveille, C.; Bartsch, H.; Marquardt, H.; Baker, S.; Tierney, B.; Hewer, A.; Grover, P.L.; Sims, P. Biochem. Biophys. Res. Commun. 1978, 85, 1568-74.

93. Levin, W.; Buening, M.K.; Wood, A.W.; Chang, R. L.; Thakker, D.R.; Jerina, D.M.; Conney, A.H. Cancer Res. 1979, 39, 3549-53.

94. Harvey, R.G.; Cortez, C.; Jacobs, S. J. Org. Chem. 1982, 47, 2120-5.

95. Beak, P.; Brown, R.A. J. Org. Chem. 1977, 42, 1823-4.

96. Jacobs, S.; Cortez, C.; Harvey, R.G. J.C.S. Chem. Commun. 1981, 1215-6.

97. Lee, H.; Harvey, R.G. unpublished data.

98. Hecht, S.S.; Loy, M.; Maconpot, R.R.; Hoffmann, D. Cancer Lett. 1976, 1, 147-54.

99. Hecht, S.S.; LaVoie, E.; Mazzrese, R.; Amin, S.; Bedenko, V.; Hoffmann, D. Cancer Res. 1978, 38, 2191-4.

100. Coombs, M.M.; Dixon, C.; Kissonerghis, A.M. Cancer Res. 1976, 36, 4526-9.

101. Hecht, S.S.; Rivenson, A.; Hoffmann, D. Cancer Res. 1980, 40, 1396-9.

102. Melikian, A.; Amin, S.; Hecht, S.S.; Hoffmann, D.; Pataki, J.; Harvey, R.G. Cancer Res. 1984, 44, 2524-9.

103. Amin, S.; Camanzo, J.; Huie, K.; Hecht, S.S. J. Org. Chem. 1984, 49, 381-4.

104. Hecht, S.S.; Radock, P.; Amin, S.; Huie, K.; Melikian, A.A.; Hoffmann, D.; Pataki, J.; Harvey, R.G. Carcinogenesis, in press.

105. Hoffmann, D.; Rathkamp, G.; Nesnow, S.; Wynder, E.L. J. Natl. Cancer Inst. 1972, 49, 1165-75.
106. LaVoie, E.J.; Hecht, S.S.; Bedenko, V.; Hoffmann, D. Carcinogenesis 1982, 3, 841-6.
107. Rice, J.E.; LaVoie, E.J.; Hoffmann, D. J. Org. Chem. 1983, 48, 2360-3.
108. Rastetter, W.H.; Nachbar, Jr., R.B.; Russo-Rodriguez, S.; Wattley, R.V.; Thilly, W.G.; Andon, B.M.; Jorgensen, W.L.; Ibrahim, M. J. Org. Chem. 1982, 47, 4873-8.
109. Harvey, R.G.; Lindow, D.F.; Rabideau, P.W. Tetrahedron 1972, 28, 2909-19.
110. Hecht, S.S.; LaVoie, E.; Amin, S.; Bedenko, V.; Hoffmann, D. In "Polynuclear Aromatic Hydrocarbons: Chemistry and Biological Effects"; Bjorseth, A.; Dennis, A.J., Eds.; Battelle: Columbus, OH, 1980; p. 417.
111. Amin, S.; Bedenko, V.; LaVoie, E.; Hecht, S.S.; Hoffmann, D. J. Org. Chem. 1981, 46, 2573-8.
112. Amin, S.; Hussain, N.; Brielmann, H.; Hecht, S.S. J. Org. Chem. 1984, 49, 1091-5.
113. Perin-Roussel, O.; Saguem, S.; Ekert, B.; Zajdela, F. Carcinogenesis 1983, 4, 27-32.
114. Perin-Roussel, O.; Ekert, B.; Zajdela, F. Chem.-Biol. Interact. 1981, 37, 109-22.

RECEIVED April 8, 1985

The Bay Region Theory of Polycyclic Aromatic Hydrocarbon Carcinogenesis

ROLAND E. LEHR[1], SUBODH KUMAR[1], WAYNE LEVIN[2], ALEXANDER W. WOOD[2], RICHARD L. CHANG[2], ALLAN H. CONNEY[2], HARUHIKO YAGI[3], JANE M. SAYER[3], and DONALD M. JERINA[3]

[1]Department of Chemistry, University of Oklahoma, Norman, OK 73019
[2]Department of Experimental Carcinogenesis and Metabolism, Hoffmann-La Roche, Inc., Nutley, NJ 07110
[3]Laboratory of Bioorganic Chemistry, National Institute of Arthritis, Diabetes, Digestive, and Kidney Diseases, National Institutes of Health, Bethesda, MD 20205

The hydrolysis rates, mutagenicities and tumorigenicities of a variety of diol epoxides and tetrahydroepoxides derived from polycyclic aromatic hydrocarbons (PAH) are examined. Bay-region diol epoxides and tetrahydroepoxides, as predicted by the "bay-region theory," are the most reactive and biologically active of those derivatives of a given PAH. A comparison of reactivity and mutagenicity data with a quantum chemical parameter that estimates chemical reactivity, $\Delta E_{deloc}/\beta$, reveals a generally good correlation for the hydrolysis data, provided that diol epoxides within groups of similar conformational preference are compared. A similar result is found for the mutagenicity data when the analysis is confined to four- and five-ring PAH, with the significant exception of diol epoxides with the epoxide in a sterically hindered bay region, which are considerably more mutagenic than other diol epoxides with similar $\Delta E_{deloc}/\beta$ values. Three- and six-ring tetrahydro- and diol epoxides, however, exhibit consistently lower mutagenicities than their four- and five-ring counterparts with similar conformational preferences and $\Delta E_{deloc}/\beta$ values. Additional restrictions apply to tumorigenicity. To date, only bay-region diol epoxides that have preferred conformations with pseudodiequatorial hydroxyl groups have been found to be significantly tumorigenic, and those only when they have a relatively high calculated $\Delta E_{deloc}/\beta$ value or a highly hindered bay-region epoxide.

Mutagenicity data for benz[a]- and benz[c]acridine diol epoxides and tetrahydroepoxides reveals a deactivating effect of nitrogen dependent upon its position relative to the epoxide. Benz[a]acridine bay region diol and tetrahydroepoxides have significantly reduced mutagenicities relative to their

0097–6156/85/0283–0063$06.25/0
© 1985 American Chemical Society

benz[a]anthracene counterparts, whereas the benz[c]-
acridine non-bay tetrahydroepoxide on the angular
benzo ring has significantly reduced mutagenicity.

For more than fifty years it has been recognized that members of
the polycyclic aromatic hydrocarbon (PAH) class of molecule exhibit
widely varied carcinogenicities. However, it is only relatively
recently that the chemistry and biochemistry of PAH has become
sufficiently understood to permit an assessment of the structure-
activity relationships involved in their diverse carcinogenicities.
A conceptual basis for understanding the chemical carcinogenesis of
PAH was laid in the 1950's by the Millers (1). Their work with
aromatic amides led them to hypothesize that many organic molecules
become carcinogenic only after "metabolic activation" to reactive,
electrophilic species capable of binding with cellular macromolecules. The identification of these reactive, "ultimate" carcinogenic forms of PAH proved difficult, as a number of exciting possibilities emerged, but were eventually discarded. Nonetheless, by
1976, evidence was accumulating that, for benzo[a]pyrene (BaP), an
important "ultimate" carcinogenic form was BaP 7,8-diol-9,10
-epoxide, produced by three steps of metabolism:

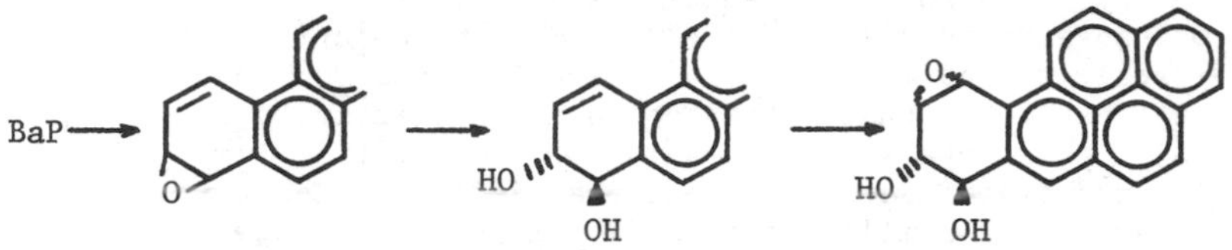

This important discovery was made possible by significant contributions from a number of research laboratories. Discussions of
these contributions are included in a number of articles (2-6).

The bay region theory

Although BaP 7,8-diol-9,10-epoxides provided a possible clue to
explaining the carcinogenesis of PAH, it was initially unknown
whether metabolic activation to diol epoxides would prove to be
generally important and whether any structural feature in the BaP
7,8-diol-9,10-epoxide could be identified that would provide a
guide to predicting which of the various diol epoxides derivable
from a PAH would be most biologically active. Jerina and Daly, by
interpreting substituent effects upon the carcinogenicity of
benz[a]anthracene (BA) derivatives and considering observed reactivities at benzylic positions on tetrahydrobenzo rings of PAH,
provided the initial postulation that it was the "bay region"
nature of the epoxide in BaP 7,8-diol-9,10-epoxide that was the
critical structural feature (3). The prototype of a "bay region"
in a PAH is the sterically hindered region between C-4 and C-5 of
phenanthrene, and bay region epoxides on tetrahydrobenzo rings of
PAH are characterized by this structural feature, though additional
aromatic rings may be fused to the aromatic nucleus. Because of
the route by which PAH are metabolically activated, there will also
be a trans, vicinal dihydroxyl substitution at the two remaining

saturated carbon atoms on the tetrahydrobenzo ring. While, as

bay
region
O 3
4 2
5
6 1
7
8 9 10

will be seen later, these hydroxyl groups have an important influence on reactivity and biological activity, for the present it
is appropriate to focus attention upon the epoxide moiety, and its
benzylic position, as the site of reactivity.

Quantum chemical calculations provided a theoretical basis for
the bay region theory. These calculations were used to estimate
the relative ease of ring-opening of the benzylic C-O epoxide bond
to form a resonance-stabilized carbocation (7,8). The precise
mechanism by which diol epoxides react with DNA and other macromolecules was unknown then and remains a topic of active research.
Nonetheless, carbocations are attractive intermediates to consider,
since they are involved in the rapid hydrolysis of the BaP bay
region diol epoxides and are highly electrophilic species consistent with the Millers' hypothesis. Further, Hulbert (9) and others
have argued that reaction via a S_N1 mechanism or transition state
with appreciable carbocation-like character provides one way of
accommodating a significant (if low) level of reactivity of an
electrophile with weakly nucleophilic macromolecules like DNA.
Thus, an estimation of the relative ease with which various benzylic epoxides on tetrahydrobenzo rings are converted to electrophilic carbocations seemed appropriate. A similar approach had
been previously applied to arylmethyl derivatives of PAH, when
those derivatives appeared to be of importance in the metabolic
activation of methylated PAH (10). The principal opportunity for
differences in activation energy for the conversion resides in the π
-electron energy changes that occur, and these can be easily estimated through perturbational molecular orbital (PMO) calculations. The PMO calculation yields a parameter, $\Delta E_{deloc}/\beta$, which
is a difference in π-electron energy between the reactant π-system
(the unsubstituted aromatic nucleus) and the π-system of the product
(an arylmethyl cation, substituted at the position on the aromatic
nucleus corresponding to that of the benzylic carbon atom of the
epoxide). Provided that these differences in π-energy are reflected in the transition state for the conversion, $\Delta E_{deloc}/\beta$ values
will provide an estimate of relative reactivity, with those
epoxides that can form more highly stabilized carbocations being
more reactive. Thus, for conversion of BA H_4-1,2-epoxide to the
benzylic carbocation (Equation 1), the PMO calculation corresponds
to the π-energy change shown in Equation 2. That is, the
π-energy difference corresponds to that between anthracene and
1-anthracenylmethyl cation. For further illustration, the analogous conversion of BA H_4-10,11-epoxide (Equation 3) is calculated
as the π-energy difference between phenanthrene and 3-phenan

-threnylmethyl cation (Equation 4). The π–systems of the aryl-
methylcarbocations are invariably stabilized relative to the π–
systems of their precursor,

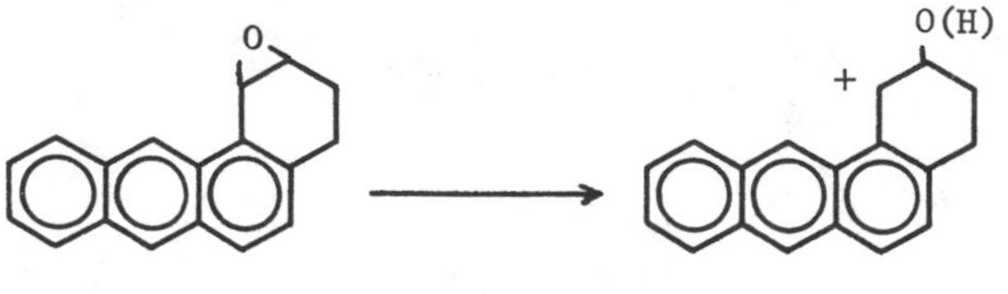

Eq. 1

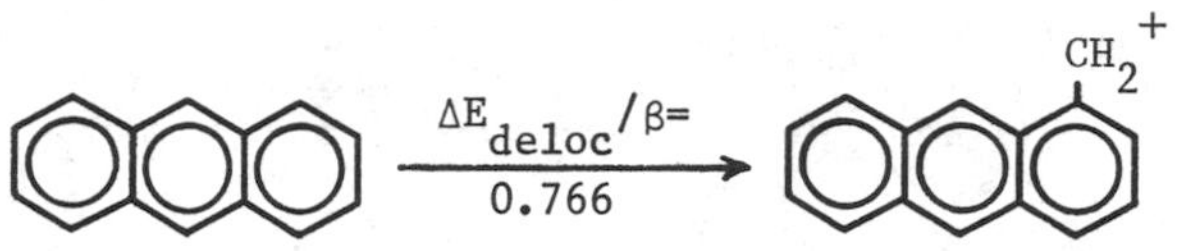

Eq. 2

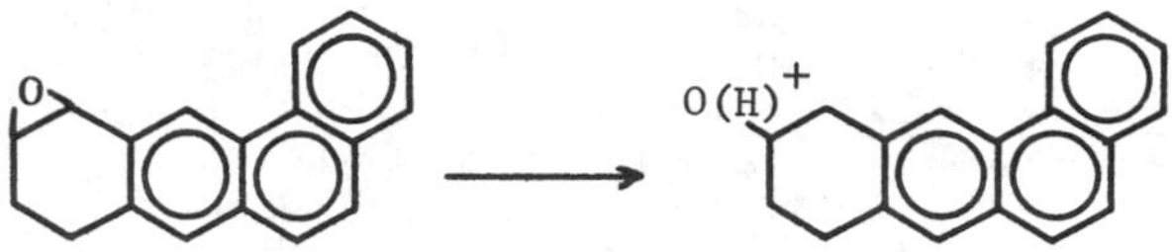

Eq. 3

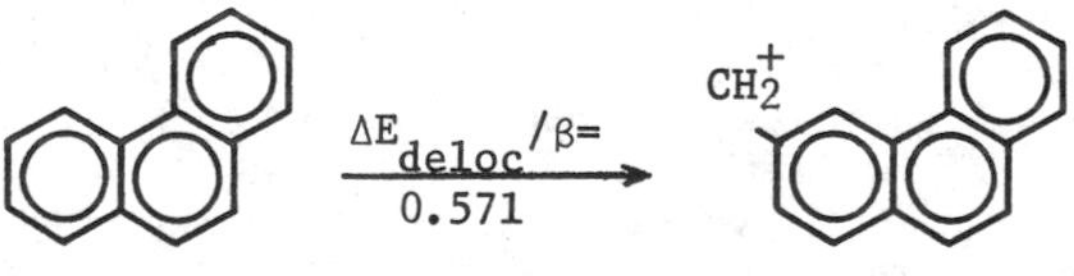

Eq. 4

with larger values of $\Delta E_{deloc}/\beta$ corresponding to greater stabilization. Notably, these calculated values vary considerably, with
significantly greater stabilization being found for carbocations
derived from bay–region tetrahydroepoxides of a given PAH than from
tetrahydroepoxides in which the epoxide is not at a bay region
position (7,8). A theoretical basis for this observation has been
offered (11). Further, the calculations of $\Delta E_{deloc}/\beta$ for benzylic
positions of different PAH revealed that more highly carcinogenic
PAH typically had higher calculated $\Delta E_{deloc}/\beta$ values at the bay-
region position than did weak or non-carcinogenic PAH (7,8). In
fact the correlation of predicted carcinogenicity for a PAH based
upon the calculated $\Delta E_{deloc}/\beta$ value at the bay-region position of
the derived tetrahydrobenzo ring is about equally "successful" as
that of the Pullmann's "K-region" calculations, since a good correlation between the parameters used in the two methods exists (8).
The structures of a variety of PAH of varying carcinogenicity that
have been investigated in our laboratories are shown in Figure 1,

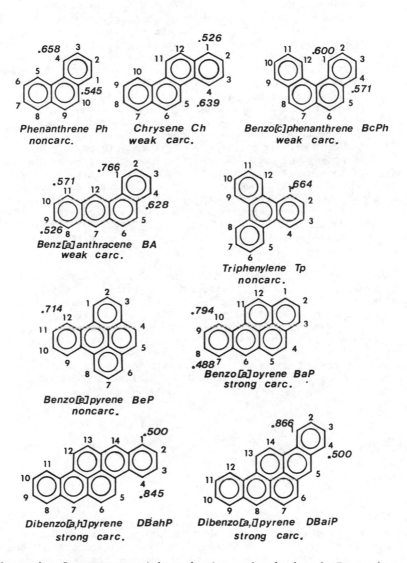

Figure 1. Structures, with numbering and calculated $\Delta E_{deloc}/\beta$ values at benzylic positions of the tetrahydro benzo–ring derivatives of PAH discussed in this chapter.

along with calculated values of $\Delta E_{deloc}/\beta$, which refer to the values calculated at the benzylic position on the tetrahydrobenzo ring derivative of the PAH. Also indicated in this Figure are the abbreviations to be used for the PAH in this article, and the numbering of the PAH. The higher calculated values of $\Delta E_{deloc}/\beta$ at the bay-region positions of each PAH are evident, and it is appropriate to note that experimental results with all the carcinogens in Figure 1 are consistent with metabolic activation to bay region diol epoxides as a major route of activation of the PAH. Indeed, for the more than ten even-alternant PAH so far studied in some depth, the experimental results strongly support "bay-region" activation. Recent reviews (12-14) have summarized these results, which came from detailed studies of the metabolism of PAH and their dihydrodiol derivatives, studies of the mutagenicity of PAH and their derivatives both with and without metabolic activation, and studies of the tumorigenicity of PAH and their derivatives. This manuscript will focus primarily upon the "ultimate" mutagens and carcinogens, the diol epoxides, as well as the related tetrahydroepoxides, with a view to exploring the structure-activity relationships that govern their chemical and biological behavior, and the extent to which quantum chemical calculations are useful in explaining those relationships.

Tetrahydroepoxides as models. Since the quantum chemical calculations apply most rigorously to the simple benzo-ring tetrahydroepoxides and since the calculations neglect influences of the hydroxyl groups in the diol epoxides, it is instructive first to examine the benzo-ring tetrahydroepoxides as simplified models for the reactive site in the diol epoxides. Most of the information about tetrahydroepoxide reactivity derives from studies of the kinetics of their hydrolysis reactions, in which cis- and trans-diols, as well as tetrahydroketones can be formed (Equation 5).

$$\text{Eq. 5}$$

The reaction follows the rate law: $k_{obsd} = k_H a_H{}^+ + k_o$, (15) and values of k_o for eight PAH tetrahydroepoxides are plotted in Figure 2 against the $\Delta E_{deloc}/\beta$ values at the benzylic position bearing the epoxide. Since the experimental conditions used for the naphthalene (Np) and phenanthrene (Ph) studies differed from those for the studies of BA, triphenylene (Tp), benzo[e]pyrene (BeP) and BaP, a direct comparison of all compounds is difficult. Interestingly, for a series of tetrahydro epoxides whose hydration products have been studied, the percentage of cis-hydration during acidic hydrolysis increases as the calculated ease of formation of the carbocation increases, a result consistent with longer-lived (more stable) carbocations as one proceeds along the series (16).

Mutagenicity of tetrahydroepoxides. The mutagenic activity, without metabolic activation, of fourteen PAH tetrahydroepoxides has been measured in S. typhimurium strain TA 100. The logarithms of the relative mutagenic activities are plotted in Figure 3 vs. the value of $\Delta E_{deloc}/\beta$ calculated for conversion of each tetrahydroepoxide to its benzylic carbocation. Several features are notable: for each case for which both bay- and non-bay region H_4-epoxides for a given PAH have been studied, the bay-region epoxide is the more mutagenic; for the four BA H_4-epoxides, the relative mutagenicity of all four isomers is correctly predicted by the $\Delta E_{deloc}/\beta$ parameter; the correlation coefficient for the fourteen data points is 0.76, and it improves to '0.87 if only the data for the ten tetra- and pentacyclic PAH are considered and if the data for dibenzopyrenes and Ph are omitted. As will be seen later, the lower than predicted mutagenicity of these latter molecules persists with their diol epoxide derivatives. The mutagenicity of PAH H_4-epoxides has also been shown to correlate with Herndon's SC ratio calculations (17).

Tumorigenicity of tetrahydroepoxides. As yet, only Ch H_4-epoxide has been directly demonstrated to be tumorigenic (18). However, indirect evidence has been found in the high tumorigenicity of 3,4-dihydro BA, 9,10-dihydro BeP and 3,4-dihydrobenz[c]acridine (19-21), each of which is a likely metabolic precursor of a bay-region $\overline{H}_4$-epoxide. In the case of 9,10-dihydro BeP, cis- and trans-9,10-dihydroxy-9,10,11,12-tetrahydro BeP were identified as products of metabolism of 9,10-dihydro BeP (22), and are the expected products of hydration of the epoxide. Diols are also formed from 7,8-dihydro BaP upon metabolism with prostaglandin endoperoxide synthase (23) or with rat liver homogenates (24).

Diol epoxides. Structural considerations. Because enzymatic hydration of arene oxides produces **trans** dihydrodiols in mammalian cells, there are two diastereomeric series of diol epoxides. In

this article, we shall refer to those diol epoxides in which the benzylic hydroxyl group is cis to the oxirane oxygen atom as belonging to the "diol epoxide-1" series and to those diol epoxides in which the benzylic hydroxyl group is trans to the oxirane oxygen atom as belonging to the "diol epoxide-2" series. Depending upon the nature and position of any substituents attached to the peri positions of the adjacent aromatic ring, certain conformational preferences have been observed, as judged by the coupling constants between the carbinol protons in the nuclear magnetic resonance spectrum. The conformational preferences have been discussed in detail recently (25). To facilitate the discussion in this paper, **Table I** groups the diol epoxides for which hydrolysis and/or muta-

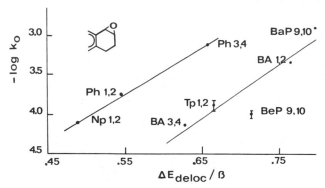

Figure 2. Log k_O versus $\Delta E_{deloc}/\beta$ plots for the spontaneous hydrolysis of tetrahydroepoxides. Rates were measured at 25 °C in 1:9 dioxane:H_2O, ionic strength 0.1 ($NaClO_4$) except in the case of the phenanthrene and naphthalene tetrahydroepoxides, whose rates were measured at 30 °C in H_2O, ionic strength 1.0 (KCl). (Adapted from Refs. 25, 26, and 27.)

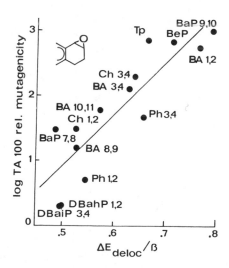

Figure 3. Plot of the logarithms of the relative mutagenicity of PAH tetrahydroepoxides toward S. typhimurium TA 100 versus $\Delta E_{deloc}/\beta$. (Adapted from Ref. 33.)

Table 1. Conformational Preferences of Diol Epoxides

Group	Hydroxyl groups	Epoxide	Compounds
		Diol Epoxides-1	
A	pseudodiaxial (slight-moderate)	nonaligned (slight-moderate)	BaP 7,8-diol-9,10-epoxide, BA 3,4-diol-1,2-epoxide, BA 8,9-diol-10,11-epoxide, BA 10,11-diol-8,9-epoxide, Ch 1,2-diol-3,4-epoxide, Ph 1,2-diol-3,4-epoxide
B	pseudodiaxial (strong)	nonaligned (strong)	BA 1,2-diol-3,4-epoxide, BeP 9,10-diol-11,12-epoxide, Tp 1,2-diol-3,4-epoxide
C	pseudodiequatorial (moderate-strong)	aligned (moderate-strong)	BcPh 3,4-diol-1,2-epoxide
		Diol Epoxides-2	
A	pseudodiequatorial (moderate-strong)	nonaligned (moderate-strong)	BaP 7,8-diol-9,10-epoxide Ba 3,4-diol-1,2-epoxide BA 8,9-diol-10,11-epoxide BA 10,11-diol-8,9-epoxide Ch 1,2-diol-3,4-epoxide Ph 1,2-diol-3,4-epoxide DBahP 1,2-diol-3,4-epoxide DBaiP 3,4-diol-1,2-epoxide BcPh 3,4-diol-1,2-epoxide
B	pseudodiaxial (strong)	aligned (strong)	BeP 9,10-diol-11,12-epoxide Tp 1,2-diol-3,4-epoxide
C	no clear preference	no clear preference	BA 1,2-diol-3,4-epoxide BP 9,10-diol-7,8-epoxide

genicity and tumorigenicity data will be compared according to
their conformational preferences. For diol epoxides-1 that do not
have the benzylic hydroxyl group in a bay region (Group A), there
is a very slight preference for the conformation with pseudoaxial
hydroxyl groups when the epoxide is in a bay region. This conformational preference becomes more pronounced when both the epoxide
and the diol groups are not in a bay region. For diol epoxides-1
with their benzylic hydroxyl group at a bay region position (Group
B), there is a strong preference for the pseudodiaxial conformation
regardless of whether the epoxide is in a bay region. For diol
epoxides-1 with the benzylic hydroxyl group at a non-bay position,
but with the epoxide at a strongly hindered bay region position
such as the fjord region in BcPh (Group C), there is a moderate-
strong preference for a pseudodiequatorial conformation. The
pseudodiaxial and pseudodiequatorial conformations of diol
epoxides-1 are shown below. For diol epoxides-2, the compounds can

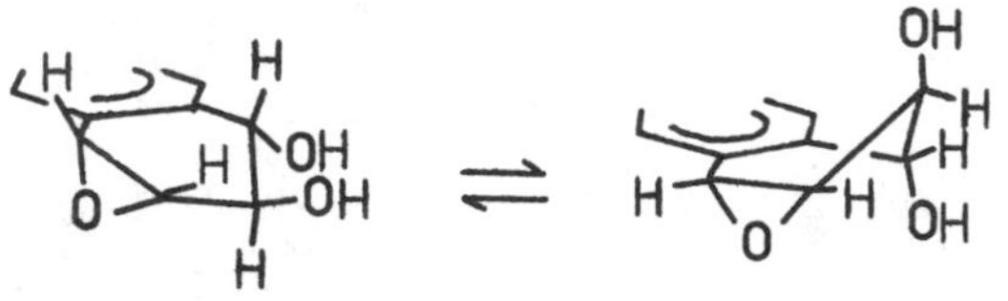

aligned   nonaligned

also be separated into three groups. When the benzylic hydroxyl
group is at a non-bay position (Group A) and the epoxide is either
at a bay or non-bay position, there is a moderate-strong preference
for pseudodiequatorial hydroxyl groups. When both the benzylic
hydroxyl group and the epoxide are in a bay-region, there is a
strong preference for the pseudodiaxial conformation for the hydroxyl groups (Group B). When the benzylic hydroxyl group is at a
bay region position and the epoxide is at a non-bay position, there
is no clear preference for either conformation (Group C). The
conformations for the diol epoxide-2 isomers are shown below.

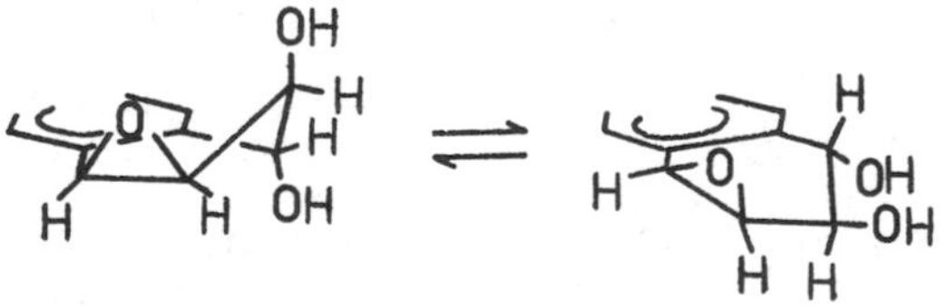

aligned   nonaligned

     A significant conformational influence on chemical reactivity
of these diol epoxides is the orientation of the benzylic C-O bond
of the epoxide relative to the aromatic ring system. In both conformations shown at the left, this bond is in better alignment with
the aromatic pi orbitals than in the conformations shown at the
right. Such an "aligned" conformation of the epoxide might be expected to lead to better overlap between the aromatic pi orbitals
and the p-orbital of a developing carbocation in epoxide cleavage

reactions relative to the "nonaligned" conformation. This is indeed the case for neutral solvolysis reactions (k_o) as will be discussed in the following section.

Reactivity of diol epoxides. By far the most thoroughly and systematically studied reaction of diol epoxides has been their hydrolysis. In Figure 4, log k_o values for eight diol epoxides-1 and ten diol epoxides-2 are plotted vs. $\Delta E_{deloc}/\beta$. Within each diastereomeric series, the conformational preference of the molecule is indicated by open circles (Group A diol epoxides, Table 1), closed circles (Group B diol epoxides, Table 1) or a square (Group C diol epoxide, Table 1). For the Group A diol epoxides, an excellent correlation of log k_o with $\Delta E_{deloc}/\beta$ is observed in both series. The steep slopes of these lines indicate (30) that considerable positive charge is accumulated at the transition state in the k_o process for these molecules and the considerable variation (>100 fold) in k_o values within each series is notable. For the diol epoxides-1, the conformationally related Group B diol epoxides derived from BA, Tp and BeP show a similar sensitivity to calculated values of $\Delta E_{deloc}/\beta$, and their reactivities are correctly ordered. Notably, these three compounds exhibit negative deviations from the line determined by the Group A diol epoxides. The BcPh diol epoxide-1, a Group C diol epoxide, is conformationally unique in this series. It exhibits reactivity expected for a Group A diol epoxide with the same $\Delta E_{deloc}/\beta$ value. For the diol epoxide-2 series, the BcPh diol epoxide is conformationally related to the Group A diol epoxides and shows the anticipated low reactivity. The Group B diol epoxides derived from Tp and BeP exhibit almost equal k_o values. The BA 1,2-diol-3,4-epoxide, with no strong conformational preference, is fairly reactive. It is possible to explain much of the reactivity difference in the k_o region by considering the orientation of the benzylic C-O bond of the epoxide ring with respect to the aromatic system in the various conformations (25). Specifically, those diol epoxides that have no marked conformational preference or those that prefer the aligned conformation (Groups A and C in the diol epoxide-1 series and Groups B and C in the diol epoxide-2 series)will generally be more reactive in k_o reactions than diol epoxides with comparable values of $\Delta E_{deloc}/\beta$ but which *strongly prefer* the nonaligned conformation (Group B in the diol epoxide-1 series and Group A in the diol epoxide-2 series). The negative deviations of the three Group B diol epoxides-1 and the small positive deviations of BA 1,2-diol-3,4-epoxide-2 and TP diol epoxide-2 from the lines in Figure 4 are illustrative of this point. The general thrust of the data in Figure 4 is that a good correlation of log k_o with $\Delta E_{deloc}/\beta$ values is observed when major conformational differences between diol epoxides are taken into consideration. Other quantum chemical parameters (17,31,32) have been shown to correlate with diol epoxide hydrolysis data about equally well (30).

Mutagenicity of diol epoxides. The intrinsic mutagenicities of ten diol epoxides-1 and twelve diol epoxides-2 toward S. typhimurium strain TA 100 and Chinese hamster V79 cells have been determined (14,28,29). For diol epoxides-1, the logarithms of the relative mutagenicities vs. the calculated values of $\Delta E_{deloc}/\beta$ are given in

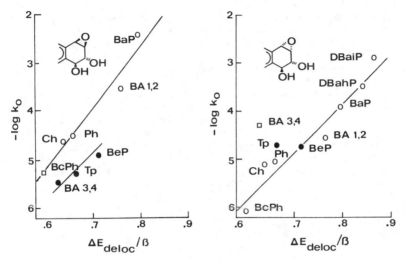

Figure 4.   Plots of −log k_O versus $\Delta E_{deloc}/\beta$ for diol epoxides.
Rates were measured at 25 °C in 1:9 dioxane:H_2O, ionic strength
0.1 ($NaClO_4$).   The numbers following the PAH abbreviation, if
given, designate the position of the epoxide ring.   If no
abbreviation is used, the diol epoxide is at a bay region.
(Adapted from Refs. 25 and 30.)

Figure 5A (for the bacteria) and Figure 5B (for Chinese hamster V79 cells). It will be noted that conformational Group A diol epoxides (open circles) of four and five rings have mutagenicities closely correlated with calculated $\Delta E_{deloc}/\beta$ values, whereas the conformationally related three-ring Ph diol epoxide is about ten times less mutagenic than would be predicted based upon the correlation of the other Group A diol epoxides. It may be recalled that the Ph 1,2- and 3,4-tetrahydroepoxides were also less mutagenic than expected based upon the correlation of the other tetrahydroepoxides (Fig. 3), but the approximately four-fold diminished activity of the tetrahydroepoxides is less than that of the diol epoxides. The relative mutagenicities of the Group B diol epoxides are closely correlated with $\Delta E_{deloc}/\beta$ in the bacteria, but not in the mammalian cells, where the mutagenicity of the BA 1,2-diol-3,4-epoxide appears to be unexpectedly high. Except for the BA 1,2-diol-3,4-epoxide value in Chinese hamster V79 cells, the mutagenicities of these diol epoxides fall well below the values expected based on the correlation of the four- and five-ring diol epoxides with hydroxyl groups not at a bay region. The conformationally unique BcPh diol epoxide exhibits a very high mutagenicity in both systems, and is the most mutagenic diol epoxide-1 toward Chinese hamster V79 cells, which belies its low calculated $\Delta E_{deloc}/\beta$ value and its low observed reactivity. Figures 6A and 6B present the log mutagenicities vs. $\Delta E_{deloc}/\beta$ plots for the diol epoxides-2. The BcPh 3,4-diol-1,2-epoxide is again highly mutagenic and much more than expected based on the correlation of the other Group A (open circles) diol epoxides. Otherwise, the five conformationally related Group A diol epoxides of four and five rings have mutagenicities correlated fairly well with $\Delta E_{deloc}/\beta$ but the Ph and dibenzopyrene diol epoxide mutagenicity values fall appreciably below the values predicted on the basis of the correlation of the five other diol epoxides. These low values are again consistent with the low values observed for tetrahydroepoxides derived from the three PAH. The group B (closed circles) diol epoxides derived from BeP and Tp show much reduced mutagenicities with respect to values expected for Group A diol epoxides with similar $\Delta E_{deloc}/\beta$ values, but their relative mutagenicities are correctly ranked. The conformationally intermediate BA 1,2-diol-3,4-epoxide exhibits reduced mutagenicity more closely related to the Group B diol epoxides.

Several observations can be made about the mutagenicity values. First, the highly-hindered BcPh bay-region diol epoxides exhibit mutagenicities far in excess of what might have been anticipated based upon their calculated $\Delta E_{deloc}/\beta$ values, and the basis for their high mutagenicities must lie in some factor other than π-electron stabilization. It is possible that, once a group of similarly hindered diol epoxides have been studied, there may be a gradation of mutagenicities that reflects relative $\Delta E_{deloc}/\beta$ values, but at a level of mutagenicity raised well above that of other diol epoxides. There is growing evidence (34,46) that the bay-region diol epoxides of 5-methylchrysene in which the methyl group shares the same bay region as the epoxide are responsible for much of the mutagenicity and carcinogenicity of 5-methylchrysene, though they are also reported to be chemically unreactive (46). Similarly, a hindered bay-region diol epoxide appears to be a

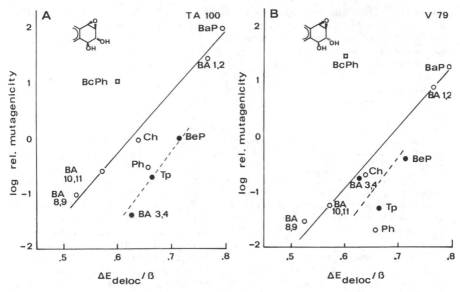

Figure 5. Plots of log rel. mutagenicity toward S. typhimurium
TA 100 and Chinese hamster V79 cells versus $\Delta E_{deloc}/\beta$ for PAH diol
epoxides-1. Numbers refer to the position of the epoxide; the
epoxide is at a bay region if the number is not given. (Adapted
from Refs. 14, 28, and 29.)

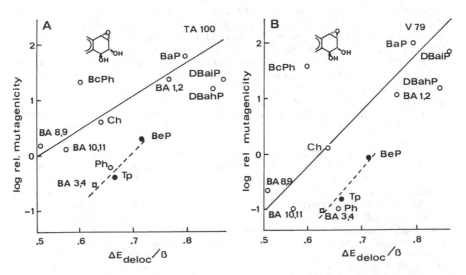

Figure 6. Plots of log rel. mutagenicity toward S. typhimurium
TA 100 and Chinese hamster V79 cells versus $\Delta E_{deloc}/\beta$ for PAH
diol epoxides-2. Numbers refer to the position of the epoxide;
the epoxide is at a bay region if the number is not given.
(Adapted from Refs. 14, 28, and 29.)

likely candidate as an ultimate carcinogen of 7,12-dimethylbenz-
[a]anthracene, and of numerous PAH with methyl groups positioned to
interact sterically with bay-region diol epoxides derived from the
PAH (35,45).

   Secondly, mutagenicity is strongly attenuated in diol epoxides
like those of BeP and Tp, in which the benzylic hydroxyl group as
well as the epoxide is at a bay region. In these cases, a pseudodiaxial conformation for the hydroxyl groups is strongly favored.
Reduced mutagenicity is also observed for BA 1,2-diol-3,4-epoxide-
2, in which the presence of the benzylic hydroxyl group at a bay
region position results in a greater than usual (ca. 50%) amount of
pseudodiaxial conformer. Recently, the 7,8-diol-9,10-epoxides
derived from 6-fluoro BaP have been shown to exist predominantly in
conformations with pseudodiaxial hydroxyl groups, and they have
mutagenicities reduced considerably relative to the corresponding
BaP 7,8-diol-9,10-epoxides (36). The fluorine atom is not responsible for the reduced mutagenicity, since the 6-F BaP H_4 9,10-
epoxide exhibits mutagenicity similar to that of BaP H_4 9,10-
epoxide. Interestingly, geminal dimethyl substitution at C-7 of
the tetrahydro ring also reduces

H_3C CH_3

the mutagenicity of 7,7-dimethyl-BaP H_4 9,10-epoxide, in which one
of the methyl groups must be axial, to less than one-tenth that of
BaP H_4 9,10-epoxide in S. typhimurium TA 100 (37).

Tumorigenicity of diol epoxides. Based on the diol epoxides tested
in skin and/or newborn mouse models (12-14) several qualitative
statements are possible. Diol epoxides-1 are generally nontumorigenic or weakly tumorigenic with one exception, the bay-region
diol epoxide-1 derived from BcPh, which is a potent carcinogen in
mouse skin (38). All tumorigenic diol epoxides-2 are bay-region
diol epoxides and can be ranked in the order: BcPh>BaP>DBahP =
DBaiP>BA> Ch. The BeP diol epoxide is very weakly tumorigenic and
the Ph and BA 1,2-diol-3,4-epoxides (39,40) are nontumorigenic on
mouse skin. These results reveal a restriction of tumorigenicity
to only those diol epoxides with preferred pseudodiequatorial conformations of their hydroxyl groups. This is the preferred conformation of only one bay region diol epoxide-1, that of BcPh, which
unlike other diol epoxide-1 isomers is highly tumorigenic. Further, of those diol epoxides-2 possessing a preferred pseudoedi-
equatorial conformation for their hydroxyl groups, BcPh 3,4-diol-
1,2-epoxide has an exceptional combination of high tumorigenicity
and low reactivity (and low $\Delta E_{deloc}/\beta$ value). Although the dibenzopyrene bay region diol epoxides-2 are potent tumorigens, their
tumorigenicity is not as high as might have been expected based on
their reactivities and $\Delta E_{deloc}/\beta$ values. The BaP, BA and Ch diol
epoxides have relative tumorigenicities that correspond to their
relative calculated and observed reactivities.

Benzacridine tetrahydroepoxides and diol epoxides. The isosteric
molecules BA, benz[a]acridine (BaAcr) and benz[c]acridine (BcAcr)
and their derivatives provide excellent probes for studying the
effect of electronic changes upon biological properties. The

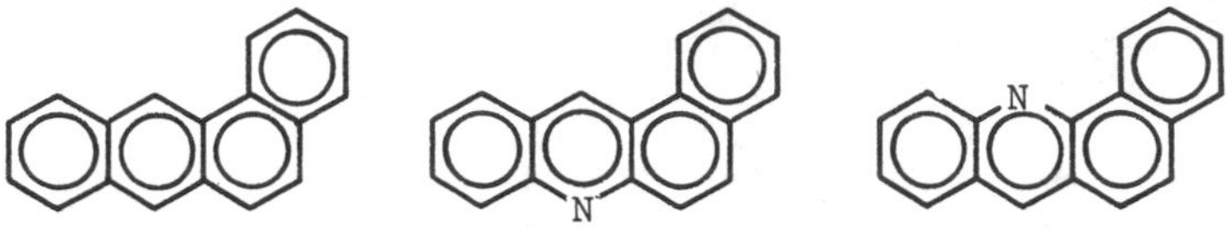

BA   BaAcr   BcAcr

tumorigenicity of a variety of BaAcr and BcAcr derivatives has been
examined, and generally BcAcr derivatives have been found to be
more active than their BaAcr counterparts (41). Recent studies
(21,29) indicate that these relative tumorigenicities may be explicable, in part, as due to a significant effect of nitrogen substitution that is dependent upon the position of nitrogen substitution relative to the epoxide on the angular tetrahydrobenzo
ring.

Effects of nitrogen substitution can be predicted by quantum
chemical calculations (42,43), or more qualitatively by examining
the charge distribution in the relevant carbocations. When a
carbocation is generated at C-1, the positive charge can be
delocalized as shown below.

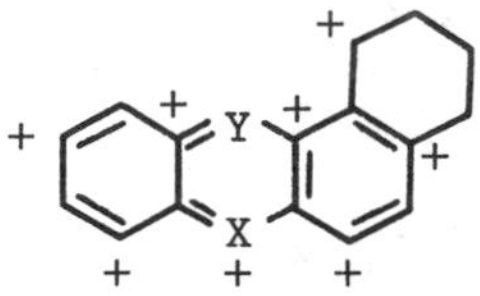

BA:X=Y=CH
BaAcr:X=N;Y=CH
BcAcr:X=CH;Y=N

It will be noted that the positive charge is distributed, in part,
to the position occupied by the electronegative nitrogen atom (X)
of the benz[a]acridine, whereas it is not possible to distribute
the charge by resonance to the nitrogen atom (Y) of BcAcr. Thus,
the C-1 bay-region carbocation derived from benz[a]acridine is
predicted to be destabilized relative to the corresponding carbocations derived from BA and BcAcr. The opposite effect is expected
for the carbocations derived from tetrahydroepoxides at C-4, the
non-bay region position on the angular ring. Here, the carbocation
derived from BcAcr is predicted to be destabilized relative to the
BaAcr carbocation.

The mutagenicities of the bay region diol epoxides derived from BA,
BaAcr and BcAcr toward S. typhimurium strain TA 100 (similar results are obtained in strain TA 98) and Chinese hamster V79 cells

are shown in Figure 7. In both systems, the BaAcr diol epoxides are

BA:X=Y=CH
BaAcr:X=N;Y=CH
BaAcr:X=CH;Y=N

significantly less mutagenic (less than one-tenth) than the corresponding BcAcr diol epoxides. The BcAcr diol epoxides are less
mutagenic than the analogous BA diol epoxides, with isomer-1 having
about 50% and 90% the mutagenicity of the analogous BA diol epoxides in bacteria and mammalian cells, respectively, and isomer-2
being about 25% and 33% as mutagenic as the analogous BA diol
epoxides in the two systems. Thus, the bay region diol epoxides of
BaAcr are much more affected by the nitrogen substitution than
those from BcAcr.

The mutagenicities of the angular benzo ring tetrahydroepoxides of BA, BaAcr and BcAcr provide a clear-cut demonstration of
the selectivity of the effect of nitrogen atom substitution for
carbon. In Figure 8 the relative mutagenicities of these epoxides
in S. typhimurium TA 98 and TA 100 are shown. Using the corresponding BA H_4-epoxides for comparison, it is clear that for the
BaAcr H_4-epoxides the mutagenicity of the H_4 1,2-epoxide is sharply
reduced (to less than half that of BA H_4 1,2-epoxide) whereas the
H_4 3,4-epoxide has 70-90% of the mutagenicity of BA H_4 3,4-epoxide.
The effect is opposite for the BcAcr H_4-epoxides. In that case,
the BcAcr H_4 1,2-epoxide is actually more mutagenic than the BA H_4
1,2-epoxide whereas BcAcr H_4 3,4-epoxide is less than 30% as mutagenic as BA H_4 3,4-epoxide. The effect of nitrogen substitution,
relative to BA, is to accentuate the differences in mutagenicity of
the BcAcr H_4-epoxides, so that the bay region isomer is 29-230
times as mutagenic as the non-bay region H_4-epoxide. However, the
lowering of the mutagenicity of the bay region H_4-epoxide of BaAcr
results in virtually identical mutagenicities for the two angular
ring H_4-epoxides of BaAcr. In fact, in strain TA 100 the mutagenicity of the non-bay region H_4-epoxide of BaAcr slightly exceeds
the mutagenicity of the bay region H_4-epoxide. This apparent
reversal is easily accommodated by the electronic considerations
that form part of the basis for the bay-region theory.

Summary and conclusions

For the alternant PAH that have been studied extensively, bay-
region diol epoxides are important metabolically activated forms.
Studies of the chemical and biological activity of a variety of
diol epoxides have provided insight into the factors related to
reactivity and biological activity. Chemical reactivity, as measured by spontaneous hydrolysis, correlated well with calculated
quantum chemical parameters that estimate π-electron stabilization
upon conversion of the epoxide to a benzylic carbocation, provided

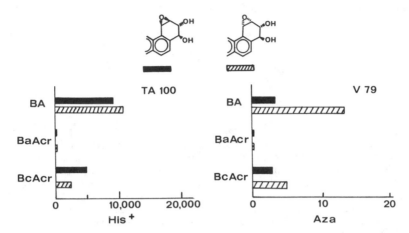

Figure 7. Mutagenicities of BA, BaAcr, and BcAcr bay region diol epoxides in S. typhimurium TA 100 and Chinese hamster V79 cells. (Adapted from Ref. 29).

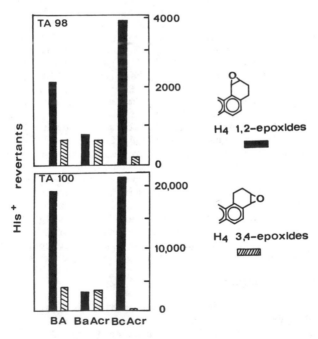

Figure 8. Mutagenicities of BA, BaAcr, and BcAcr angular ring tetrahydroepoxides in S. typhimurium TA 98 and TA 100. (Adapted from Ref. 29 and unpublished results from these laboratories.)

that diol epoxides of similar conformational preferences are compared. Mutagenicities are similarly fairly well correlated if conformational preferences are taken into account, with two striking exceptions: the BcPh diol epoxides, in which the bay region epoxide moiety is strongly sterically hindered and three- and six-ring PAH diol epoxides, which show consistently lower mutagenicities than expected based upon the mutagenicities of four- and five-ring PAH with similar conformational preferences. The unique behavior of the BcPh diol epoxides suggests that some factor not related to electronic considerations is of major importance to its activity. Mutagenicities of diol epoxides which strongly favor the pseudodiaxial conformations for their hydroxyl groups, though correctly ranked internally by $\Delta E_{deloc}/\beta$ values, are much lower than the mutagenicities of diol epoxides with similar $\Delta E_{deloc}/\beta$ values without this structural feature. The relative mutagenicities of angular ring tetrahydroepoxides of BA, BaAcr and BcAcr demonstrate the importance of the position of nitrogen substitution relative to the epoxide moiety, and the substantially lowered mutagenicities of the bay-region BaAcr tetrahydroepoxide and of the BcAcr non-bay region tetrahydroepoxide can be related to electronic destabilization by nitrogen when the positive charge in the carbocation derived from the epoxide is delocalized onto nitrogen.

Significant tumorigenicity is confined to only those diol epoxides which have preferred conformations with pseudodiequatorial hydroxyl groups and i) a sterically hindered bay region or ii) a fairly high reactivity. Also, though not reviewed here, the absolute configuration at the four chiral carbon atoms in the diol epoxide is extremely important. Thus, for all PAH except BcPh so far investigated (44), only one of the stereoisomeric bay region diol epoxides with trans-hydroxyl groups contributes significantly to the mutagenicity and tumorigenicity.

The effects of structure on the mutagenicity and tumorigenicity of PAH diol epoxides appear to follow consistent patterns, and should enable fairly accurate predictions of these properties for given diol epoxides. The prediction of relative carcinogenicity of the parent PAH, however, is dependent as well upon an ability to predict the extent of its metabolism to the bay region diol epoxide of correct absolute configuration. Significant progress has been made in understanding the metabolism of PAH (14,44), but a discussion of this topic is beyond the scope of this paper. Finally, the existence of other routes of metabolic activation for PAH may emerge as studies continue, although there is no substantial evidence for such routes at present.

Acknowledgment

Partial support of this work by Grant No. CA 22985 from the National Cancer Institute to R.E.L. is gratefully acknowledged.

Literature Cited

1. Miller, E.C.; Miller, J.A. Cancer 1981, 47, 1055-1064 and references cited therein.

2. Conney, A.H. Cancer Res 1982, 42, 4875-4917.
3. Jerina, D.M.; Daly, J.W. In "Drug Metabolism-from Microbe to Man," Parke, D.V.; Smith, R.L., Eds; Taylor and Francis, Ltd.: London, 1976, 13-32.
4. Phillips, D.H. Nature, 1983, 303, 468-472.
5. Harvey, R.G. American Scientist 1982, 70, 386-393.
6. Dipple, A. Cancer Res. 1983, 43, 2422s-2425s.
7. Jerina, D.M.; Lehr, R.E.; Yagi, H.; Hernandez, O.; Dansette, P.M.: Wislocki, P.G.; Wood, A.W.; Chang, R.L.; Levin, W.; Conney, A.H. In "In Vitro Metabolic Activation in Mutagenesis Testing" de Serres, F.J.; Fouts, J.R.; Bend, J.R.; Philpot, R.M., Eds.; Elsevier/North Holland Biomedical Press: Amsterdam, 1976, 159-177.
8. Jerina, D.M.; Lehr, R.E. In "Microsomes and Drug Oxidations" Ulbrick, V.; Roots, I.; Hilderbrandt, A.; Estabrook, R.W., Eds: Pergamon Press: Elmsford, NY, 1978, 709-720.
9. Hulbert, P.B. Nature 1975, 256, 146-148 and references cited therein.
10. Dipple, A.; Lawley, P.D.; Brookes, P. Eur. J. Cancer 1968, 4, 493.
11. Lowe, J.P.; Silverman, B.D. J. Am. Chem. Soc., 1981,103, 2852-2855.
12. Nordqvist, M.; Thakker, D.R.; Yagi, H.; Lehr, R.E.; Wood, A.W.; Levin, W.; Conney, A.H.; Jerina, D.M. In "Molecular Basis of Environmental Toxicity" Bhatnagar, R.S., Ed; Ann Arbor Science: Ann Arbor, MI, 1980, 329-357.
13. Thakker, D.R.; Yagi. H.; Nordqvist, M.; Lehr, R.E.; Levin, W.; Wood, A.W.; Chang, R.L.; Conney, A.H.; Jerina, D.M. In "Chemical Induction of Cancer" Arcos, J.C.; Woo, Y.T.; Argus, M.F., Eds; Academic Press: New York, 1982, 727-747.
14. Levin, W.; Wood, A.W.; Chang, R.; Ryan, D.; Thomas, P.; Yagi, H.; Thakker, D.R.; Vyas, K.; Boyd, C.; Chu, S.-Y. In "Drug Metabolism Reviews," 1982, 13, 555-580.
15. Whalen, D.L.; Ross, A.M.; Montemarano, J.A.; Thakker, D.R.; Yagi, H.; Jerina, D.M. J. Am. Chem. Soc., 1979, 101, 5086-5088.
16. Sayer, J.M.; Yagi, H.; Silverton, J.V.; Friedman, S.L.; Whalen, D.L.; Jerina, D.M. J. Am. Chem. Soc., 1982, 104, 1972-1978.
17. Herndon, W.C. Tetrahedron Lett., 1981, 22, 983.
18. Buening, M.K.; Levin, W.; Karle, J.M.; Yagi, H.; Jerina, D.M.; Conney, A.H. Cancer Res. 1979, 39, 5063-5068.
19. Levin, W.; Thakker, D.R.; Wood, A.W.; Chang, R.L.; Lehr, R.E.; Jerina, D.M.; Conney, A.H. Cancer Research 1978, 38, 1705-1710.
20. Buening, M.K.; Levin, W.; Wood, A.W.; Chang, R.L.; Lehr, R.E.; Taylor, C.W.; Yagi, H.; Jerina, D.M.; Conney, A.H. Cancer Research 1980, 40, 203-206.
21. Levin, W.; Wood, A.W.; Chang, R.L.; Kumar, S.; Yagi, H.; Jerina, D.M.; Lehr, R.E.; Conney, A.H. Cancer Research 1983, 43, 4625-4628.
22. Wood, A.W.; Levin, W.; Thakker, D.R.; Yagi, H.; Chang, R.L.; Ryan, D.E.; Thomas, P.E.; Dansette, P.M.; Whittaker, N.; Turijman, S.; Lehr, R.E.; Kumar, S.; Jerina, D.M.; Conney, A.H. J. Biol. Chem. 1979, 254, 4408-4415.
23. Reed, G.A.; Marnett, L.J. J. Biol. Chem. 1982, 257, 11368-11376.

24. Waterfall, J.F.; Sims, P. Biochem. J. 1972, 128, 265-277.
25. Sayer, J.M.; Whalen, D.L.; Friedman, S.L.; Paik, A.; Yagi, H.; Vyas, K.P.; Jerina, D.M. J. Am. Chem. Soc. 1984, 106, 226-233.
26. Becker, A.R.; Janusz, J.M.; Bruice, T.C. J. Am. Chem. Soc. 1979, 101, 5679-5687.
27. Rogers, D.E.; Bruice, T.C. J. Am. Chem. Soc. 1979, 101, 4713-4719.
28. Lehr, R.E.; Yagi, H.; Thakker, D.R.; Levin, W.; Wood, A.W.; Conney, A.H.; Jerina, D.M. In "Carcinogenesis. Vol. 3: Polynuclear Hydrocarbons" Jones, P.W.; Freudenthal, R.I., Eds; Raven Press: New York, 1978, 231-241.
29. Wood, A.W.; Chang, R.L.; Levin, W.; Ryan, D.E.; Thomas, P.E.; Lehr, R.E.; Kumar, S.; Schaefer-Ridder, M.; Engelhardt, U.; Yagi, H.; Jerina, D.M.; Conney, A.H. Cancer Research 1983, 43, 1656-1662.
30. Sayer, J.M.; Lehr, R.E.; Whalen, D.L.; Yagi, H.; Jerina, D.M. Tetrahedron Lett. 1982, 23, 4431-4434.
31. Smith, I.A.; Berger, G.D.; Seybold, P.G.; Serve, M.P. Cancer Res. 1978, 38, 2968.
32. Loew, G.H.; Sudhindra, B.S.; Ferrell, Jr., J.E. Chem.-Biol. Interact. 1979, 26, 75.
33. Wood, A.W.; Levin, W.; Chang, R.L.; Yagi, H.; Thakker, D.R.; Lehr, R.E.; Jerina, D.M.; Conney, A.H. In "Polynuclear Aromatic Hydrocarbons, 3rd International Symposium on Chemistry and Biology-Carcinogenesis and Mutagenesis" Jones, P.W.; Leber, P., Eds; Ann Arbor Science: Ann Arbor, MI, 1979, 531-551.
34. Melikian, A.A.; La Voie, E.J.; Hecht, S.S.; Hoffmann, D. Carcinogenesis 1983, 4, 843-849.
35. Di Giovanni, J.G.; Diamond, L.; Harvey, R.G.; Slaga, T.J. Carcinogenesis 1983, 4, 403-407.
36. Thakker, D.R.; Yagi, H.; Sayer, J.M.; Kapur, U.; Levin, W.; Chang, R.L.; Wood, A.W.; Conney, A.H.; Jerina, D.M. J. Biol. Chem. 1984, 259, 1249-1256.
37. Unpublished result, these laboratories.
38. Levin, W.; Wood, A.W.; Chang, R.L.; Ittah, Y.; Croisy-Delcey, M.; Yagi, H.; Jerina, D.M.; Conney, A.H. Cancer Res. 1980, 40, 3910-3914.
39. Slaga, T.J.; Huberman, E.; Selkirk, J.K.; Harvey, R.G. and Braken, W.M. Cancer Res. 1978, 38, 1699-1704.
40. Slaga, T.J.; Gleason, G.L.; Mills, G.; Ewald, L.; Fu, P.P.; Lee, H.M.; Harvey, R.G. Cancer Res. 1980, 40, 1981-1984.
41. Lacassagne, A.; Buu-Hoi, N.P.; Daudel, R.; Zajdela, F. Adv. Cancer Res. 1956, 4, 315.
42. Lehr, R.E.; Jerina, D.M. Tetrahedron Lett. 1983, 24, 27-30.
43. Smith, I.A.; Seybold, P.G. J. Heterocyclic Chem. 1979, 16, 421.
44. Jerina, D.M.; Yagi, H.; Thakker, D.R.; Sayer, J.M.; van Bladeren, P.J.; Lehr, R.E.; Whalen, D.L.; Levin, W.; Chang, R.L.; Wood, A.W.; Conney, A.H. In "Foreign Compound Metabolism" Caldwell, J.; Paulson, G.D., Eds; Taylor and Francis, Ltd.: London, 1984 (in press).
45. Levin, W.; Wood, A.W.; Chang, R.L.; Newman, M.S.; Thakker, D.R.; Conney, A.H.; Jerina, D.M. Cancer Lett. 1983, 20, 139-146.

46. Melikian, A.; Amin, S.; Hecht, S.S.; Hoffmann, D.; Pataki, J.;
     Harvey, R.G. In "Proceedings of the 75th Annual Meeting of the
     American Association for Cancer Research" 1984, Abstract 361,
     p. 92.

RECEIVED April 9, 1985

Effects of Methyl and Fluorine Substitution on the Metabolic Activation and Tumorigenicity of Polycyclic Aromatic Hydrocarbons

STEPHEN S. HECHT, SHANTU AMIN, ASSIEH A. MELIKIAN, EDMOND J. LaVOIE, and DIETRICH HOFFMANN

Naylor Dana Institute for Disease Prevention, American Health Foundation, Valhalla, NY 10595

The effects of methyl and fluorine substitution on the metabolic activation and tumorigenicity of polycyclic aromatic hydrocarbons (PAH) are reviewed. The structural requirements favoring tumorigenicity of methylated PAH are a bay region methyl group and a free peri position, both adjacent to an unsubstituted angular ring. The enhancing effect of a bay region methyl group on PAH tumorigenicity appears to be due to the relatively high reactivity with DNA and exceptional tumorigenicity of a dihydrodiol epoxide metabolite having a methyl group and epoxide ring in the same bay region. The inhibiting effect of a peri methyl substituent on tumorigenicity can be due to either the diaxial conformation of the trans dihydrodiol at the adjacent double bond, or to inhibition of dihydrodiol formation. Substitution of fluorine in the angular ring of PAH can prevent bay region dihydrodiol epoxide formation and can thereby diminish tumorigenicity. Fluorine substitution at the peri position adjacent to the angular ring also inhibits tumorigenicity, by mechanisms similar to those observed in peri-methyl substituted PAH.

Human exposure to complex mixtures of polycyclic aromatic hydrocarbons (PAH) occurs through inhalation of tobacco smoke and polluted indoor or outdoor air, through ingestion of certain foods and polluted water, and by dermal contact with soots, tars, and oils (1). Methylated PAH are always components of these mixtures and in some cases, as in tobacco smoke and in emissions from certain fuel processes, their concentrations can be in the same range as some unsubstituted PAH. The estimated emission of methylated PAH from mobile sources in the U.S. in 1979 was approximately 1700 metric tons (2). The occurrence of methylated and unsubstituted PAH has been recently reviewed (1,2). In addition to their environmental occurrence, methylated PAH are among the most important model compounds in experimental carcinogenesis. 7,12-Dimethylbenz[a]anthracene, one of

0097-6156/85/0283-0085$06.00/0
© 1985 American Chemical Society

the most potent known PAH carcinogens, is extensively used for the
induction of breast tumors in Sprague-Dawley rats. 7,12-Dimethyl-
benz[a]anthracene is an excellent example of a strong carcinogen,
formed by methyl substitution of an essentially inactive parent compound, benz[a]anthracene. Many examples are available of striking
differences in tumorigenic activity between methyl substituted PAH
and their parent compounds or between different methyl isomers in the
same series (3). In this review, we will discuss the structural
features that favor tumorigenicity among the methylated PAH and will
consider the mechanistic basis for these observations.

Fluorinated PAH have been used exclusively as model compounds to
probe the mechanisms of PAH metabolic activation. It has generally
been assumed that substitution of a small, electronegative fluorine
atom at a particular position in a PAH ring system will block enzymatic oxidation at that position. Thus, if a fluorinated PAH were
less tumorigenic than its parent compound, it could be concluded that
the carbon bearing the fluorine atom was involved in metabolic activation of the parent PAH. This "fluorine-probe approach" has been
useful and in many cases the tumorigenicity data have been supported
by metabolism studies. The effects of fluorine substitution on PAH
tumorigenicity and metabolic activation will be reviewed. In some
ways, the effects of fluorine or methyl substitution on PAH tumorigenicity are similar.

Structural Requirements Favoring Tumorigenicity of Methylated PAH

In 1979, based on our work on methylchrysenes and on literature data
on other methylated PAH, we suggested that the structural requirements favoring tumorigenicity of methylated PAH were a bay region

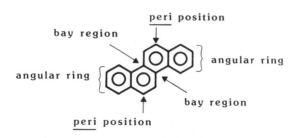

methyl group and a free peri position, both adjacent to an unsubstituted angular ring (4). The bay regions, peri positions, and angular
rings of chrysene are indicated. Examples of highly tumorigenic
methylated PAH having a bay region methyl group adjacent to an unsubstituted angular ring are illustrated in Figure 1. All of these compounds are more tumorigenic than the corresponding unsubstituted
PAH. 1,4-Dimethylphenanthrene and 4,10-dimethylphenanthrene are more
tumorigenic than several other dimethylphenanthrenes (6). However,
4-methylphenanthrene, which has a bay region methyl group adjacent to
an unsubstituted angular ring, is inactive (5). This exception will
be discussed further below. 7,12-Dimethylbenz[a]anthracene is the
most potent of the dimethylbenz[a]anthracenes (3). 12-Methylbenz[a]-
anthracene is more tumorigenic than benz[a]anthracene and 1-,2-,3-,

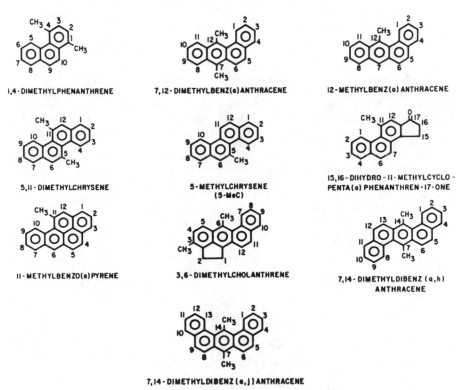

Figure 1.

Highly tumorigenic methylated PAH having a bay region methyl group adjacent to an unsubstituted angular ring. All compounds are more tumorigenic than their parent PAH. With the exception of 12-methylbenz[a]anthracene, all compounds shown are also more tumorigenic than any other assayed methyl or dimethyl isomers. References: 1,4-dimethylphenanthrene (5,6); 7,12-dimethylbenz[a]-anthracene (3); 12-methylbenz[a]anthracene (3,7); 5,11-dimethyl-chrysene (4); 5-methylchrysene (8); 15,16-dihydro-11-methylcyclo-penta[a]phenanthren-17-one (9); 11-methylbenzo[a]pyrene (10); 3,6-dimethylcholanthrene (11,12); 7,14-dimethyldibenz[a,h]anthra-cene (11,13); 7,14-dimethyldibenz[a,j]anthracene (11).

4-,5-,9-,10-, and 11-methylbenz[a]anthracene (3,7). However, it is less tumorigenic than 7-methylbenz[a]anthracene and has activity similar to those of 6-methylbenz[a]anthracene and 8-methylbenz[a]-anthracene (3,7). 5,11-Dimethylchrysene is the most potent dimethyl-chrysene tested, with tumorigenic activity exceeding that of 5-methylchrysene (5-MeC) (4). 5-MeC is more tumorigenic than chrysene or any of the other methylchrysenes (8). 15,16-Dihydro-11-methyl-cyclopenta[a]phenanthrene-17-one is more tumorigenic than the parent compound or other monomethyl isomers in this system (9). 11-Methyl-benzo[a]pyrene is more tumorigenic than benzo[a]pyrene or any of the other monomethylbenzo[a]pyrenes (10). 3,6-Dimethylcholanthrene is more tumorigenic than cholanthrene or 3-methylcholanthrene (11,12). 7,14-Dimethyldibenz[a,h]anthracene and 7,14-dimethyldibenz[a,j]-anthracene are more tumorigenic than the parent hydrocarbons or the 7-methyl analogues (11,13). These results firmly establish the enhancing effect on tumorigenicity of a bay region methyl group adjacent to an unsubstituted angular ring.

However, the peri position adjacent to the angular ring must also be unsubstituted for maximum activity to be observed. Examples of inhibition of tumorigenesis by methyl substitution at the peri position are illustrated in Figure 2. The comparative carcinogenic activities on mouse skin of 5,12-dimethylchrysene, 5,11-dimethylchry-sene, and 5-MeC provide a clear example, as illustrated in Figure 3. The requirement for a free peri position can also be seen by the inactivity of 9,14-dibenz[a,c]anthracene, which has a bay region methyl group but no free peri position adjacent to the angular ring (11).

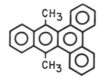

9,14-dimethyldibenz(a,c)anthracene

As discussed in previous chapters, the metabolic activation of methylated PAH involves formation of dihydrodiol epoxides in the angular ring, as in the unsubstituted PAH. This process can be par-tially inhibited by substitution in the angular ring. Metabolic studies on 7-methylbenzo[a]pyrene have shown that 7,8-dihydro-7,8-dihydroxy-7-methylbenzo[a]pyrene is formed, but to a lesser extent than in the metabolism of benzo[a]pyrene (20,21). Therefore, it is not surprising that 7-methylbenzo[a]pyrene, as well as 8-,9-,and 10-methylbenzo[a]pyrene are less tumorigenic than benzo[a]pyrene. In the benz[a]anthracene series, 1,7,12-trimethylbenz[a]anthracene and 2,7,12-trimethylbenz[a]anthracene are less active than 7,12-dimethyl-benz[a]anthracene (22), probably as a result of inhibition of 3,4-dihydrodiol-1,2-epoxide formation. While there are many examples of methylated PAH which have tumorigenic activities in agreement with the structural requirements discussed above, there are also some notable exceptions. These include 6-,7-, and 8-methylbenz[a]anthra-cene. None of these compounds have bay region methyl groups yet all

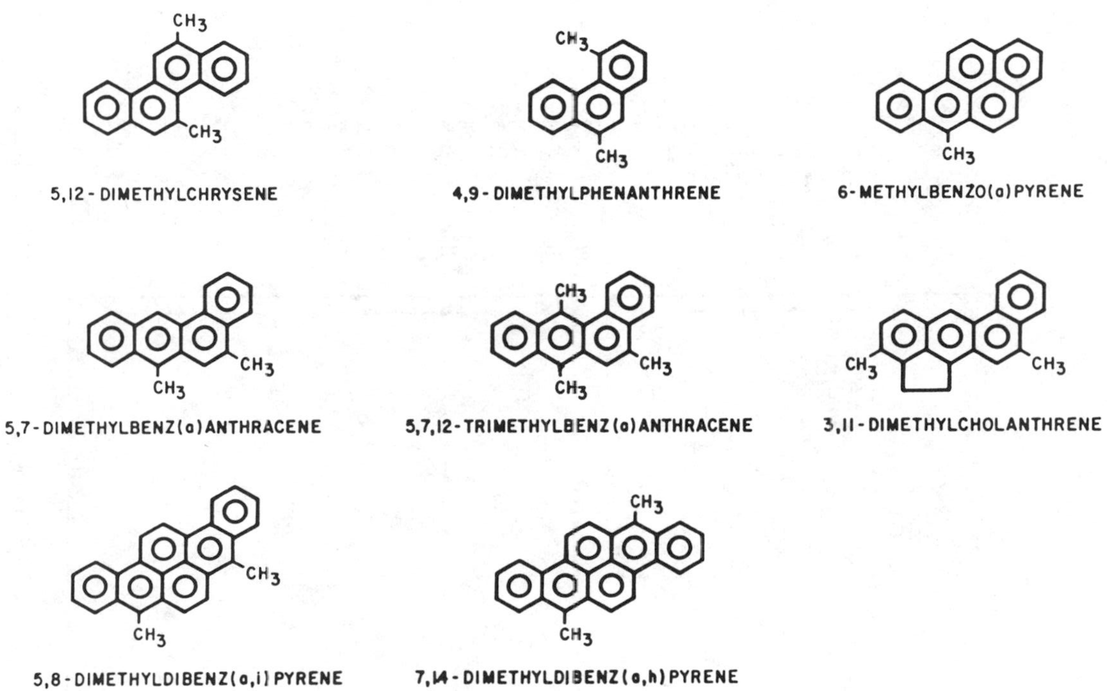

Figure 2.
Inhibition of tumorigenicity by peri-methyl substitution. The methylated PAH shown are less tumorigenic than other methyl isomers in the same series or than their parent compounds. References: 5,12-dimethylchrysene (14); 4,9-dimethylphenanthrene (6); 6-methylbenzo[a]pyrene (10,15); 5,7-dimethylbenz[a]anthracene (16); 5,7,12-trimethylbenz[a]anthracene (17); 3,11-dimethylcholanthrene (18); 5,8-dimethyldibenz[a,i]pyrene (19); 7,14-dimethyldibenz[a,h]pyrene (19).

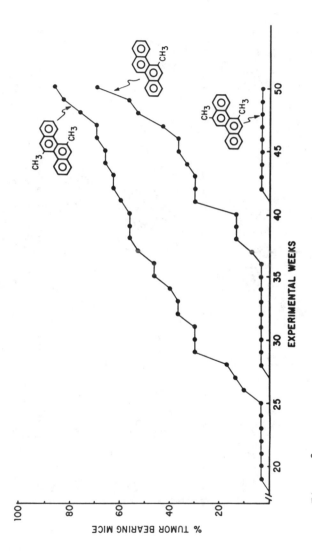

Figure 3.
Comparative tumorigenicity on mouse skin of 5-MeC, 5,11-dimethyl-chrysene, and 5,12-dimethylchrysene. Each compound (5 μg) was applied to mouse skin 3 times weekly for 50 weeks.

are strong tumorigens.   7-Methylbenz[a]anthracene is the strongest
tumorigen among the monomethylbenz[a]anthracenes (3,7).   The reasons
for the unique activity of 7-methylbenz[a]anthracene are not known.
4-Methylphenanthrene fulfills the structural requirements but, as in
the case of the other monomethylphenanthrenes, is inactive as a
tumor initiator on mouse skin (5).   This seems to be due to facile
metabolic detoxification by formation of the 9,10-dihydrodiol, a process which is blocked in the tumorigenic isomers 1,4- and 4,10-di-
methylphenanthrene (5,6).   Among the methylated benzo[c]phenan-
threnes, the 3-,4-,5-, and 6-methyl isomers are the most tumorigenic.   The 1-methyl isomer, in which the methyl group is present in
a 4-sided "fjord", is only weakly active like the parent hydrocarbon
(23).

Mechanistic Basis for the Enhancing Effect on Tumorigenicity of a Bay Region Methyl Group

5-MeC is an excellent model compound for studying the enhancing
effect on tumorigenicity of a bay region methyl group because it has
two bay regions, one of which has the methyl group adjacent to an
unsubstituted angular ring.   Bioassays of 5-MeC metabolites for tumor
initiating activity on mouse skin demonstrated that 1,-2-,3-,7-,8-
and 9-hydroxy-5-MeC were less tumorigenic than was 5-MeC.   5-Hydroxy-
methylchrysene had activity comparable to that of 5-MeC (24).   Among
the dihydrodiol metabolites, 1,2-dihydro-1,2-dihydroxy-5-MeC (5-MeC-
1,2-diol, Figure 4) was more tumorigenic than was 5-MeC whereas
5-MeC-7,8-diol was less tumorigenic than 5-MeC.   5-MeC-9,10-diol was
inactive (25).   Both 5-MeC-1,2-diol and 5-MeC-7,8-diol could form bay
region dihydrodiol epoxides and, by analogy to unsubstituted PAH,
these metabolites might be expected to be ultimate carcinogens.
However, the relatively high activity of 5-MeC-1,2-diol compared to
5-MeC and 5-MeC-7,8-diol suggested that the methyl group in the bay
region had a special effect on dihydrodiol epoxide tumorigenicity.
This was found to be the case when the tumorigenic activities of
anti-1,2-dihydroxy-3,4-epoxy-1,2,3,4-tetrahydro-5-MeC   (anti-DE-I,
Figure 4), syn-DE-I, and anti-DE-II were compared in newborn mice and
on mouse skin.   The results of the newborn mouse experiment which are
summarized in Table I clearly show that anti-DE-I has exceptional
tumorigenic activity and is a major ultimate carcinogen of 5-MeC.
Anti-DE-I is the only example of an ultimate carcinogen of a methylated PAH to be established in a bioassay.   Anti-DE-I was also more
tumorigenic than anti-DE-II on mouse skin, but was less active than
5-MeC (Table II).   These results demonstrate that a bay region methyl
group can enhance the tumorigenicity of a dihydrodiol epoxide metabolite as in the case of anti-DE-I.   Thus, the enhancing effect of the
bay region methyl group on PAH tumorigenicity would appear to result
from the unique activity of a bay region dihydrodiol epoxide metabolite, having the methyl group and epoxide ring in the same bay
region.
    Experiments on the metabolic activation of 5-MeC in mouse skin
are in agreement with the pivotal role of anti-DE-I in expressing its
tumorigenicity.   The major DNA adduct formed in mouse skin treated
with [^{3}H]5-MeC has the structure indicated in Figure 5, resulting
from addition of the exocyclic amino group of deoxyguanosine to car-

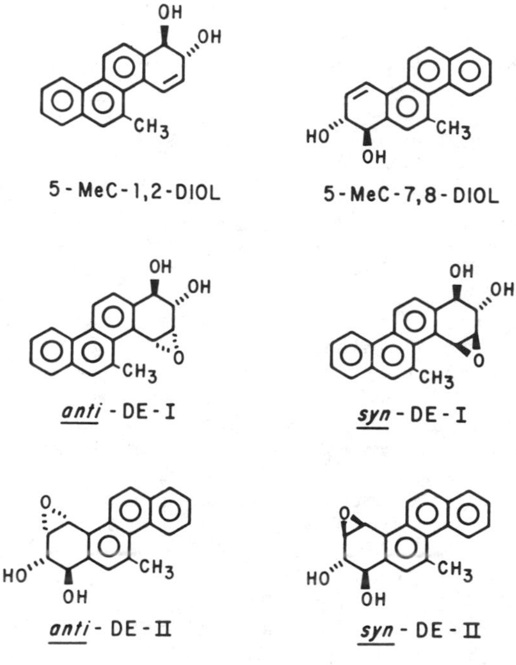

Figure 4.
Structures of 5–MeC bay region dihydrodiol epoxides and their precursor dihydrodiols.

Table I. Tumorigenicity of 5-MeC Metabolites in Newborn Mice

Compound	No of mice injected	Effective no of mice	Pulmonary tumors		Hepatic tumors	
			% tumor bearing animals	tumors/animal	% tumor bearing animals	tumors/animal
5-MeC	100	Female 48	21	0.25	12	0.29
		Male 35	20	0.26	23	0.43
		Total 83	21	0.25	17	0.34
5-MeC-1,2-diol	100	Female 43	12	0.14	7	0.23
		Male 44	11	0.18	25	0.52
		Total 87	12	0.16	16	0.38
5-MeC-7,8-diol	100	Female 45	18	0.24	11	0.49
		Male 46	13	0.13	2	0.02
		Total 91	15	0.19	7	0.25
Anti-DE-I	100	Female 48	81	5.6	4	0.13
		Male 38	82	3.3	34	2.6
		Total 86	81	4.6	19	1.2
Syn-DE-I	100	Female 41	29	0.34	7	0.46
		Male 49	6	0.06	14	0.20
		Total 90	17	0.17	11	0.37
Anti-DE-II	100	Female 50	6	0.06	4	0.04
		Male 49	18	0.18	2	0.02
		Total 99	12	0.12	3	0.03
DMSO	100	Female 41	7	0.07	2	0.02
		Male 48	4	0.04	2	0.04
		Total 89	6	0.06	2	0.03

Note: Ha/ICR mice were given i.p. injections of each compound (total dose, 56 nmol) in DMSO on the 1st, 8th, and 15th days of life. Mice were weaned at age 21 days, separated by sex, and sacrificed at age 35 weeks.

Table II. Tumor Initiating Activity of 5-MeC Metabolites on Mouse Skin

	Dose (nmol)	Percent Tumor Bearing Animals		Tumors Per Animal	
		15 weeks[a]	25 weeks	15 weeks	25 weeks
5-MeC	100	50	90	1.2	5.2
	33	45	80	1.1	3.9
5-MeC-1,2-diol	100	85	100	4.3	12.7
	33	70	85	2.4	9.9
5-MeC-7,8-diol	100	10	75	0.1	1.3
	33	5	30	0.1	0.3
Anti-DE-I	100	60	80	1.8	4.4
	33	35	65	0.4	1.3
Anti-DE-II	100	0	0	0	0
	33	0	5	0	0.1
Acetone	-	0	10	0	0.1

Note: Groups of 20 female CD-1 mice (age 50-55 days) were shaved and treated with a single dose of each compound in 0.1 ml acetone. Ten days later, each group was treated 3 times weekly with 2.5 µg of tetradecanoyl phorbol acetate in 0.1 ml acetone, for 25 weeks.

[a] Weeks of treatment with tetradecanoyl phorbol acetate.

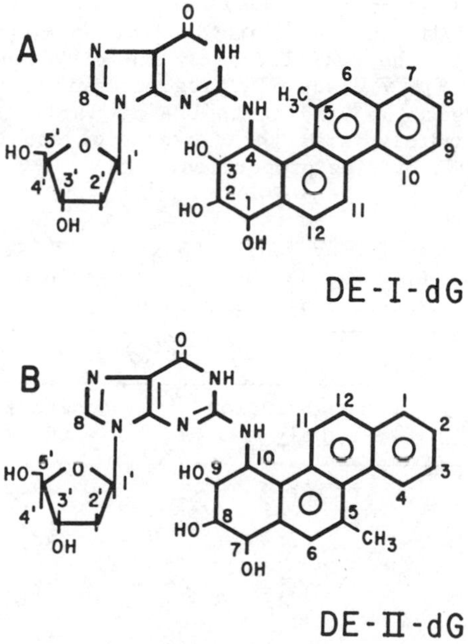

DE-I-dG

DE-II-dG

Figure 5.
Structures of the major adducts formed upon reaction of (A)anti-DE-I and (B)anti-DE-II with DNA in vitro.

bon 4 of anti-DE-I. A structurally similar major adduct is formed
from anti-DE-II (26). However, anti-DE-I adducts exceed anti-DE-II
adducts by 2-3 fold in mouse skin, 4-48 hr after treatment with [³H]-
5-MeC (27). The predominance of anti-DE-I adducts over anti-DE-II
adducts in mouse skin is not due to differences in extents of formation of 5-MeC-1,2-diol and 5-MeC-7,8-diol since the levels of these
metabolites in mouse epidermis are the same from 0.33-4 hr after
treatment with [³H]5-MeC (27).

The rates of hydrolysis and binding to DNA of anti-DE-I, syn-DE-
I, anti-DE-II, syn-DE-II, and anti-1,2-dihydroxy-3,4-epoxy-1,2,3,4-
tetrahydrochrysene (anti-chrysene-DE) were studied in order to relate
the chemical reactivity of these dihydrodiol epoxides to their biological activities. The half-lives of the dihydrodiol epoxides in
cacodylate buffer at pH 7.0 and 37°C are summarized in Table III and
their relative extents of binding to DNA in Table IV. It is clear
that the rates of hydrolysis of the dihydrodiol epoxides do not correlate with their DNA binding properties.

Table III. Half-Lives of Dihydrodiol Epoxides of 5-MeC and Chrysene
       at pH 7.0 and 37 °C in the Absence and Presence of Native and
                    Denatured Calf Thymus DNA

Compound	$t^{1}/_{2}$ (minutes)		
	Buffer Solution Only	Denatured DNA	Native DNA
Anti-DE-I	59	24	3.5
Syn-DE-I	62	48	22
Anti-DE-II	17.5	9	2
Syn-DE-II	5.4	4.9	2.8
Anti-chrysene-DE	104	77	21

When the hydrolyses were carried out in the presence of denatured
DNA, a rate enhancement of 1.1 to 2.5 fold was observed while in the
presence of native DNA the enhancement was 2 to 17 fold (see Table
III). The ratios of the rates of hydrolysis in the presence of
native DNA to the rates of hydrolysis in the presence of denatured
DNA did correlate with the extents of binding to DNA as illustrated
in Figure 6. Our interpretation of these results is that intercalation of the dihydrodiol epoxide in DNA precedes reaction, as has been
observed with benzo[a]pyrene-7,8-dihydrodiol-9,10-epoxides (28-30).
This may be the key factor in determining extents of binding of dihydrodiol epoxides to DNA in vitro. It is of interest that the
greatest rate enhancement and highest extent of binding were observed
for anti-DE-I, which also binds to DNA to a greater extent than anti-
DE-II in vivo and is the most tumorigenic of the methylated bay
region dihydrodiol epoxides tested. While these results suggest that

Table IV. Relative Extents of Binding of Dihydrodiol Epoxides of
5-MeC and Chrysene to Native Calf-Thymus DNA at pH 7.0 and 37 °C

Compound	Relative Extents of Binding
Anti-DE-I	4.9
Syn-DE-I	2.0
Anti-DE-II	2.6
Syn-DE-II	1
Anti-chrysene-DE	2.4

extents of DNA binding of the dihydrodiol epoxides are a determinant of tumorigenic activity, the exceptional tumorigenicity of anti-DE-I can probably not be explained on this basis alone.

The results of these studies demonstrate that the enhancing effect of a bay region methyl group on tumorigenicity, as in 5-MeC, is due to the exceptional tumorigenicity and relatively high reactivity with DNA of a dihydrodiol epoxide metabolite, anti-DE-I, having a methyl group and an epoxide ring in the same bay region. Among the monomethylchrysenes, only 5-MeC can form such a metabolite. This partially explains its unique activity. Studies on the metabolic activation of 15,16-dihydro-11-methylcyclopenta[a]phenanthrene-17-one and 7,12-dimethylbenz[a]anthracene have shown that bay region dihydrodiol epoxides are likely ultimate carcinogens (31,32). It appears likely that the enhancing effect of a bay region methyl group on tumorigenicity in these systems is also a result of the exceptional tumorigenicity of these dihydrodiol epoxide metabolites having a methyl group and epoxide ring in the same bay region. In the case of 7,12-dimethylbenz[a]anthracene, it has been proposed that the bay region syn-dihydrodiol epoxide may be important in its metabolic activation (33). The low tumorigenicity of syn-DE-I suggests, however, that this may not be the case and indicates the importance of steric factors in determining dihydrodiol epoxide tumorigenicity, as reported for unsubstituted PAH.

An important structural feature of the methylated PAH with a bay region methyl group is their non-planarity. Steric hindrance between the methyl group and the adjacent bay region hydrogen causes distortion and deviation from planarity in 7,12-dimethylbenz[a]anthracene and 5-MeC (34,35). It has been suggested that non-planarity may play a role in the high tumorigenicity of these compounds. In view of the results discussed above, it would appear that this effect would have to operate at the level of the dihydrodiol epoxide metabolites. It would be useful to determine the x-ray crystal structures of anti-DE-I and anti-DE-II in order to establish whether differences in planarity of these metabolites, if any, could contribute to their differing extents of reaction with DNA.

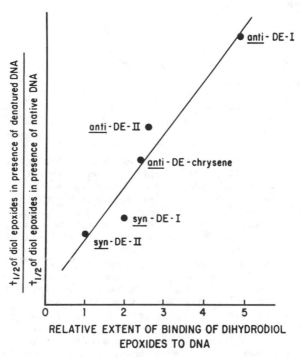

Figure 6.
  Plot of the ratios of the half-lives of dihydrodiol epoxides in
  the presence of denatured DNA to those in the presence of native
  DNA vs. extents of DNA binding in vitro.

Mechanistic Basis for the Inhibitory Effect on Tumorigenicity of a Peri Methyl Group

Unhindered angular ring trans-dihydrodiols such as trans-7,8-dihydro-7,8-dihydroxybenzo[a]pyrene exist preferentially in the pseudo di-equatorial conformation and can usually be oxidized enzymatically to the corresponding bay region dihydrodiol epoxides. However, a peri methyl group causes crowding and as a result the conformation of a trans-dihydrodiol in the adjacent angular ring will be preferentially diaxial. This phenomenon has been noted by several groups and has been previously reviewed (36, 37). It has been suggested that the inhibitory effect of a peri methyl group on tumorigenicity is due to the relative difficulty of enzymatic conversion of diaxial dihydro-diols to their corresponding dihydrodiol epoxides (36,37). Some experimental evidence supports this suggestion although detailed metabolic studies on compounds such as 6-methylbenzo[a]pyrene, 3,11-dimethylcholanthrene, and 5,7,12-trimethylbenz[a]anthracene have not been reported.

Whereas peri methyl substitution does not block dihydrodiol formation in the adjacent ring in the benz[a]anthracene system (38,39), it apparently does so in the chrysene system. 7,8-Dihydro-7,8-dihydroxy-5,12-dimethylchrysene was a major metabolite of 5,12-dimethylchrysene in rat and mouse hepatic 9000 x g supernatant, but 1,2-dihydro-1,2-dihydroxy-5,12-dimethylchrysene could not be de-tected. Similarly, the ratio of 7-hydroxy-5,12-dimethylchrysene to 1-hydroxy-5,12-dimethylchrysene was about 100 to 1 in liver super-natants from 3-methylcholanthrene pretreated mice and rats. In contrast, 1,2-dihydro-1,2-dihydroxy-5,11-dimethylchrysene was a major metabolite of 5,11-dimethylchrysene (40). These results suggest that the low tumorigenicity of 5,12-dimethylchrysene is due to inhibition of formation of its likely major proximate carcinogen, 1,2-dihydro-1,2-dihydroxy-5,12-dimethylchrysene.

Effects of Fluorine Substitution on the Tumorigenicity of PAH

Table V summarizes literature on the tumorigenic activities of fluor-inated PAH, tested either as tumor initiators or complete carcinogens on mouse skin. In general, the results are consistent with the hy-pothesis that fluorine substitution could block the formation of angular ring bay region dihydrodiol epoxides. Thus, decreased tumor-igenicity was observed upon substitution of fluorine in the angular rings of 7-methylbenz[a]anthracene, 7,12-dimethylbenz[a]anthracene, benzo[a]pyrene, dibenzo[a,i]pyrene, and dibenzo[a,h]pyrene. In the case of 5-methylchrysene and 5-hydroxymethylchrysene which each have 2 angular rings and 2 bay regions, decreased tumorigenicity was observed only upon fluorine substitution in the 1-4 ring which is the major site of metabolic activation as discussed above. Decreased tumorigenicity was also observed upon fluorine substitution at the peri-positions adjacent to the angular rings involved in metabolic activation, as seen with the methylated PAH. Increases in tumorigen-icity were observed upon substitution of fluorine at the 7- and 12-positions of benz[a]anthracene derivatives, although the increases were much less than observed upon substitution of methyl groups at those positions. Whereas the effects of fluorine substitution in the angular rings and peri-positions are reasonably well understood as

Table V. Tumorigenicity of Fluorinated PAH on Mouse Skin[a]

	+	NC	−
Benz[a]anthracene	7F (16) 12F 7,12diF	4F (41) 5F	
7-methylbenz[a]anthracene[b]	12F (16)	6F (41) 9F 10F	3F (16,41–43) 4F 5F
12-methylbenz[a]anthracene	7F (16)		5F (16)
7,12-dimethylbenz[a]anthracene[c]		11F (17,44)	1F (17,44) 2F 5F
5-methylchrysene	6F (45)	6F (45,46) 7F 9F 11F	1F (45,46) 3F 12F
5-hydroxymethylchrysene		7F (24)	3F (24)
benzo[a]pyrene[d]			6F (47–49) 7F 8F 9F 10F
dibenzo[a,i]pyrene[e]			2F (50,51) 3F 2,10-diF
dibenzo[a,h]pyrene			3,10-diF (52)

[a] +; more active than parent hydrocarbon; NC, no change; −, less active.

[b] When tested by s.c injection in rats, 6F was + (41,53); 2F, 3F, 5F, 9F, 10F were − (22,41–43). When tested by s.c. injection in mice, 3F, 6F were + (41); 5F, 9F, 10F were − (41–43).

[c] When tested by s.c. injection in rats 1F, 2F, 4F were − and 8F, 11F were NC (22,53).

[d] When tested for induction of lung adenomas in mice or by s.c. injection in rats, 6F was NC (49).

[e] When tested by s.c. injection in mice 3F and 2,10-diF were − (54).

described below, the reasons for the increases in tumorigenicity upon substitution at the 7- and 12-positions of benz[a]anthracene are obscure.

Effects of Florine Substitution on the Metabolic Activation of PAH

Metabolism studies on fluorinated derivatives of 5-MeC, 5-hydroxymethylchrysene, benzo[a]pyrene and dibenzo[a,i]pyrene have all shown that fluorine effectively blocks oxidation of a PAH at the formal double bond to which the fluorine is attached (24,45,47,50). Only one example of metabolic loss of fluorine has been reported; 6-fluorobenzo[a]pyrene was partially converted to the 1,6- and 3,6-quinones (48). Since dihydrodiol epoxide formation requires successive oxidation in the same ring, a single fluorine atom in an angular ring will inhibit this pathway of metabolic activation. Thus, the DNA binding (55) and carcinogenicity of the angular ring fluorinated compounds are lower than those of the corresponding hydrocarbons (24,45,47,50).

The effects of fluorine substitution at the peri-positions adjacent to angular rings of PAH are not as straightforward. It has been shown that the trans-dihydrodiols adjacent to peri fluorine substituents adopt the pseudo-diaxial conformation, as in the case of peri-methyl substitution. Thus, NMR experiments demonstrated that the 5,6- and 8,9-dihydrodiols of 7-fluorobenz[a]anthracene and the 7,8-dihydrodiol of 6-fluorobenzo[a]pyrene exist in the pseudo-diaxial conformation (48,56). It was suggested that the inhibitory effect of a peri-fluorine substituent on tumorigenicity, as observed for 5F-7-methylbenz[a]anthracene, 5F-12-methylbenz[a]anthracene, 5F-7,12-dimethylbenz[a]anthracene, 6F-benzo[a]pyrene, and 12F-5-methylchrysene, might be due either to a low rate of conversion of the diaxial dihydrodiols to the corresponding bay region dihydrodiol epoxides or to the inherently lower tumorigenicity of the dihydrodiol epoxide metabolite (56). For 6F-benzo[a]pyrene, this explanation appears to be correct (48). Thus, the 7,8-dihydrodiol is formed metabolically at similar rates from both 6F-benzo[a]pyrene and benzo[a]pyrene. Both dihydrodiols have the same absolute configuration, but the 7,8-dihydrodiol of 6F-benzo[a]pyrene is diaxial and is not appreciably mutagenic toward Chinese hamster V79 cells, in contrast to benzo[a]pyrene-7,8-dihydrodiol.

In contrast, the fluorine atom at the peri-position of 12F-5-methylchrysene influences dihydrodiol formation in the adjacent angular ring. Whereas the ratio of 5-MeC-7,8-diol to 5-MeC-1,2-diol in mouse epidermis was 1:1, 2 hr after topical application of [^{3}H]5-MeC, the ratio of 12F-5-methylchrysene-7,8-diol to 12F-5-methylchrysene-1,2-diol was 68:1. In contrast to 5-MeC, the metabolites formed from 12F-5-methylchrysene in mouse skin resulted almost exclusively from oxidation at the 7,8-bond (57). Thus, metabolic switching to the less tumorigenic 7,8-dihydrodiol appears to be the basis for the lower tumorigenicity of 12F-5-methylchrysene compared to 5-MeC.

Prospects for Further Research

With some exceptions, the relationship between structure and activity of various methylated and fluorinated PAH is now reasonably well understood. Important exceptions are 6-,7-, and 8-methylbenz[a]-

anthracene and 7,12-dimethylbenz[a]anthracene. There is presently no satisfactory explanation for the potent tumorigenicity of these compounds compared to other monomethyl or dimethylbenz[a]anthracenes. Nevertheless, current knowledge of structure-tumorigenicity relationships should allow accurate prediction of the tumorigenic activities on mouse skin of untested methylated PAH. Extension of methylated PAH testing to other bioassay systems would be desirable. Little is known, for example, about the carcinogenicity of methylated PAH administered by inhalation or in the diet. In addition, the carcinogenic activities of mixtures of PAH should be more extensively investigated since human exposure is always to mixtures.

Whereas bay region dihydrodiol epoxides appear to be major ultimate carcinogens of a number of methylated PAH, the stereochemical aspects of dihydrodiol epoxide reactions with DNA and tumorigenicity require further investigation. Among the unsubstituted PAH, it is known that the absolute configuration of dihydrodiol epoxide metabolites is a key feature in their tumorigenic activities. Whether such stereochemical subtleties operate for the bay region dihydrodiol epoxides having a methyl group and epoxide ring in the same bay region is unknown, but appears likely based on the differences in tumorigenicity between anti-DE-I and syn-DE-I.

Since the general features of methylated PAH metabolic activation are known, it should now be possible to design effective chemopreventive strategies. A key to this approach is a better understanding of the ability of the organism to detoxify dihydrodiol epoxide metabolites by conjugation with glutathione. If glutathione conjugates of methylated PAH dihydrodiol epoxides are formed, it may be possible to enhance their rates of formation by various pretreatments. In addition, it will be important to identify naturally occurring or synthetic compounds that can prevent dihydrodiol epoxide formation or reaction with DNA in vivo.

Finally, it is important that further research be carried out on the identification and mechanism of action of environmental cocarcinogens and tumor promoters which can enhance the carcinogenicity of PAH. Human exposure to at least trace amounts of PAH is unavoidable. The probability of eventual tumor development may be controlled primarily by repeated exposure to cocarcinogens or promoters.

Acknowledgments

Our research on methylated PAH is supported by Grant CA-32242 from The National Cancer Institute.

Literature Cited

1. International Agency for Research on Cancer. "IARC Monographs on the Evaluation of the Carcinogenic Risk of Chemicals to Humans, Volume 32"; IARC: Lyon, France, 1983; pp. 33-53.
2. Committee on Pyrene and Selected Analogues, Board on Toxicology and Environmental Health Hazards, National Research Council. "Polycyclic Aromatic Hydrocarbons: Evaluation of Sources and Effects"; National Academy Press; Washington, D.C., 1983.
3. Dipple, A. In "Chemical Carcinogens"; Searle, C.E., Ed.; American Chemical Society: Washington, D.C., 1976, pp. 245-314.

4.   Hecht, S.S.; Amin, S.; Rivenson, A.; Hoffmann, D. Cancer Lett.
     8, 65-70, 1979.
5.   LaVoie, E.J.; Tulley-Freiler, L.; Bedenko, V.; Hoffmann, D.
     Cancer Res. 1981, 41, 3441-7.
6.   LaVoie, E.J.; Bedenko, V.; Tulley-Freiler, L.; Hoffmann, D.
     Cancer Res. 1982, 42, 4045-9.
7.   Wislocki, P.G.; Fiorentini, K.M.; Fu, P.P.; Yang, S.K.; Lu,
     A.Y.H. Carcinogenesis 1982, 3, 215-7.
8.   Hecht, S.S.; Bondinell, W.E.; Hoffmann, D.   J. Natl. Cancer
     Inst. 1974, 53, 1121-33.
9.   Coombs, M.M.; Bhatt, T.S.; Croft, C.J. Cancer Res. 1973, 33,
     832-7.
10. Iyer, R.P.; Lyga, J.W.; Secrist, J.A., III; Daub, G.H.; Slaga,
     T.J. Cancer Res. 1980, 40, 1073-6.
11. DiGiovanni, J.; Diamond, L.; Harvey, R.G.; Slaga, T.J. Carcino-
     genesis 1983, 4, 403-7.
12. Levin, W.; Wood, A.W.; Chang, R.L.; Newman, M.S.; Thakker, D.R.;
     Conney, A.H.; Jerina, D.M. Cancer Lett. 1983, 20, 139-46.
13. Heidelberger, C.; Baumann, M.E.; Greisbach, L.; Ghobar, A.;
     Vaughan, T.M. Cancer Res. 1962, 22, 78-83.
14. Hecht, S.S.; Hirota, N.; Loy, M.; Hoffmann, D.   Cancer Res.
     1978, 38, 1694-6.
15. Cavalieri, E.; Roth, R.; Grandjean, C.; Althoff, J.; Patil, K.;
     Liakus, S.; Marsh S. Chem-Biol Interactions 1978, 22, 53-67.
16. Wood, A.W.; Levin, W.; Chang, R.L.; Conney, A.H.; Slaga, T.J.;
     O'Malley, R.F.; Newman, M.S.; Buhler, D.R.; Jerina, D.M. J.
     Natl. Cancer Inst. 1982, 69, 725-8.
17. Slaga, T.J.; Huberman, E.; DiGiovanni, J.; Gleason, G.; Harvey,
     R.G. Cancer Lett. 1979, 6, 213-20.
18. Slaga, T.J.; Gleason, G.L.; Hardin, L. Cancer Lett. 1979, 7,
     97-102.
19. Lacassagne, A.; Buu-Hoi, N.P.; Zajdela, F. C.R. Acad Sci. 1958,
     246, 1477-80.
20. Wong, T.K.; Chiu, P-L; Fu, P.P.; Yang, S.K.   Chem-Biol Inter-
     actions 1981, 36, 153-66.
21. Kinoshita, T.; Konieczny, M.; Santella, R.; Jeffrey, A.M.
     Cancer Res. 1982, 42, 4032-6.
22. Harvey, R.G.   In "Safe Handling of Chemical Carcinogens,
     Mutagens, Teratogens and Highly Toxic Substances, Volume 2";
     Walters, D.B., Ed.; Ann Arbor Science: Ann Arbor, MI, 1980, pp.
     439-468.
23. Stevenson, J.L.; von Haam, E. Amer. Ind. Hyg. Assoc. J. 1965,
     26, 475-8.
24. Amin, S.; Juchatz, A.; Furuya, K.; Hecht, S.S.   Carcinogenesis
     1981, 2, 1027-32.
25. Hecht, S.S.; Rivenson, A.; Hoffmann, D. Cancer Res. 1980, 40,
     1396-99.
26. Melikian, A.A.; Amin, S.; Hecht, S.S.; Hoffmann, D.; Pataki, J.;
     Harvey, R.G. Cancer Res., 1984, 44, 2524-9.
27. Melikian, A.A.; LaVoie, E.J.; Hecht, S.S.; Hoffmann, D. Carcin-
     ogenesis 1983, 4, 843-49.
28. Geacintov, N.E.; Yoshida, H.; Ibanez, V.; Harvey, R.G. Bio-
     chem. Biophys. Res. Commun. 1981, 100, 1569-77.
29. Meehan, T.; Gamper, H.; Becker, J.H.   J. Biol. Chem. 1982,
     257-10479-85.

30. MacLeod, M.C.; Selkirk, J.K. Carcinogenesis 1982, 3, 287-92.
31. Coombs, M.M.; Kissonerghis, A.M.; Allen, A.J.; Vose, C.W. Cancer Res. 1979, 39, 4160-5.
32. Moschel, R.C.; Baird, W.M.; Dipple, A. Biochem. Biophys Res. Commun. 1977, 76, 1092-8.
33. Sawicki, J.T.; Moschel, R.C.; Dipple, A. Cancer Res. 1983, 43, 3212-8.
34. Iball, J. Nature 1964, 201, 916-7.
35. Kashino, S.; Zacharias, D.E.; Prout, C.K.; Carrell, H.L.; Glusker, J.P.; Hecht, S.S.; Harvey, R.G. Acta Cryst. C., 1984. In press.
36. Yang, S.K.; Chou, M.W.; Fu, P.P. In "Carcinogenesis: Fundamental Mechanisms and Environmental Effects"; Pullmann, B.; Ts'o, P.O.P.; Gelboin, H., Eds.; D. Reidel: London, 1980; pp. 143-56.
37. Slaga, T.J.; Iyer, R.P.; Lyga, W.; Secrist, A., III; Daub, G.H.; Harvey, R.G. In "Polynuclear Aromatic Hydrocarbons: Chemistry and Biological Effects"; Bjorseth, A.; Dennis, A.J., Eds.; Batelle Press: Columbus, Ohio, 1980, pp. 753-69.
38. Yang, S.K.; Chou, M.W.; Fu, P.P. In "Polynuclear Aromatic Hydrocarbons: Chemistry and Biological Effects"; Bjorseth, A.; Dennis, A.J., Eds.; Batelle Press: Columbus, Ohio, 1980, pp. 645-62.
39. Yang, S.K.; Chou, M.W.; Fu, P.P. In "Polynuclear Aromatic Hydrocarbons: Chemical Analysis and Biological Fate"; Cooke, M.; Dennis, A.J., Eds.; Batelle Press: Columbus, Ohio, 1981, pp. 253-64.
40. Amin, S.; Camanzo, J.; Hecht, S.S. Carcinogenesis 1982, 3, 1159-63.
41. Miller, J.A.; Miller, E.C. Cancer Res. 1963, 23, 229-39.
42. Miller, E.C.; Miller, J.A. Cancer Res. 1960, 20, 133-7.
43. Newman, M.S. In "Polynuclear Aromatic Hydrocarbons: Chemistry, Metabolism and Carcinogenesis"; Freudenthal, R.; Jones, P.W., Eds.; Raven Press: New York, 1976, pp. 203-7.
44. Huberman, E.; Slaga, T.J. Cancer Res. 1979, 39, 411-14.
45. Hecht, S.S.; LaVoie, E.; Mazzarese, R.; Hirota, N.; Ohmori, T.; Hoffmann, D. J. Natl. Cancer Inst. 1979, 63, 855-61.
46. Hecht, S.S.; Hirota, N.; Loy, M.; Hoffmann, D. Cancer Res. 1978, 38, 1694-98.
47. Buhler, D.R.; Unlu, F.; Thakker, D.R.; Slaga, T.J.; Newman, M.S.; Levin, W.; Conney, A.H., Jerina, D.M. Cancer Res. 1982, 42, 4779-83.
48. Buhler, D.R.; Unlu, F.; Thakker, D.R.; Slaga, T.J.; Conney, A.H.; Wood, A.W.; Chang, R.L.; Levin, W.; Jerina, D.M. Cancer Res. 1983, 43, 1541-9.
49. Buening, M.K.; Levin, W.; Wood, A.W.; Chang, R.L.; Agranat, I.; Rabinovitz, M.; Buhler, D.R.; Mah, H.D.; Hernandez, O.; Simpson, R.B.; Jerina, D.M.; Conney, A.H.; Miller, E.C.; Miller, J.A. J. Natl. Cancer Inst. 1983, 71, 309-15.
50. Hecht, S.S.; LaVoie, E.J.; Bedenko, V.; Pingaro, L.; Katayama, S.; Hoffmann, D.; Sardella, D.J.; Boger, E.; Lehr, R.E. Cancer Res. 1981, 41, 4341-45.
51. Chang, R.L., Levin, W.; Wood, A.W.; Lehr, R.E.; Kumar, S.; Yagi, H.; Jerina, D.M.; Conney, A.H. Cancer Res. 1982, 42, 25-9.

52. Sardella, D.J.; Boger, E.; Ghoshal, P.K. In "Polynuclear Aromatic Hydrocarbons: Chemical Analysis and Biological Fate"; Cooke, M.; Dennis, A.J., Eds.; Battelle Press: Columbus, Ohio, 1981, pp. 529-38.

53. Harvey, R.G.; Dunne, F.B. Nature 1978, 273, 566-8.

54. Boger, E.; O'Malley, R.F.; Sardella, D.J. J. Fluorine Chem. 1976, 8, 513-25.

55. Daniel, F.B.; Joyce, N.J. J. Natl. Cancer Inst. 1983, 70, 111-8.

56. Chiu, P.L.; Fu, P.P.; Yang, S.K. Biochem. Biophys. Res. Commun. 1982, 106, 1405-11.

57. Amin, S.; Camanzo, J.; Hecht, S.S. Cancer Res. 1984, 44, 3772-8.

RECEIVED March 5, 1985

Mechanisms of Interaction of Polycyclic Aromatic Diol Epoxides with DNA and Structures of the Adducts

NICHOLAS E. GEACINTOV

Chemistry Department, New York University, New York, NY 10003

Spectroscopic studies on complexes derived from the binding of benzo(a)pyrene-7,8-diol-9,10-epoxide (BaPDE) to DNA indicate that the conformations of the adducts can be broadly classified into two types: site I which displays most of the properties of inter-calative adducts, and site II which is characterized by an orientation of the planar aromatic residues tilted closer to the axis of the helix. Both the syn and the anti diastereomers of BaPDE form unstable type I physical intercalation complexes and undergo speci-fic and general acid catalysis to form tetraols (>90%) and covalent adducts (<10%). The biologically highly active (+) anti-BaPDE enantiomer undergoes a marked reorientation upon covalent binding to form almost exclusively site II adducts, while the less active (-) anti-BaPDE enantiomer, as well as racemic syn-BaPDE, give mixtures of site I and site II adducts. The presence of site II type of adducts appears to be correlated with high tumorigenic and mutagenic activi-ties in this family of stereoisomeric diol epoxides.

The reaction of metabolically generated polycyclic aromatic diol epoxides with DNA in vivo is believed to be an important and criti-cal event in chemical carcinogenesis (1,2). In recent years, much attention has been devoted to studies of diol epoxide-nucleic acid interactions in aqueous model systems. The most widely studied reactive intermediate is benzo(a)pyrene-7,8-diol-9,10-epoxide (BaPDE), which is the ultimate biologically active metabolite of the well known and ubiquitous environmental pollutant benzo(a)pyrene. There are four different stereoisomers of BaPDE (Figure 1) which are characterized by differences in biological activities, and reactivi-ties with DNA (2-4). In this review, emphasis is placed on studies of reaction mechanisms of BPDE and related compounds with DNA, and the structures of the adducts formed.

0097–6156/85/0283–0107$06.00/0
© 1985 American Chemical Society

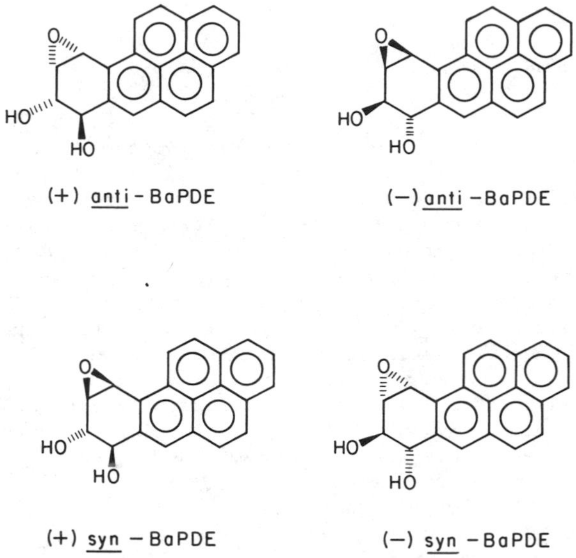

(+) <u>anti</u> - BaPDE   (−) <u>anti</u> - BaPDE

(+) <u>syn</u> − BaPDE   (−) <u>syn</u> − BaPDE

Figure 1.   Stereoisomers of benzo(<u>a</u>)pyrene-7,8-diol-9,10-oxide.

Principles of Experimental Methods

The pyrene-like aromatic chromophore of BaPDE is characterized by a prominent and characteristic absorption spectrum in the ~310-360 nm spectral region, and a fluorescence emission in the ~370-460 nm range. These properties are sensitive to the local microenvironment of the pyrenyl chromophore, and spectroscopic techniques are thus useful in studies of the structures of the DNA adducts and in monitoring the reaction pathways of BaPDE and its hydrolysis products in DNA solutions.

Linear Dichroism. In this technique, the DNA molecules are aligned either by an applied electric field pulse or in a flow gradient. The orientation of the aromatic residues of the metabolite model compounds bound to the DNA (either covalently or non-covalently) relative to the orientation of the DNA bases is probed utilizing linearly polarized light. The linear dichroism ΔA can be either negative or positive, and is defined as

$$\Delta A = A_{//} - A_{\perp} = (3/2) \, A \, (3 \cos^2\theta - 1) \, f \qquad (1)$$

where $A_{//}$ and $A_{\perp}$ are the absorbance components measured by the linearly polarized light oriented parallel and perpendicular with respect to the direction of the applied electric field, or flow direction. The orientation angle θ of the transition moment vectors is measured relative to this direction, A is the isotropic absorbance of the sample, and f is a wavelength independent function which defines the degree of orientation of the DNA. The axis of the DNA double helix tends to align parallel to the direction of the flow or electric field. Since, in the Watson-Crick model $\theta = 90°$, ΔA is negative in sign below ~310 nm. Depending on the orientation of the in-plane, long axis transition moment of the pyrene chromophore, ΔA can be either negative or positive in the 310 - 360 nm region.

When there is only one oriented chromophore present, it is important to note that the magnitude of ΔA is proportional to A. In such cases, the linear dichroism spectra resemble the absorption spectra (5,6), i.e. $|\Delta A| \propto A$.

Two types of DNA binding sites. Two different spectroscopically distinct types of binding sites have been identified utilizing absorption, fluorescence and linear dichroism data on non-covalent (6), and covalent (7) pyrene-like metabolite model compound-DNA complexes.

Site I is characterized by a relatively large red shift of ~10 nm in the absorption maxima (relative to the aqueous solution spectra), exhibiting maxima at ~337 and ~354 nm, and a negative ΔA spectrum; all of these properties are consistent with an intercalation-complex geometry in which the planar pyrene ring-system is nearly parallel to the planes of the DNA bases.

Site II is characterized by a relatively small 2-3 nm red shift in the absorption spectrum and a positive ΔA spectrum. In this conformation, the planes of the pyrene moeities tend to align parallel rather than perpendicular to the axis of the DNA helix.

This classification has recently been adopted by other workers as well(8-10).

Formation of Physical Intercalative Diol Epoxide-DNA Complexes

Stopped flow kinetic measurements indicate that when two aqueous solutions, one containing BaPDE and the other DNA, are mixed rapidly, a non-covalent site I-type complex is formed within 5 ms or less (11).
      The linear dichroism spectra of such non-covalent complexes can be measured by the flow orientation method, and typical ΔA spectra obtained with racemic anti- and syn-BaPDE are shown in Figure 2. These experiments were carried out at pH 9.2 in order to minimize the degradation of the diol epoxide molecules during the measurement (less than 2 minutes). The structures and the negative sign of the ΔA spectra suggest that both compounds form intercalative physical complexes. The ΔA spectra of the covalent adducts are also shown for comparison (after allowing the reactions to go to completion and extracting the tetraol hydrolysis products of BaPDE with ether); for anti-BaPDE ΔA is positive, as found previously (5), but for the syn adducts ΔA is negative, as shown already by Undeman et al (10). Thus, considerable re-orientation of the pyrenyl moeity is occurring in the case of (+) anti-BaPDE as a result of the covalent binding reaction, while with (+) syn-BaPDE the conformational changes, if any, appear to be minor. The kinetics of such linear dichroism changes have recently been studied utilizing the enantiomers (+) and (-) anti-BaPDE (12).
      Analogous results have been recently obtained with trans-1,2-dihydroxy-anti-3,4-epoxy-1,2,3,4-tetrahydro-5-methyl chrysene (13) and the epoxide 1-oxyranylpyrene (14). Thus, the formation of non-covalent intercalative site I complexes appears to be a general phenomenon which governs the interaction of polycyclic aromatic epoxides with DNA (15-17).

Reaction Pathways of BaPDE in Aqueous DNA Solutions

The experimentally observed pseudo-first order rate constant k is increased in the presence of DNA (18,19). This enhanced reactivity is a result of the formation of physical BaPDE-DNA complexes; the dependence of k on DNA concentration coincides with the binding isotherm for the formation of site I physical intercalative complexes (20). Typically, over $\sim$90% of the BaPDE molecules are converted to tetraols, while only a minor fraction bind covalently to the DNA bases (18,21-23). The dependence of k on temperature (21,24), pH (21,23-25), salt concentration (16,20,21,25), and concentration of different buffers (23) has been investigated. In 5 mM sodium cacodylate buffer solutions the formation of tetraols and covalent adducts appear to be parallel pseudo-first order reactions characterized by the same rate constant k, but different ratios of products (21,24). Similar results are obtained with other buffers (23). The formation of carbonium ions by specific and general acid catalysis has been assumed to be the rate-determining step for both tetraol and covalent adduct formation (21,24).
      The experimental observations in cacodylate buffer solutions are consistent with a mechanism involving a kinetically common intermediate according to the following reaction scheme:

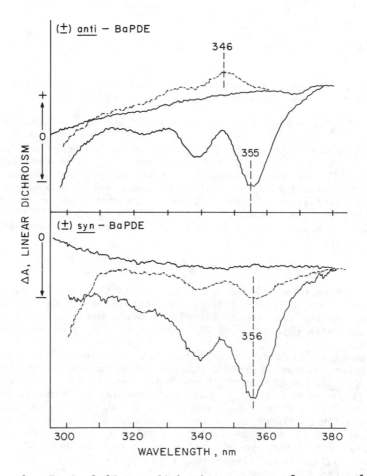

Figure 2.   Typical linear dichroism spectra of non-covalent (solid lines) and covalent (dashed lines) DNA complexes (data of M. Shahbaz).

$$\text{BaPDE + DNA} \xrightleftharpoons[k_2]{k_1} \text{[BaPDE...DNA]} \xrightarrow{k_3} \text{[BaPDE}^+\text{...DNA]} \qquad (2)$$

with k_h pathway from BaPDE + DNA to Tetraols, k_T pathway from [BaPDE...DNA] to Tetraols, and k_C pathway from [BaPDE$^+$...DNA] to Covalent adducts.

As long as the rate constants $k_1, k_2 \gg k_3$, the pseudo-first order rate constant k for the reaction of BaPDE is (20,21):

$$k = (1-X_b)k_h + k_3 X_b, \qquad\qquad X_b = K[DNA]/(1 + K[DNA]) \quad (3)$$

where X_b is the fraction of diol epoxide molecules bound to DNA, and $K = k_1/k_2$ is the equilibrium binding constant. The two terms in equation (3) represent weighted contributions of the reactions of free BaPDE molecules, and of molecules complexed with DNA, and this equation provides an adequate fit to the experimental data with K = 12,000 M^{-1} in 5 mM sodium cacodylate solution at pH 7, 25°C(20).

The fraction of diol epoxide molecules which bind covalently to DNA (f_{cov}) is (21):

$$f_{cov} = [k_C/(k_C+k_T)][k_3 X_b/k] \qquad\qquad\qquad (4)$$

The first term on the right-hand side is the fraction of carbonium ions which decay by forming covalent bonds, while the second term denotes the fraction of all diol epoxide molecules which react while complexed physically to DNA, rather than as free molecules in solution.

Equation (4) demonstrates that the relationship between the association constant K, which is sensitive to the ionic strength (16,17,21,25), and the level of covalent binding, f_{cov}, is a complex one. It is known that f_{cov} decreases upon the addition of NaCl or MgCl$_2$, and this effect has been taken as evidence that physical intercalation complexes play a role in the covalent binding reaction (17,22,26). While this conclusion may still be correct, such evidence is insufficient since it has been shown that not only K, but also k_3 (21,25), and the branching ratio k_C/k_T (21) in Equation (4) depend on the salt concentration.

Physical Intercalation Complexes, Covalent Binding and Hydrolysis

The possible existence of two types of binding sites for physical BaPDE-DNA complexes has stimulated various proposals regarding the relative importance of each in the two reaction pathways of tetraol formation and covalent binding. In order to facilitate the following discussion, the various possible reaction pathways are summarized in Figure 3. In addition, we have added the possibility that there may be a rapid exchange between physical binding sites (k_E, k_{-E}), a factor which seems to have been neglected up till now.

Physical binding studies (8,9,27) suggest that physical complex formation with DNA by intercalation appears to be sequence-specific. Thus, BaPT and pyrene intercalate much more strongly in poly(dA-

dT):poly(dA-dT) than in poly(dG-dC):poly(dG-dC) and in the homopolymers poly(dG):poly(dC) and poly(dA):poly(dT). This preference for
dA-dT sequences has prompted Chen to suggest that the covalent
binding of BaPDE to guanine proceeds at external site II physical
complexes, rather than at intercalative site I complexes, while
BaPDE molecules intercalated at dA:dT rich sequences preferentially
undergo hydrolysis (8). In figure 3, Chen's hypothesis corresponds
to $k_3^I(T) \gg k_3^{II}(T)$, and $k_3^{II}(C) \gg k_3^I(C)$.

Meehan and Bond (23) on the other hand, have taken an opposite
view, namely that $k_3^{II}(C) \ll k_3^I(C)$, while $k_3^{II}(T) \gg k_3^I(T)$. Thus, in
this view, the hydrolysis occurs at external binding sites, while
covalent binding occurs at intercalation sites. Furthermore, they
reject the common intermediate model (Equation 2) on the basis of
their belief that the rates of reaction for tetraol formation and
adduct formation and the ratio of the products should be the same in
such a model. While these rates of reaction are the same and the
product ratios are observed to be different, this is fully consistent for a set of parallel pseudo-first order reactions involving a
common intermediate (29) as pointed out above. Thus, the data of
Meehan and Bond does not demonstrate the validity of the two-domain
model (23).

The reaction schemes of Chen (8) and of Meehan and Bond (23),
neglect the possibility that exchange between the two physical
binding sites (Figure 3) may be occurring on time scales which are
much faster than those characterizing the chemical reaction pathways
of BaPDE. Thus, while it is still possible that the reactions may
be occurring at physically different binding sites, kinetically only
one common precursor for these reactions may be distinguishable
($k_1, k_2, k_E, k_{-E} \gg k_3$ in Figure 3). Utilizing a kinetic flow dichroism method, we have established that there is a distinct kinetic relationship between the disappearance of physically bound (+)-
anti-BaPDE molelcules at type I intercalative binding sites, and the
appearance of covalent adducts at sites II. However, because of the
foregoing arguments involving the rapid exchange of physically bound
molecules between sites I and sites II, the exact nature of the
microcomplexes which are involved in the covalent and hydrolysis
reactions remains to be elucidated.

Structure of the Covalent Adducts

There is considerable disagreement between different researchers on
the conformations of covalent adducts derived from the binding of
BaPDE to DNA. It has been reported that the covalent binding of
BaPDE to closed circular DNA results in the unwinding of the DNA
helix (26,29). Since analogous unwinding effects are produced by
non-covalently intercalated acridine dyes, these effects have been
attributed to the formation of covalent intercalative BaPDE complexes (29). However, these conclusions can be criticized since
other types of conformations, or effects other than covalent-inter-
calative binding of BaPDE may give rise to the unwinding of the
double helix (26,30). Furthermore, the absorption, linear dichroism, and fluorescence properties of the covalent (+)-anti-BaPDE-
DNA complexes are not consistent with those of classical intercalation complexes, as is shown below.

Linear Dichroism. The ΔA spectra of covalent adducts derived from the binding of racemic anti-BaPDE and of the enantiomer (+)-anti-BaPDE to DNA are positive in sign and similar in shape (5,31); this is expected since the (+) enantiomer binds more extensively to DNA than the (-) enantiomer (15). These covalent adducts are therefore of the site II type.

Subsequent studies of adducts derived from the covalent binding of other epoxide model compounds with pyrenyl chromophores (Figure 4) to DNA (7,14,32-33) show that the occurrence of positive ΔA spectra is the exception rather than the rule. A detailed analysis of the linear dichroism spectra of covalent adducts obtained with the compounds shown in Figure 4, demonstrates that there is a marked heterogeneity of adducts. Qualitatively the linear dichroism spectra can be accounted for in terms of superpositions of positive ΔA spectra due to site II binding, and negative ΔA spectra due to site I binding, with the latter dominating in 9,10-BaPE, 7,8-BaPE (unpublished), BePDE, BePE and 1-OP covalent DNA adducts. Other studies on adducts derived from the binding of benzo(a)pyrene-9,10-diol-7,8-oxide to DNA (22) suggest that site I complexes also dominate in this case.

The site I adducts are characterized by a near-parallel (within 25°) average orientation of the planar pyrene residue with the planes of the DNA bases, and a relatively strong interaction between the π-electrons of the pyrene residues and the DNA bases.

Hogan et al (34), who studied the electric linear dichroism of (+) anti-BaPDE bound covalently to small DNA fragments (~145 base pairs), showed that the linear dichroism within the DNA absorption band decreased with an increasing level of binding. These results indicate that kinks are produced in the DNA molecule upon interaction with the diol epoxide molecules. They suggested that the covalently bound BaPDE moeities reside at these kinks in wedge-shaped intercalation complexes. This model is reasonable; however, such a structure is more consistent with the negative, red-shifted linear dichroism spectra of site I binding sites (31) than with the major site II type of binding site observed with the covalent (+)-anti-BaPDE-DNA adducts studied by Hogan et al. The red-shift of only 2-3 nm displayed by the covalently bound residues appears to be too small for an intercalative geometry; the larger red shift, and the negative linear dichroism displayed by adducts bound at sites I, appear more consistent with this model.

While it is reasonable to assume that kinks are formed at the site of the covalent binding of BaPDE (34), it is also possible that such bends arise elsewhere on the double helix, due to the known formation of nicks and single-strand breaks (35,36). Therefore, the existence of kinks in the DNA helix at site I or site II binding sites should be further investigated, before such a model can be adopted definitively.

It is interesting to note that the absence of the two OH groups at the 7 and 8 positions of 9,10-BaPE (7), and the 7,8-carbon atoms in 1-OP (14), lead to a partial loss of stereoselective effects in the covalent binding of these molecules to DNA; in the adducts derived from these two molecules, site I adducts dominate. In contrast, in the case of the diol epoxide (+)anti-BaPDE, site II complexes account for over 90% of the binding(31).

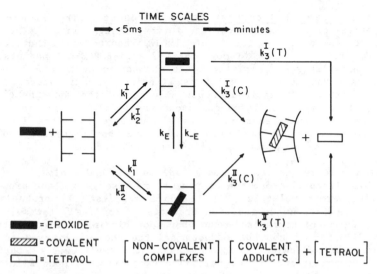

Figure 3. Possible reaction schemes of BaPDE in aqueous solutions containing DNA.

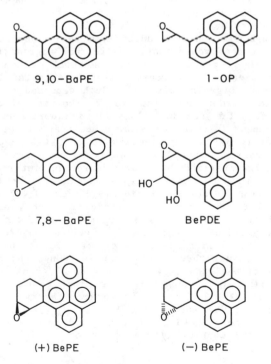

9,10-BaPE

1-OP

7,8-BaPE

BePDE

(+) BePE

(−) BePE

Figure 4. Structures of model compounds which upon covalent binding to DNA give predominantly site I adducts.

Fluorescence and Heterogeneity of Adducts. The fluorescence
properties of adducts can provide further insight into the heterogeneity of the adducts. Prusik et al (37) reported that there are two
fluorescent components in covalent (+)-anti-BaPDE-DNA adducts. One
component with a 75% amplitude, was characterized by a 8.2 ns lifetime, and the other by a 125 ns lifetime (in air-saturated solutions) with a 25% amplitude. Upon dilution of the DNA, the relative
amplitude of the long-lived component was found to increase. Recent
measurements by Undeman et al (10) indicate that the long-lived
component is a minor one (42 ns, 6%), and that there are two other
short-lived decay components (1.6 ns, 52% and 7.0 ns, 42% amplitude).

Our more recent measurements (38) confirm Undeman's finding that
the long-lived fluorescence components represents a minor proportion
of fluorescent molecules; the earlier samples (37) probably contained fluorescent degradation products, most likely tetraols. The
fluorescence of tetraols is quenched upon binding physically to DNA;
thus dilution of the DNA adducts shifted the equilibrium to free,
fluorescent molecules, thus increasing the amplitudes of the long-
lived component and the overall fluorescence yield upon dilution of
the samples (37). Such DNA concentration effects on the fluorescence of covalent adducts were not observed by Undeman et al (39)
and by Hogan et al (34). Since the fluorescence lifetimes of the
major portion of covalently bound molecules is very small, the
quenching of this fluorescence by intermolecular DNA-DNA interactions, as previously proposed (37), will be difficult to observe
even if the adducts are externally located.

Even though previously studied samples of covalent BaPDE-DNA
adducts may have been contaminated by tetraol degradation products,
a long-lived fluorescence component is still present even after
repeated extractions of the aqueous solutions containing the covalent adducts with organic solvents. Such repeated organic solvent
extractions are known to be effective for the removal of physically
bound tetraols. The residual long-lived fluorescence component,
even though present in relatively low concentration, gives rise to a
disproportionately large contribution to the steady-state fluorescence yield because of its long lifetime (38). This fluorescence
component is characterized by a low degree of interaction of the
pyrene chromophore with the DNA bases and thus probably represents
an externally bound adduct as previously discussed (37).

We find that the fluorescence yield of freshly prepared covalent (+)-anti-BaPDE-DNA adducts in oxygen-free solutions is 66+2
lower than the yield of the tetraol 7,8,9,10-tetrahydroxytetrahydro-
benzo(a)pyrene (BaPT) in the absence of DNA. Since the fluorescence
lifetime of BaPT under these conditions is 200ns, the mean fluorescence lifetime of the adducts (see reference 37) can be estimated to
have a lower limit of ~3ns, which is close to the mean value of
0.52x1.6 + 0.42x4.0 = 2.7 ns estimated from the two short fluorescence components of Undeman et al (10).

The fluorescence of BaPT (6) and of other related compounds (40)
seems to be nearly completely quenched upon physical intercalation
into DNA.

Fluorescence Quenching and Accessibility of Adducts. The solvent
accessibility of fluorophores bound to DNA can be probed by magnetic

resonance methods (41), and by fluorescence quenching techniques.
If F_o is the yield in the absence of quenchers, F the yield in
the presence of quenchers (concentration Q), and τ_o is the fluorescence lifetime when [Q] = 0, the relative yield as a function of
[Q] is given by the Stern-Volmer equation:

$$F_o/F = 1 + \tau_o K [Q] \qquad (5)$$

K is the bimolecular encounter rate, multiplied by a probability
factor which determines the efficiency of quenching per collisional
encounter. This constant is dependent on the degree of solvent
exposure of the adducts. For freely accessible BaPT molecules in
aqueous solutions $K = 9 \times 10^9 M^{-1} s^{-1}$ in the case of molecular oxygen
and $7.5 \times 10^8 M^{-1} s^{-1}$ in the case of acrylamide (38). The accessibility of the pyrenyl chromophores in the DNA adducts has been investigated utilizing oxygen (10,37), or acrylamide (22,34) as quenchers. In the case of oxygen it is known that the accessibility of
intercalated chromophores is reduced by a factor of 20-30 (42,43).
When the dye ethidium bromide is intercalated between the bases of
DNA, it appears to be completely inaccessible to acrylamide and its
fluorescence yield is unaffected at acrylamide concentrations as
high as 0.8M (22). Utilizing a high pressure cell to measure F_o/F
for covalent (+)-anti-BaPDE-DNA adducts for oxygen concentrations up
to 90 mM, Undeman et al (10) concluded that the adducts display at
best a low accessibility to molecular oxygen. On the basis of
acrylamide quenching data, Hogan et al (34) concluded that the
fluorescent pyrenyl moeities in the covalent adducts are at least 9-
fold less accessible than in nuclease-digested complexes. The
wedge-shaped intercalation model for the covalent complexes rests,
in part, on these fluorescence quenching results. However, MacLeod
et al (22) compared the acrylamide quenching of the fluorescence of
ethidium bromide-DNA intercalation complexes with those of the covalent (+)-BaPDE-DNA adducts, and they concluded that the pyrenyl
moeity is not intercalated.
The above results were obtained utilizing measurements of F_o/F
as a function of [Q] according to Equation 5. It should be emphasized that a low extent of fluorescence quenching can be interpreted
in one of two ways: (1) low accessibility of fluorescent adducts
(small K), or (2) low values of fluorescence lifetimes τ_o, even
though the adducts may be totally accessible. If the fluorescence is
heterogeneous, the Stern-Volmer plots of F_o/F versus [Q] are not
straight lines, but will exhibit downward curvature with increasing
quencher concentration (10,34); this effect can also be caused by a
heterogeneity of lifetimes, rather than by a heterogeneity of acces-
sibilities, a possibility which was not considered by Hogan et al
(34).
In view of these uncertainties we have re-investigated the
fluorescence properties of covalent (+)-anti-BaPDE-DNA adducts (38).
The heterogeneity of the fluorescence is clearly apparent in Figure
5a which is a plot of the ratio Fexc/A (fluorescence excitation
spectrum divided by the absorption spectrum); the absorption spectrum A is also shown in this figure for comparison. If one single
species were present, F_{exc}/A should be independent of wavelength, as
seems to be the case above $\sim$354 nm. The maxima at 312, 325 and 341
nm indicate that a species with a higher fluorescence yield is

present; since these maxima do not coincide with the absorption
maxima, the relative concentration of this species must be small and
corresponds to the minor long lifetime component discussed in the
previous section.

The wavelength dependence of F_o/F at an oxygen concentration of
1.3mM and an acrylamide concentration of 0.335M are compared in
Figure 5b. The maxima correspond to those in the F_{exc}/A spectrum in
Figure 5a, indicating that the minor fluorescence species is more
readily quenched (possibly due to a large τ_o) than the major species, whose fluorescence dominates at excitation wavelengths above
356. However, even above 356 nm, the acrylamide quenching Stern-
Volmer plots exhibit marked downward curvatures (data not shown),
indicating that the fluorescence is heterogeneous in this wavelength
range of excitation as well (38). From such acrylamide quenching
plots we find that the smallest measured values of the factor $K\tau_o$
lie in the range of 0.5-1.0 M^{-1}; these values are consistent with
the results of Hogan et al (34). Since τ_o is less than 3 ns, and
could be even as low as 1.6 ns (10), we conclude that the pyrenyl
chromophores are either totally accessible to acrylamide, or, at
best, display a decreased accessibility by a factor of only 2-3
relative to free BaPT molecules. Similar results are obtained upon
a detailed analysis of the oxygen quenching (38).

In summary, a detailed examination of the fluorescence
quenching properties, and taking into account the low values of
fluorescence lifetimes τ_o, indicates that the pyrenyl chromophores
in covalent BaPDE-DNA adducts are accessible to quenching molecules. This conclusion represents a further argument against classical intercalation structures for these covalent complexes. Furthermore, it is likely that these adducts are characterized by many
species each differing slightly from the other because of its microenvironment; such a heterogeneity has been proposed previously
(5), and gives rise to a broadening of the absorption bands and,
possibly, to a heterogeneity of fluorescence lifetimes τ_o.

These conclusions are, at this point, limited to the site II
complexes which dominate in covalent (+)-BaPDE-DNA adducts. The
accessibility of site I type complexes, which we have termed "quasi-
intercalative" (7), is presently being investigated.

Theoretical modeling of adduct conformations

Several groups of researchers have considered adduct conformations
from a theoretical point of view. Lin et al (44) have proposed that
physical intercalation of BaPDE precedes the formation of covalent
BaPDE-DNA adducts, as suggested by Meehan and Straub (15) on more
qualitative grounds. The steric properties of physical DNA intercalation complexes obtained with the two enantiomers of anti-BaPDE
(45,46), and the transition-state alkylation geometries with N2 of
guanine (48) have been examined. Space-filling molecular model
building (48), molecular mechanics and computer graphic studies
suggest that the formation of the major N2 guanine adduct does not
result in a significant structural perturbation of the DNA structure
(49,50). Hingerty and Broyde have performed minimized conformational potential energy calculations for a BaPDE adduct with the
dinucleotide dCpdG (51). Miller and his co-workers have considered

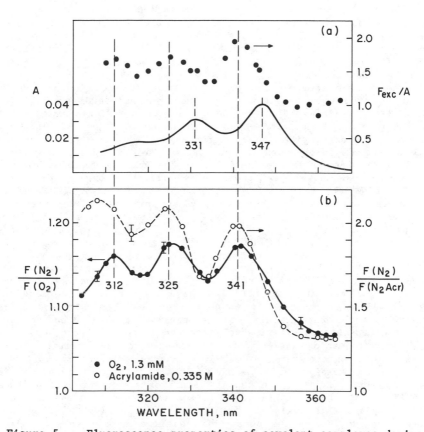

Figure 5.   Fluorescence properties of covalent complexes derived from the covalent binding of (+)-anti-BaPDE to DNA.
(a) Absorption spectrum (A, solid line) and fluorescence excitation spectrum (F_{exc}) divided by A.
(b)   Fluorescence quenching ratio as a function of excitation wavelength. Closed circles : comparison of fluorescence yields in nitrogen-saturated [F(N_2)] and oxygen-saturated [F(O_2)] aqueous solutions.   Open circles: comparison of yields of nitrogen-saturated, and nitrogen-saturated solution containing 0.335 M acrylamide [F(N_2Acr)].   (Adapted from Ref. 38.)

different adduct geometries (52), and the reader is referred to Miller's paper for greater detail (53).

Stereoselective Covalent Binding of BaPDE Enantiomers to DNA

For racemic syn-BaPDE, and (+) and (-) anti-BaPDE, the linear dichroism spectra of the non-covalent intercalative complexes indicate that there is little, if any, difference in the conformations of these physical complexes (Figure 2, 12). However, there are striking differences in the conformations of the covalent syn and anti-BaPDE-DNA adducts (Figure 2, reference 10).

The differences in the adduct structures derived from the binding of the two enantiomers (+) and (-) anti-BaPDE to DNA are particularly striking. The absorbance and linear dichroism spectra for the (+) and the (-) DNA adducts are shown in Figure 6. The (+) adducts are characterized by a single absorption band at 346 nm. A similar maximum is observed in the case of the (-) adduct; however, a prominent shoulder is also apparent at 354 nm. A detailed analysis of these spectra suggest that the (+) adduct is characterized by a major type II binding site (similar to the binding of racemic anti-BaPDE, since the (+) enantiomer binds more extensively to DNA than the (-) enantiomer) with a ∼10% contribution of site I (note the prominent tail beyond 350 nm in the absorption spectrum in Figure 6, top left panel). The (-) adducts appear to be mixtures of site I and site II adducts. These conclusions are fully confirmed by the linear dichroism spectra of these adducts, which are also shown in Figure 6. The ΔA spectrum is positive in the case of the (+) adduct, as expected for site II; the ΔA spectrum in the case of the (-) adduct is positive at those wavelengths where site II dominates (330 and 346 nm), while ΔA is negative at 335-340 and 354 nm where the red-shifted site I absorption bands are most prominent.

The orientation angles for site II in the (+) adducts were found to be in the range of 15-30° (average orientation of the pyrenyl long, in plane axis, relative to the average orientation of the DNA helix); however, in the (-) adducts the orientation of the site II adducts appear to be more tilted with respect to the axis of the helix, and this angle is within the range of 37-45° (31). The orientation angles of the quasi-intercalative site I complexes in the (-) adducts lie in the range of 61-79°; these angles are different from those expected for classical intercalation complexes (31).

Similar experimental results on the linear dichroism of covalent adducts derived from the covalent binding of the two enantiomers of anti-BaPDE to DNA have also been published recently(54).

Conformations of covalent adducts and biological activity

Out of the four BaPDE stereoisomers shown in Figure 1, (+)-anti-BaPDE displays by far the highest biological activity either as a tumorigen, or a mutagen in mammalian cells (1-4). It is also the only enantiomer which gives rise to dominant site II adducts with the long axis of the pyrenyl plane tilted close to the average orientation of the DNA double helix. Therefore, it appears that the occurrence of site II aducts is correlated with high biological activity in this family of benzo(a)pyrene diol epoxide stereoisomers. The quasi-intercalative site I adducts, which may involve the

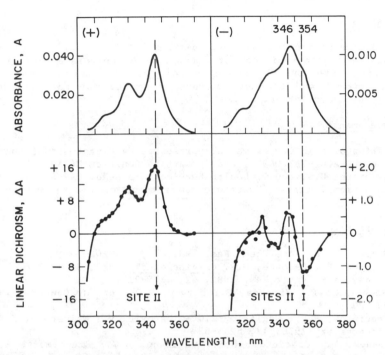

Figure 6.   Absorbance (A) and linear dichroism (ΔA)spectra of covalent adducts derived from the binding of (+)<u>anti</u>-BaPDE (left panels) and (-)<u>anti</u>-BaPDE (right panels) to double-stranded calf thymus DNA. Extent of covalent modification: 0.19% (+) and 0.06%(-) of the bases, nucleotide concentration: 7.3×10^{-4} M.   The DNA was oriented by the pulsed electric field method (<u>31</u>).

binding of BaPDE to sites other than the exocyclic amino group of guanine (2), appear to be correlated with a lower biological activity. It remains to be seen whether similar correlations can be found with other families of stereoisomeric diol epoxide molecules.

Conclusions

The existence of isomeric polycyclic aromatic diol epoxide compounds provides rich opportunities for attempting to correlate biological activities with the physico-chemical reaction mechanisms, and conformational and biochemical properties of the covalent DNA adducts which are formed.

Acknowledgments

The writing of this review was supported by the Department of Energy (Contract DE-ACO2-78EVO4959) and in part by Contract E(11-1)2366, as well as by the U.S. Public Health Service, Grant No. CA20851 awarded by the National Cancer Institute, Department of Health and Human Services.

Literature Cited

1.   Harvey, R. G. Acc. Chem. Res. 1981, 14, 218-26.
2.   Brookes, P.; Osborne, M.R. Carcinogenesis 1983, 3, 1223-6.
3.   Conney, A.H.Cancer Res. 1982, 42, 4875-917.
4.   Newbold, R.F.; Brookes, P.; Harvey, R.G. Int. J. Cancer 1979, 24, 203-9.
5.   Geacintov, N.E.; Gagliano, A.G.; Ivanovic, V.; Weinstein, I.B. Biochemistry 1978, 17, 5256-62.
6.   Ibanez, V.; Geacintov, N.E.; Gagliano, A.G.; Brandimarte S.; Harvey, R.G. J. Am. Chem. Soc. 1980, 102, 5661-66.
7.   Geacintov, N.E.; Gagliano, A.G.; Ibanez, V.; Harvey, R.G. Carcinogenesis 1982, 3, 242-53.
8.   Chen, F.M. Nucleic Acids Res. 1983, 11, 7232-50.
9.   Chen, F.M.; Carcinogenesis 1984, 5, 753-8.
10. Undeman, O.; Lycksell, P.O.; Graslund, A.; Astlind, T.; Ehrenberg, A.; Jernstrom, B.; Tjerneld, F.; Norden, B. Cancer Res. 1983, 43, 1851-60.
11. Geacintov, N.E.; Yoshida, H.; Ibanez, V.; Harvey, R.G. Biochem. Biophys. Res. Commun. 1981, 100, 1569-77.
12. Geacintov, N.E.; Hoshida, H.; Ibanez, V.; Jacobs, S.A.; Harvey, R.G. Biochem. Biophys. Res. Commun. 1984, 122, 33-9.
13. Kim, M.H.; Geacintov, N.E.; Pope, M.; Harvey, R.G. to be published.
14. Kim, M.H.; Geacintov, N.E.; Pope, M.; Harvey, R.G. Biochemistry 1984, 23, 5433-5439.
15. Meehan, T.; Straub, K. Nature 1979, 277, 410-12.
16. MacLeod, M.C.: Selkirk, J.K. Carcinogenesis 1982, 3, 287-92.
17. Meehan, T.; Gamper, H.; Becker, J.F. J. Biol. Chem. 1982, 257, 10479-85.
18. Geacintov, N.E.; Ibanez, V.; Gagliano, A.G.; Yoshida, H.; Harvey, R.G. Biochem. Biophys. Res. Commun. 1980, 92, 1335-42.
19. Kootstra, A.; Haas, B.L.; Slaga, T.J. Biochem. Biophys. Res. commun. 1980, 94, 1432-38.

20. Geacintov, N.E.; Yoshida, H.; Ibanez, V.; Harvey, R.G. Biochemistry 1982, 21, 1864-69.
21. Geacintov, N.E.; Hibshoosh, H.; Ibanez, V.; Benjamin, m.J.; Harvey, R.G. Biophys. Chem. 1984, 20, 121-133.
22. MacLeod, M.C.; Mansfield, B.K.; Selkirk, J.K. Carcinogenesis 1982, 3, 1031-37.
23. Meehan, T.; Bond, D.M. Proc. Natl. Acad. Sci. (U.S.A.) 1984, 81, 2635-41.
24. Geacintov, N.E.; Ibanez, V.; Benjamin, M.J.; Hibshoosh, H.; Harvey, R.G. In "Polynuclear Aromatic Hydrocarbons: Formation, Metabolism and Measurement"; Cooke, M.; Dennis, A.J., Eds; Battelle Press, Columbus, Ohio, 1983, 554-570.
25. Michaud, D.P.; Gupta, S.C.; Whalen, D.L.; Sayer J.M.; Jerina, D.M. Chem. Biol. Interactions 1983, 44, 41-9.
26. Gamper, H.B.; Straub, K.; Calvin, M.; Bartholomew, J.C. Proc. Natl. Acad. Sci. 1980, 77, 2000-4.
27. Yang, N.C., Hrinyo, T.P.; Petrich, J.W.; Hang, D.D.H. Biochem. Biophys. Res. Commun. 1983, 114, 8-13.
28. Pearson, R.G.; Frost, A.A. "Kinetics and Mechanisms", Wiley, New York, 1965, 2nd edition, pp. 160-162.
29. Drinkwater, N.R., Miller, J.A., Miller, E.C., Yang, N.C., Cancer Res. 1978, 38, 3247-55.
30. Pulkrabek, P.; Leffler, S.; Grunberger, D.; Weinstein, I.B. Biochemistry 1979, 18, 5128-34.
31. Geacintov, N.E.; Ibanez, V.; Gagliano, A.G.; Jacobs, S.A.; Harvey, R.G. J. Biomol. Struct. Dynamics 1984, 1, 1473-84.
32. Gagliano, A.G.; Geacintov, N.E.; Ibanez, V.; Harvey, R.G.; Lee, H.M. Carcinogenesis 1982, 3, 969-76.
33. Geacintov, N.E.; Gagliano, A.G.; Ibanez, V.; Lee, H.; Jacobs, S.A.; Harvey, R.G. J. Biomol. Struct. Dynamics 1983, 1, 913-23.
34. Hogan, M.E.; Dattagupta, N.; Whitlock, J.P. Jr. J. Biol. Chem. 1981, 256, 4505-13.
35. Agarwal, K.L.; Hrinyo, T.P.; Yang, N.C. Biochem. Biophys. Res. Commun. 1984, 114, 14-19.
36. Gamper, H.B.; Bartholomew, J.C.; Calvin, M. Biochemistry, 1980, 19, 3948-56.
37. Prusik, T.; Geacintov, N.E.; Tobiasz, C.; Ivanovic, V.; Weinstein, I.B. Photochem. Photobiol. 1979, 29, 223-32.
38. Zinger, D.; Geacintov, N.E.; Harvey, R.G.; to be published.
39. Undeman, O.; Sahlin, M.; Graslund, A.; Ehrenberg, A.; Dock, L.; Jernstrom, B. Biochem. Biophys. Res. Commun. 1980, 94, 458-65.
40. Shahbaz, M.; Harvey, R.G.; Prakash, A.S., Boal, T.R., Zegar, I.S.; LeBreton, P.R. Biochem. Biophys. Res. Commun. 1983, 112, 1-7.
41. Lefkowitz, S.M.; Brenner, H.C. Biochemistry 1982, 21, 3735-41.
42. Lakowitz, J.R.; Weber, G. Biochemistry 1973, 12, 4161-70.
43. Poulos, A.T.; Kuzmin, V.; Geacintov, N.E. J. Biochem. Biophys. Meth. 1982, 6, 269-81.
44. Lin, J.H., LeBreton, P.R.; Shipman, L.L. J. Phys. Chem. 1980, 84, 642-49.
45. Nakata, Y., Malhotra, D., Hopfinger, J.A.; Bickers, D.R. J. Pharm. Sci. 1983, 72, 809-11.
46. Subbiah, A., Islam, S.A.; Neidle, S. Carcinogenesis 1983, 4, 211-15.
47. Kikuchi, O., Pearlstein, R., Hopfinger, A.J.; Bickers, D.R. J. Pharm. Sci. 1983, 72, 800-08.

48. Kadlubar, F.F. Chem. Biol. Interactions 1980, 31, 255-63.
49. Beland, F.A. Chem. Biol. Interactions 1978, 22,329-39.
50. Aggarwal, A.K.; Islam, S.A.; Neidle, S. J. Biomol. Struct.
     Dynamics 1983, 1, 873-81.
51. Hingerty, B.; Broyde, S. J. Biomol. Struct. Dynamics 1983, 1,
     905-12.
52. Taylor, E.R.; Miller, K.J.; Bleyer, A.J. J. Biomol. Struct.
     Dynamics 1983, 1, 883-904.
53. Miller, K.J., 1985, this Symposium.
54. Jernstrom, B.; Lycksell, P.O.; Graslund, A.; Norden, B.
     Carcinogenesis 1984, 5, 1129-35.

RECEIVED March 5, 1985

X-ray Analyses of Polycyclic Hydrocarbon Metabolite Structures

JENNY P. GLUSKER

Fox Chase Cancer Center, Institute for Cancer Research, Philadelphia, PA 19111

The methods of X-ray diffraction analysis of crystal structures can be used to investigate steric effects in carcinogenic polycyclic aromatic hydrocarbons (PAHs) and their activated metabolites. For example, the presence of a methyl group adjacent to the bay region of a PAH greatly enhances its carcinogenicity; X-ray studies have shown the extent of the steric distortions of the bay region, in-plane and out-of-plane, caused by this methylation. In most PAHs the conformation of the molecule is determined by nonbonded H$\cdots$H interactions. If hydrogen atoms would approach each too closely if the molecule were planar, then bond angle and torsion angle changes are made to accommodate this strain. In addition, in more saturated ring systems such as PAH metabolites, this type of nonbonded interaction between hydrogen atoms may force a hydroxyl or other substituent to have an axial rather than an equatorial conformation. This has been analyzed for metabolites such as diols and diol epoxides.

The mechanism of carcinogenesis by PAHs is believed to involve alkylation of an informational macromolecule in a critical, but at present unknown, manner. Such an interaction with a protein has been modelled by alkylation of a peptide; this showed a conformational change occurred on alkylation. It has not yet been possible to study the structure of DNA alkylated by an activated carcinogen; this is because DNA is a fiber and the structural order in it is not sufficient for a crystal structure determination. However the crystal structures of some alkylated portions of nucleic acids are described, particularly some nucleosides alkylated by chloromethyl derivatives of DMBA. In crystals of these alkylation products the PAH portion of the adduct shows a tendency to lie between the bases of other nucleoside

0097–6156/85/0283–0125$16.00/0
© 1985 American Chemical Society

molecules in the crystal structure, although the more
buckled part of the PAH does not take part in this
stacking. There also appears to be an interaction
between the oxygen of the ribose sugar and part of
the PAH (the methyl group in a 12-methylbenz[a]-
anthracenyl derivative). Some preliminary studies in
computer modelling experiments show that, at present,
the X-ray data on adducts are consistent with both
current models of interaction after covalent attachment - partial intercalation of the aromatic portion
of the PAH between the bases of DNA or the situation
where the aromatic group lies in a groove of DNA.

The carcinogenicity of polycyclic aromatic hydrocarbons (PAHs) was
first deduced by Sir Percivall Pott in London in 1775 (1) when he
noted an appreciable incidence of scrotal cancer in young chimney
sweeps. He correctly noted that this affliction was caused by an
accumulation of soot on the skin. The work of his grandson, Henry
Earle (2), and of Curling (3, 4) led to the ideas that, since
scrotal cancer did not affect all chimney sweeps, another factor,
such as an inherited predisposition, might also play a part in its
occurrence, and that the disease could have an appreciable latent
period. The fact that coal tar could, by itself, cause skin tumors
was shown by Yamagiwa and Ichikawa (5). Shortly after that Passey
(6) extracted a single chemical from soot and it was demonstrated
that it was a carcinogen. Finally the chemical formulae of many of
the active carcinogenic components of coal tar and soot were
established by the experiments of Kennaway, Heiger, Mayneord, Cook
and Hewitt (7). They showed that the carcinogens from soot and coal
tar are PAHs with formulae such as I (benzo[a]pyrene), II
(7,12-dimethylbenz[a]anthracene), III (3-methylcholanthrene), IV
(benz[a]anthracene) and V (dibenz[a,h]anthracene). It was PAHs like
these that were responsible for the problems of the chimney sweeps.
In fact, carcinogenic PAHs continue to be a problem and are environmental hazards found wherever pyrolysis of organic matter occurs, as
in cooking and in cigarette smoking (8-10).

V

The fate of such apparently innocuous hydrophobic chemicals in
the body has been the subject of much study (11). These compounds
are fairly insoluble in water, but they are soluble in fats and oils.
The body, in an effort to solubilize, and therefore excrete, these
foreign compounds, epoxidizes and hydroxylates them, often forming
conjugates with amino acids and peptides such as glutathione (12).
Such metabolism of PAHs proceeds via quinones, alcohols, epoxides
and peptide conjugates to give products that can be eliminated from
the body. But occasionally, by a process known as "activation," a
diol epoxide, with functional groups in very specific positions in
the molecule, is formed (13). These diol epoxides are formed by the
action of the enzyme cytochrome P-450 and epoxide hydrase. Initially an epoxide is formed and this is hydrated by epoxide hydrase
to a trans-diol. Further epoxidation gives the diol epoxide; this
is an alkylating agent that can interact with a cellular macromolecule such as DNA and distort it in some manner after alkylating it.
This is believed to be the beginning of the carcinogenic process
caused by certain PAHs.

The three-dimensional structures of these PAHs and their metabolites, of their activated products and of adducts with portions of
DNA will be the subject of this chapter. While the formulae I - V
appear to be only two-dimensional (that is, flat), it will be shown
that the presence of methyl groups, diol epoxide groupings, etc. in
certain areas of such molecules results in three-dimensional
molecular structures that may be buckled and strained. Some of the
three-dimensional structures of such PAHs and their metabolites will
be described. In addition, the possible structural distortions that
may be caused in DNA by specific alkylations by activated PAHs will
be discussed in the light of structural studies of adducts of PAH
alkylating agents with nucleosides. From such information it is
hoped eventually to be able to determine what makes a PAH carcinogenic, and in what way the distortions in DNA, resulting from alkylation by an activated carcinogenic PAH, make DNA behave differently
from normal so that a carcinogenic process is initiated.

Geometry of PAHs

The techniques of X-ray diffraction analyses of crystals of
compounds of interest can be used to determine, with high precision,
the three-dimensional arrangement of atoms, ions and molecules in
such crystals (14); in each case the result is referred to as the
"crystal structure." X-ray diffraction by crystals was discovered
by von Laue, Friedrich and Knipping (15) and the technique was
applied by the Braggs to the determination of the structures of

simple crystals such as those of sodium chloride and diamond (16, 17), and by Lonsdale to the structure of crystals of hexamethyl-benzene (18). This latter study was the first evidence that the benzene ring is a symmetrical hexagon of carbon atoms with each C-C bond length the same within experimental error. The structures drawn by Kekulé with alternating single and double bonds are not found, even at low temperatures where any interconversion of isomers would be expected to be slower. Thus the benzene molecule is truly "aromatic" with six equivalent C-C bonds. Diffraction studies of benzene (19, 20) (which have necessarily been done at low tempera-tures since benzene is liquid at room temperatures) have shown that the C-C bond length is 1.392(2) Å, in agreement with results from spectroscopic data (21).

When two or more benzene rings are fused together to give naphthalene, anthracene, etc., X-ray diffraction studies show that some localization of double bonds occurs (22-24); this affects the chemical reactivities of different regions in the molecule. The experimentally measured bond lengths in PAHs are those that would be expected from a consideration of the various types of resonance hybrids (25) that are possible.

The units by which crystallographers describe interatomic distances are Ångstrom units (Å = 10^{-8} cm.). Normal values for carbon-carbon interatomic distances are 1.34 Å for a double bond (as in ethylene) and 1.54 Å (as for diamond) for a single bond. In a truly aromatic compound (such as benzene) the C-C bond length, as mentioned above, is 1.39 Å. C-C-C angles are 109.5° for a tetrahe-dral carbon atom (sp^3) and 120.0° for a trigonal carbon atom (sp^2). As shown in Figure 1 for BP and DMBA, measured C-C bond lengths, interbond angles and torsion are variable around these values.

Most carcinogenic PAHs contain the phenanthrene grouping in them (see Figure 2) and there are two important areas in these mole-cules that are often referred to in studies of PAH carcinogenesis. One is the "K-region" which corresponds to the very active double bond between atoms 9 and 10 in phenanthrene (26). In BP (I) the "K-region" is the 4,5 bond and in DMBA (II) it is the 5,6 bond. Bonds in the "K-region" are usually short (1.34-1.35 Å), near in values to that for a pure double bond, while most other C-C bonds in PAHs are in the range 1.36-1.44 Å. The second area of interest is the "bay region" (27) which corresponds to the hindered region between the 4- and 5- positions of phenanthrene (the area between C10 and C11 of BP (I) or C1 and C12 of DMBA (II), for example). Some bay- and K-regions are illustrated in Figure 2. While it was originally thought that the condition for carcinogenicity of a PAH is the presence in it of a K-region, it is now clear that the pres-ence of a bay-region with a methyl group attached to one side of it is a better criterion (27). Interestingly the correlations that were originally found between the presence of a K-region and carcino-genicity derive from the fact that most carcinogenic PAHs containing a phenanthrene-like group will also have a bay-region; Pullman (28) said that "bay-region could be a kind of 'back door' to the K-region".

Early work on the structure of carcinogenic PAHs was done by John Iball at the University of Dundee, Scotland (29); he had suggested the use of the "Iball index" (30) as a measure of the

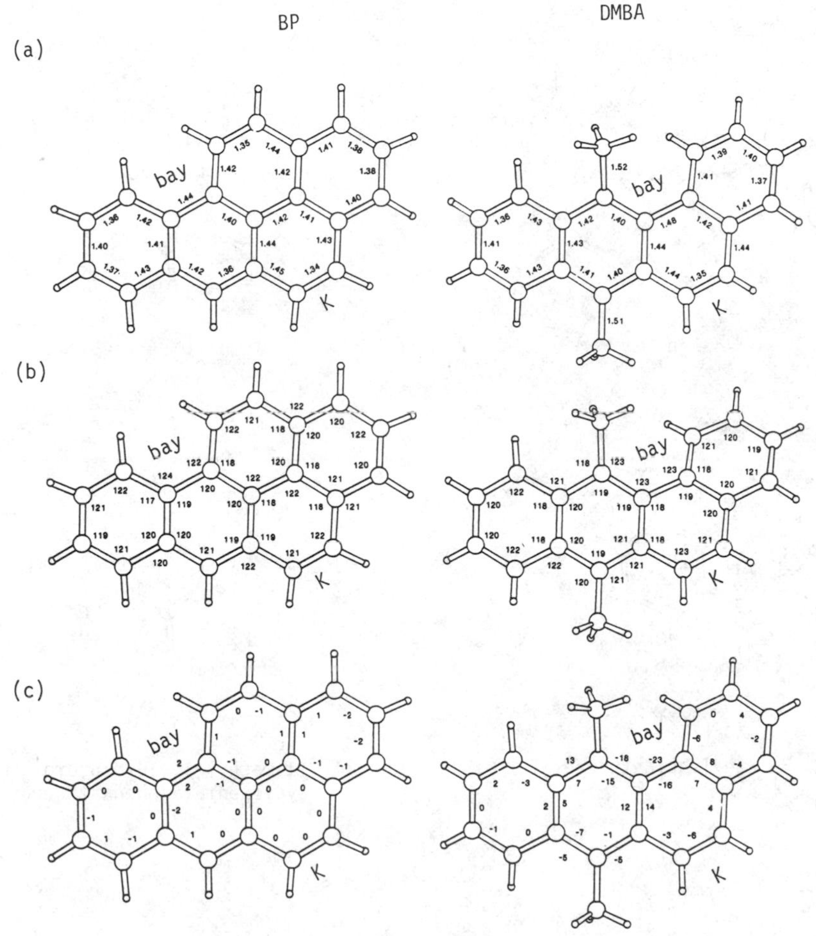

Figure 1. (a) Bond lengths, (b) interbond angles and (c) torsion angles for BP (left) and DMBA (right). Estimated standard deviations are 0.002 Å for distances and 0.2 for angles. If the distribution of errors is normal there is a 99% chance that a measurement will differ by less than 2.7 e.s.d. from the mean value of that quantity. A torsion angle is the angle of twist of a bond. In a series of four bonded atoms, A-B-C-D, the torsion angle about the B-C bond is defined as the angle of rotation about that bond required to make the projection of the line A-B coincide with the projection of the line C-D, when viewed along the B-C direction. In this and all subsequent diagrams the larger circles are carbon atoms, the smaller circles are hydrogen atoms. In later diagrams oxygen atoms are stippled and nitrogen atoms are black.

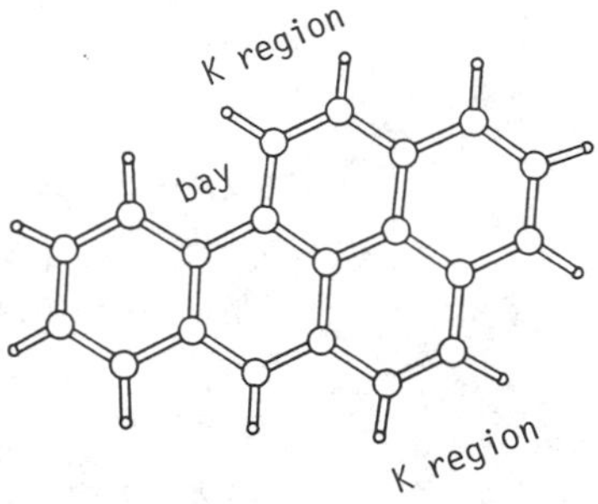

I BP

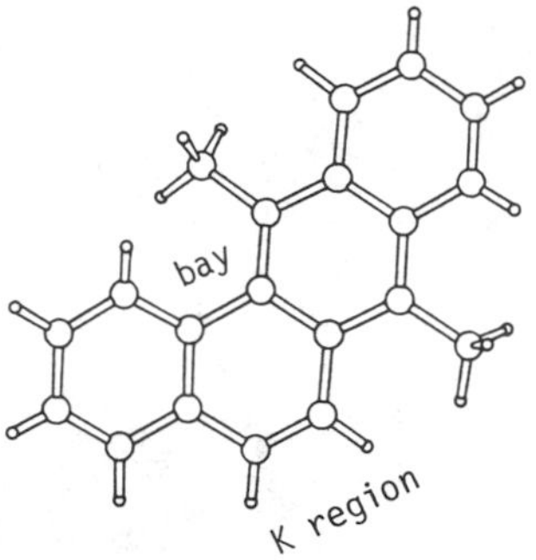

II DMBA

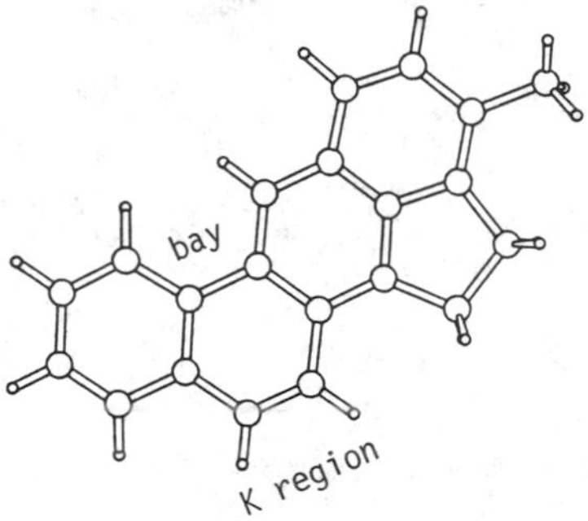

III 3-methylcholanthrene

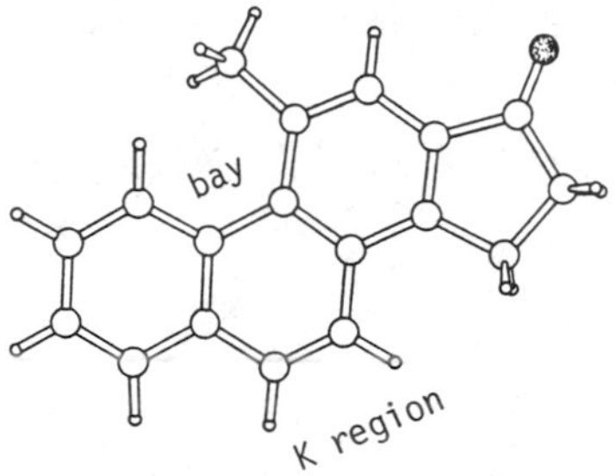

VII 11-methyl-15,16-dihydro-
     cyclopenta[a]phenanthrene

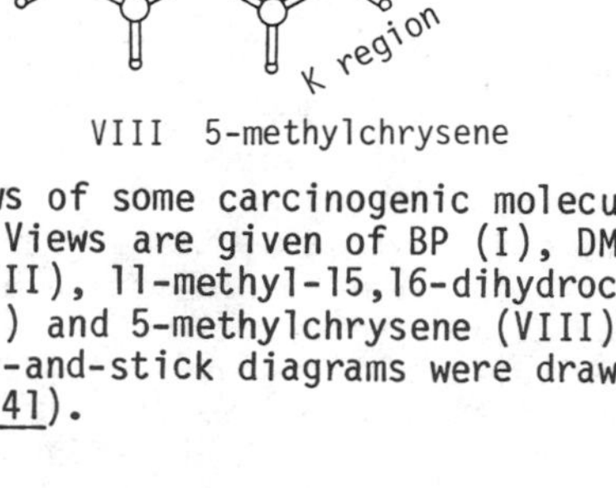

VIII 5-methylchrysene

Figure 2. Views of some carcinogenic molecules showing K- and
bay- regions. Views are given of BP (I), DMBA (II), 3-methyl-
cholanthrene (III), 11-methyl-15,16-dihydrocyclopenta[a]phen-
anthracene (VII) and 5-methylchrysene (VIII). These and all
subsequent ball-and-stick diagrams were drawn using the computer
program VIEW (141).

relative carcinogenicity of a given compound. This index involved
the percentage of cancer in mice that have survived beyond the
shortest time of the latent period versus the average latent period.
Iball studied chrysene (31) and 1,2-cyclopentenophenanthrene (32)
since he was interested in the structural similarities between
sterols, bile acids, sex hormones, toad poisons and carcinogenic
PAHs. Iball also studied 3-methylcholanthrene (MC)(III) (33, 34)
and benzo[a]pyrene (BP) (I) (35, 36). 7,12-Dimethylbenz[a]anthra-
cene (DMBA) (II) was studied by Sayre and Friedlander (37), refined
by Iball (38) and remeasured at low temperature by Zacharias and
Glusker (39). Interestingly, when the structure was first reported
the result was assumed to be erroneous because the molecule was so
buckled; it was, of course, correct (40). Other PAHs studied by
crystallographic techniques include the weak carcinogen
benz[a]anthracene (BA) (IV) (41) and the carcinogenic PAH dibenz-
[a,h]anthracene (DBA) (V) (42-45).
    It is of interest to compare the shapes and sizes of various
molecules with appreciable carcinogenic activity. For example,
aflatoxin B₁ (VI), found in moldy peanuts and grain, is one of
the most powerful carcinogens known. Its crystal structure was
determined by van Soest and Peerdeman (46-48).

VI

The mode of action of this carcinogen is believed to involve
epoxidation of a double bond (49), as indicated in Figure 3; in this
Figure the similarities of the shapes, and particularly of the sites
of activation of aflatoxin, BP and DMBA are demonstrated. Aflatoxin
has functional groups at each end of the molecule, unlike an activated PAH which has functional groups only at one end of the molecule.
As shown in Figure 3, the sizes of BP and DMBA are not only similar
to those of aflatoxin, but also to those of base pairs in DNA as
noted by Haddow (50) and of steroids, such as estradiol (51), as
noted by Huggins and Yang (52). Such structural similarities between steroids and carcinogenic PAH have intrigued scientists since
Cook and Hazelwood converted the steroid, deoxycholic acid, to the
carcinogen, 3-methylcholanthrene. in 1933 (53). Since then Coombs
has demonstrated the carcinogenicity of 11-methyl-15,16-dihydro-
cyclopenta[a]phenanthrene (VII) which has an aromatized steroid-like
nucleus (54). The significance of these structural similarities is
not clear at this time.
    In this chapter we shall consider distortions from planarity in
PAH systems. These distortions are introduced by substituent methyl
and other groups that are too bulky to fit in the available space
and at the same time allow the substituted PAH to remain planar

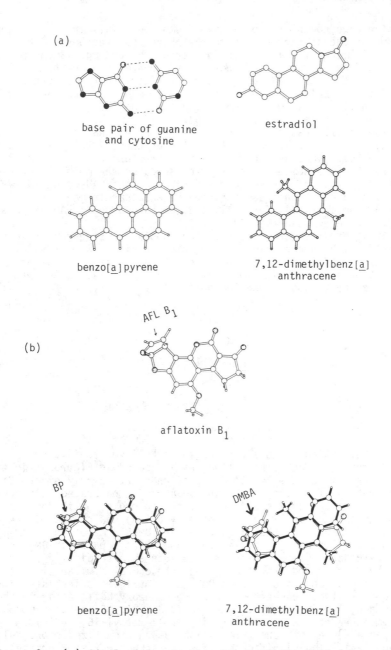

Figure 3. (a) Similarities in the shapes of BP, DMBA, a base pair of guanine and cytosine, and the steroid, estradiol. In (b) the shapes of B^D and DMBA (filled bonds) are compared with that of aflatoxin B₁ (open bonds) and the sites of activation of each to an epoxide is indicated by the arrows.

VII

(although in methyl-substituted PAHs at least two of the methyl
hydrogen atoms must, of course, be out of this plane). Most of this
type of strain comes from repulsions between hydrogen atoms that are
not bonded to adjacent carbon atoms. The minimum allowable non-
bonded H...H distance appears, from structural studies, to lie in
the region of 1.6 Å. X-ray crystallographic results show graphic-
ally how the strain that occurs on methyl substitution adjacent to
the bay-region is accommodated within the molecule; an inspection of
Figure 1 is recommended. In the bay-region the C-C bond length may
be increased to 1.48 Å, and it is usual to see distortions of interbond angles from the normally expected values of 118-122°. However,
in spite of this strain the sum of the angles around an sp^2 carbon
atom will remain at 360°, that is, the carbon retains an essentially
planar arrangement of atoms around it.

The buckling of the molecule from steric hindrance arises from
torsion about the bonds rather than from distortions from planarity
of sp^2-hybridized carbon atoms. Therefore an excellent measure of
strain in PAHs is provided by the C-C-C-C torsion angles, that is
the angles of twist of various C-C bonds. When a PAH is completely
planar all of the torsion angles are either 0° or 180°. However in
the bay regions of some methylated PAHs, values as high as 23° in
DMBA (37-39) and 36° in 1,12-dimethylbenz[a]anthracene (55) have
been reported. For example, in BP (I), there is slight overcrowding
between hydrogen atoms on C10 and C11 in the bay region. This overcrowding is relieved by a very slight twist of 2° about the C17-C18
bond so that these two hydrogen atoms on C10 and C11 lie 0.04 Å on
either side of the molecular plane (see Figure 1). Apart from this
the molecule is flat. However, as shown in Figure 4, DMBA is much
more buckled; this is a result of the overcrowding between C1 and
C19 (where C19 is the carbon atom attached to C12). Here the strain
is relieved by a twist of 23° about the C13-C14 bond in the bay
region, as well as twists about other bonds in the bay-region as
shown in Figure 4. The angle between the planes of the two outer
rings of DMBA becomes 24.0° as a result of this twisting. Similar
effects are seen in 5-methylchrysene (VIII), the only carcinogenic
monomethylchrysene, and its derivatives (56-58). Here the ring
system is more amenable to in-plane distortions and therefore torsion angles in the bay region are not as great as for DMBA (as shown
in a comparative way in Figure 5).

Thus we have a picture of a bay region that becomes distorted
on substitution with a methyl group, a substitution that may cause
in-plane or out-of-plane distortions. These are illustrated in
Figure 5 for 5,12-dimethylchrysene (in-plane distributions) and DMBA

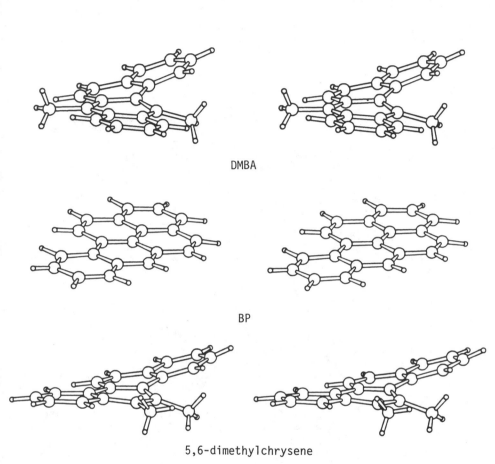

DMBA

BP

5,6-dimethylchrysene

Figure 4.   Views of DMBA, BP, 5,6-dimethylchrysene, 5,12-dimethyl-
chrysene and 5-methylchrysene. These illustrate the distortions
that occur as a result of steric effects. These and many subsequent representations of molecular structure are stereoviews and
may be viewed with stereoglasses; alternatively the reader can
focus his eyes on the two images until an image between them
begins to form and then allow his eyes to relax until the central
image becomes three-dimensional. This process calls for patience
and may take a minute or so. The reader who does not wish to do
this may simply inspect one of the two diagrams for each
structure.

                                        Continued on next page.

5,12-dimethylchrysene

5-methylchrysene

Figure 4. Continued.

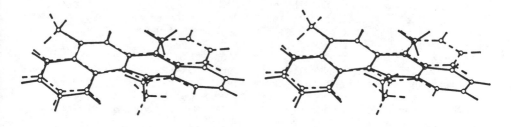

Figure 5. Bay region geometry. The result of docking the bay-region of DMBA (filled bonds) on to that of 5,12-dimethylchrysene (broken bonds) is shown (142). In DMBA the 12-methyl group is forced out of the plane of the rest of the ring system. In 5,12-dimethylchrysene (used because more accurate coordinates are available than those of 5-methylchrysene) the distortions are in-plane. Note the differing orientations of the fourth ring (upper right) that indicates different types of flexibility in the ring systems of DMBA and 5-methylchrysene.

VIII

(out-of-plane distortions). Note how differently the methyl groups (on positions 5 and 12 respectively) lie and also the differing location of the outer rings. Thus it is found that a carcinogenic PAH may be a buckled molecule, not the flat two-dimensional molecule depicted in the text books. While BP is carcinogenic, its 11-methyl derivative with a methyl group in the bay region is even more carcinogenic (59).

Intercalation

This type of buckling is of particular interest when we consider the interaction of such carcinogens with biological macromolecules such as DNA. DNA has a double helical structure with bases at distances 3.4 Å apart, hydrogen-bonded in planes perpendicular to the helix axis. By a stretching or twisting of the helix it is possible to extend the vertical distance between the bases to 6.8-7.0 Å; a planar molecule, with a thickness of about 3.5-3.7 Å, can then slip between the bases. This is a common interaction between flat aromatic molecules such as proflavine or 9-aminoacridine and DNA, and was first proposed by Lerman (60). The flat molecule becomes completely enveloped within the hydrophobic area of the nucleic acid, that is, between the hydrogen-bonded bases (61, 62). Pauling (63) lists the van der Waals radius (the distance of closest approach of another molecule) of hydrogen as 1.2 Å, 2.0 Å as the radius of a methyl group and 1.85 Å as the half-thickness of an aromatic molecule. Thus the vertical distances between bases would have to be further increased if a methyl group, which is not flat and which has no π-electron system, is to be intercalated. In practice, while very small flat molecules such as water can intercalate between bases (64), so far no cases where the more bulky methyl group intercalates have been found.

Several complexes that involve intercalation of an acridine in a portion of a nucleic acid have been studied by X-ray crystallographic techniques. These include complexes of dinucleoside phosphates with ethidium bromide, 9-aminoacridine, acridine orange, proflavine and ellipticine (65-69). A representation of the geometry of an intercalated proflavine molecule is illustrated in Figure 6 (b); this is a view of the crystal structure of proflavine intercalated in a dinucleoside phosphate, cytidylyl-(3'-5') guanosine (CpG) (70, 71). For comparison an example of the situation before such intercalation is also illustrated in Figure 6 (a) by three adjacent base pairs found in the crystal structure of a polynucleotide (72, 73). In this latter structure the vertical distance (parallel to the helix axis) between the bases is approximately

(a)

(b)

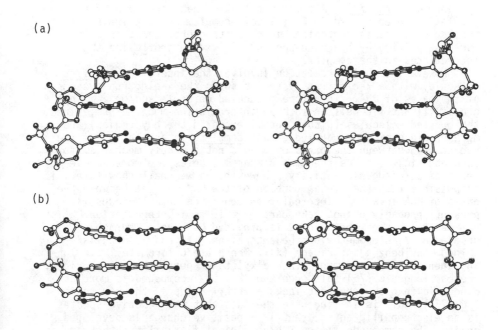

Figure 6. (a) Views of three base pairs in the crystal structure of a longer polynucleotide. No molecule is intercalated in this structure. (b) A view of the intercalation complex of proflavine in the self-complementary complex of cytidylyl-(3'-5')guanosine.

3.4 Å, compared with a value of approximately 6.8 Å in the intercalated complex. Since CpG is self-complementary, two molecules may hydrogen bond to each other in a head-to-tail manner forming two sets of Watson-Crick base pairs, each involving hydrogen bonds between cytosine and guanine. When proflavine is intercalated, the phosphodiester backbone is extended by increasing two torsion angles [those of P-O5'-C5'-C4' and O1'-C1'-N9-C8 in guanosine] by about 60° (70, 71). A similar situation has been found for proflavine intercalated in CpA; a self-complementary pair is formed because cytosine is protonated in this structure, and this allows for base pairing by hydrogen bonding (74). The proflavine then intercalates in the complex.

However, such intercalation involves the insertion of a flat molecule, with a π-electron system, between the π-electron systems of the bases of DNA. Therefore a buckled molecule and/or a methyl group will not fit well in such an intercalation mode. Thus, since the buckled molecules (with a methyl group in the bay region) are more carcinogenic, we concluded that complete intercalation of the hydrocarbon between the bases of DNA is not a likely mechanism for carcinogenicity, since the less planar molecules are more active in terms of carcinogenic activity. However, as we shall show later, it is possible that the planar portion of the PAH may lie between the bases of DNA in a semi-intercalation mode; the methyl groups, if present, probably do not take part in this semi-intercalation.

An example of these ideas is provided by crystallographic studies of 7-chloromethylbenz[a]anthracene (IX) and 12-methyl-7-chloromethylbenz[a]anthracene (X). These are chloromethyl alkylating agents that, like many other alkylating agents such as epoxides, are carcinogenic (75) (although the 12-methyl compound is more carcinogenic than the one that lacks a methyl group in the bay region (76, 77)); this is in line with the idea that the reactive species is an electrophilic agent, that is a positive charge is developed on it. The compounds IX and X have minimal flexibility; they are as planar as possible, but steric factors cause them to buckle to varying degrees. X-ray diffraction studies (78-80) showed that the compound IX, lacking a bay-region methyl group, is approximately

IX   X

planar, apart from the hydrogen and chlorine atoms of the chloromethyl group. This is illustrated in Figure 7(a). However, for reasons already described (the steric repulsion between non-bonded hydrogen atoms), when there is an additional methyl group at C12 (as in X) the molecule becomes markedly buckled as a result of interactions between the hydrogen atom on C1 and those of the methyl group attached to C12. Since the more buckled ring system is found for the more carcinogenic compound it must be concluded that planarity of the ring systems in halomethylbenz[a]anthracenes is not a requirement for carcinogenicity.

Thus complete intercalation of the aromatic PAH between the bases of DNA, in the manner described above for flat molecule such as proflavine, did not seem to be a likely mechanism for the carcinogenic action of these compounds. Since alkylation and intercalation are not simultaneously possible for steric reasons, and since one molecule is wedge-shaped and the other is flatter, it was considered more likely that the action of these compounds arose from their alkylating ability; they could alkylate a base of DNA and then, since the bulky aromatic hydrophobic group would possibly not remain protruding into the hydrophilic environment, it is possible that the aromatic PAH group could then lie in one of the grooves of DNA. This is illustrated in Figure 7(b) and (c).

K-region Derivatives

Originally it was thought, as mentioned above, that the site of action of activated carcinogenic PAHs was the K-region (26), and, indeed, those PAHs with epoxide groups in the K-region are carcinogenic. However, our main interest is in studying metabolites of these PAHs that are intermediates in carcinogenesis by the PAH, that is, the diol epoxides. Before the notion of the diol epoxide as the active agent, some crystallographic studies were done on K-region derivatives even though it is now clear that the metabolites of K-region oxides do not match those formed from the parent PAH in vivo (81). The structural studies do, however, throw some light on steric factors that are operative for the bay-region derivatives.

At first three K-region oxides were studied, those of DMBA (XI), BP (XII) and phenanthrene, the non-carcinogenic parent (XIII); the structures are shown in Figure 8. These epoxides, which were considered very reactive, were found to remain stable both in air and in the X-ray beam when in the crystalline state (82, 83).

XI   XII   XIII

The epoxide ring is an approximately symmetrical three-membered ring that lies almost perpendicular to the plane of ring containing the double bond that has undergone an addition reaction to give the epoxide, and the hydrogen atoms stick out above the ring. In general the bond lengths in an epoxide ring vary from 1.43 to 1.46 Å and the interbond angles are 60°. An averaged geometry of an epoxide ring is shown in Figure 9. In Figure 10 we show, from an

(a)

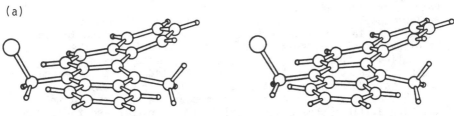

7-chloromethyl-12-methylbenz[a]anthracene

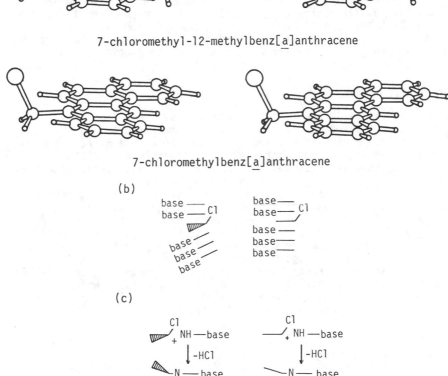

7-chloromethylbenz[a]anthracene

(b)

(c)

Figure 7. (a) Views of 7-chloromethyl-12-methylbenz[a]anthracene showing the extent of buckling of the molecule and of 7-chloromethylbenz[a]anthracene which is much more planar. (b) Diagram illustrating the effect of intercalation of the chloromethyl PAH in DNA. (c) Diagram illustrating the effect of alkylation of the extracyclic amino groups of the bases of DNA. Since the more buckled molecule is more carcinogenic it is concluded that alkylation rather than intercalation is the significant mode of interaction in this carcinogenic process.

(a)

(b)

(c)

(d)

(e)

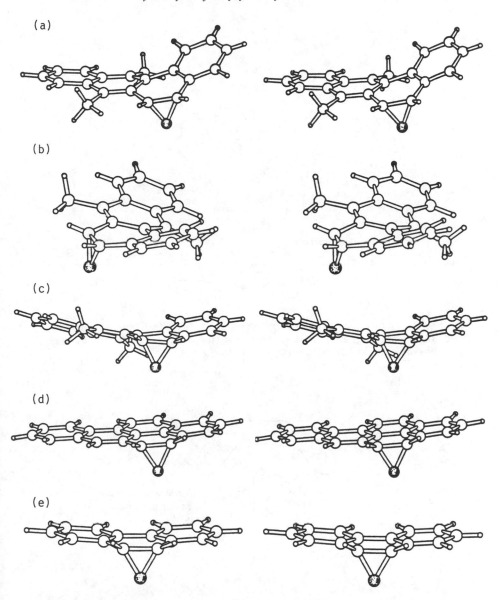

Figure 8.   The structures of the K-region oxides of DMBA (XI) (a, b and c), BP (XII) (d) and phenanthrene (XIII) (e). Views (c), (d) and (e) are directly onto the plane of the epoxide group. Note the asymmetry around the epoxide ring and the non-planarity of the aromatic ring system in the K-region oxide of DMBA. Also note that the 7-methyl and the 5,6-oxide groups in (a) are in the relative positions for a diol epoxide of a bay-region methylated diol epoxide. Perhaps the role of the bay-region methyl group in such diol epoxides is to cause such a distortion in the oxide ring and so affect its activity.

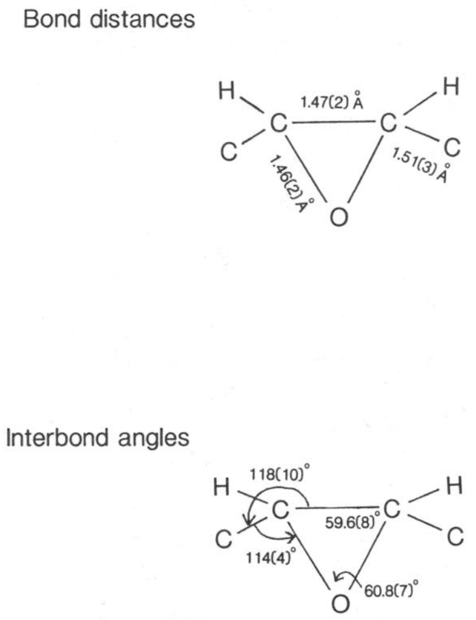

Bond distances

Interbond angles

Figure 9. Averaged geometry of an epoxide showing bond distances and interbond angles. Values in parentheses are estimated standard deviations of these average values.

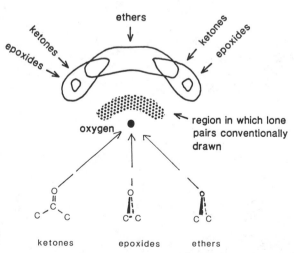

Figure 10. Contours of the positions of hydrogen bonded groups around ketone, epoxide and ether groups. These contours are obtained from scatterplots of such hydrogen bonding groups in many crystal structures. Note that hydrogen bonding groups approach ether oxygen atoms in a narrower range of positions than they do for ketones or epoxides.

analysis of all epoxides studied by X-ray diffraction methods to
date, the arrangement of groups that hydrogen bond to an epoxide;
such data are important for any model building. An analysis of
these data showed that hydrogen bonding groups tended to lie along
the directions usually ascribed to the lone pairs of the oxygen atom
(84). In this directionality of hydrogen bonding epoxides resemble
ketones rather than ethers, as shown in Figure 10.
    The addition of an epoxide group to a PAH, especially if there
is a nearby methyl group, increases the observed buckling of the
ring system. For example the angle between the outer rings of BP is
increased on epoxidation at the K-region from 1° to 5°, while the
angle between the outer rings of DMBA is, with K-region epoxidation,
increased from 24° to 35°. The epoxide bond C-O bond lengths are
1.461(6) and 1.459(6) Å for phenanthrene oxide (the same within
experimental error), and 1.481(4) and 1.478(4) Å for BP oxide (again
symmetrical). Those for the K-region oxide of DMBA are 1.445(3) and
1.457(3) Å; these are unequal, possibly for steric reasons since the
methyl group on C7 causes steric problems in this area, as shown in
Figure 11(a). It would be expected that the longer C-O bond, that
is, C6-O, would be cleaved more readily than C5-O, and this is, in
fact, the case (see Figure 11c).
    Further structural studies on a 5,6-cis-diol of DMBA (XIV) (85)
showed the extent of the steric effect that the 7-methyl group was
having on the K-region. It was found that the 6-hydroxyl group
(nearest the 7-methyl group) is axial while the 5-hydroxyl group
(further from the 7-methyl group) is equatorial, as shown in Figure
11(b). If the 6-hydroxyl group were equatorial it would "bump" into
the 7-methyl group. Since the ring bearing the diol group is more
flexible than the ring bearing the 7-methyl group the strain is
accommodated by distortions at C5 and C6 as shown in Figure 11(b).

Another cis-diol (XV) derived from BP was obtained by the action of
osmium tetroxide on 4,5-dihydroxybenzo[a]pyrene (86). The hydroxyl
group at C5a is axial and that at C6 is equatorial, illustrating the
relative rigidity of the BP ring system and the flexibility at C6 of
the ring bearing the diol groups.

Trans-Diols and Diol Epoxides

Carcinogenesis by PAHs is now believed to proceed via "activation"
to a diol epoxide that is capable alkylating either DNA (13, 87,
88), or, possibly, another information-containing cellular macromolecule (such as a nuclear protein that interacts with DNA). The

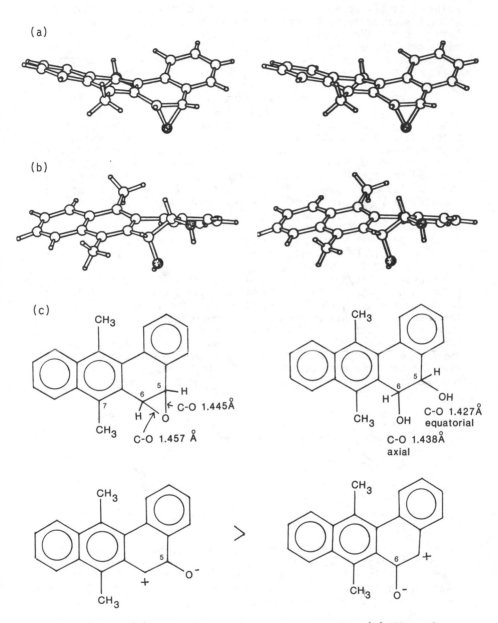

Figure 11. (a) View of K-region oxide of DMBA. (b) View of 5,6-cis diol of DMBA. (c) Some conclusions from the structural data in (a) and (b). The prediction is that the C-O bond at C6 is broken more readily than that at C5.

bay-region hypothesis (27) suggested that a diol epoxide, with an
epoxide group located adjacent to the bay-region, may be the
activated metabolite of a carcinogenic PAH that is involved in alkylating the "critical target." We can now ask the questions, how
much distortion is there in a bay-region diol epoxide, and can any
information on its chemical reactivity and the stereochemistry of
its interaction with DNA be obtained?

Our understanding of the importance of steric factors was
initiated by structure determinations by X-ray diffraction techniques of two trans-diols of benz[a]anthracene (89), XVI and XVII,
shown in Figure 12. The crystal structures showed diequatorial

XVI   XVII

conformations for the hydroxyl groups in the unhindered compound
(XVI), and diaxial conformations in the hindered compound (XVII)
with a hydroxyl group adjacent to the bay-region. In XVII the
hydroxyl group would bump into the hydrogen atom on C12 if the
conformation were diequatorial; this steric effect would be further
enhanced if there were a methyl group on C12. But what are the
conformations in solution? NMR studies showed that in XVII the
hydroxyl groups were still 100% diaxial in solution, but that for
the less hindered copound, the hydroxyl groups were 30% diaxial and
70% diequatorial in solution (89). Thus, similar conclusions, derived from a consideration of steric interactions between non-bonded
hydrogen atoms, apply to both crystal structure (i.e., solid state)
and solution data.

While the trans diols just described were racemates it has been
possible to use X-ray crystallographic methods to determine the
absolute configuration of the isomer of BP diol epoxide that has the
highest activity (90). This is the method of choice for the
determination of absolute configuration, and can be done using the
anomalous scattering effect of a fairly heavy atom (in this case a
bromoderivative was studied). This method of anomalous scattering
analysis was originally used to establish the absolute configuration
of tartaric acid (91) and of many other chiral molecules. The
determination of the X-ray molecular structure and the absolute configuration determination of (+)-trans-8-bromo-7-menthyloxyacetoxy-
7,8,9,10-tetrahydrobenzo[a]pyrene showed that it has the (7S,8S)
configuration (90). This is shown in Figure 13; subsequent chemical
reactions to give an epoxide establish that (+)-benzo[a]pyrene
7,8-oxide has the (7R,8S) configuration. This information then
could be used, with chemical correlations, to establish the absolute
configuration of the most active diol epoxide of BP (XVIII).

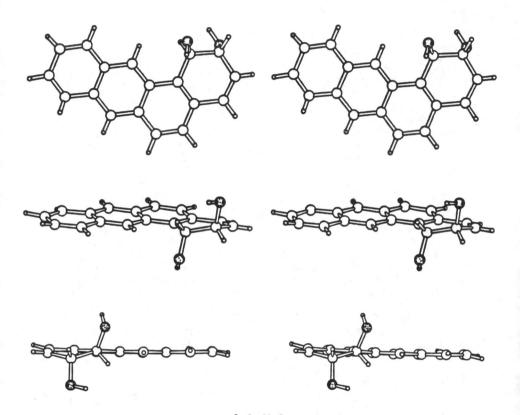

1,2-diol

Figure 12.   The structures of two trans diols of benz[a]anthracene
showing the diequatorial conformation of the unhindered 10,11-diol
and the diaxial conformation of the hindered 1,2-diol.   These
trends persist in solution where the 10,11-diol exists as an
equilibrium of 30% axial and 70% equatorial conformers (that is,
the ring is flexible); on the other hand the 1,2-diol is 100%
diaxial even in solution.   If the 1-hydroxyl group were equatorial
it would "bump" into the hydrogen atom on C12.

                                        Continued on next page.

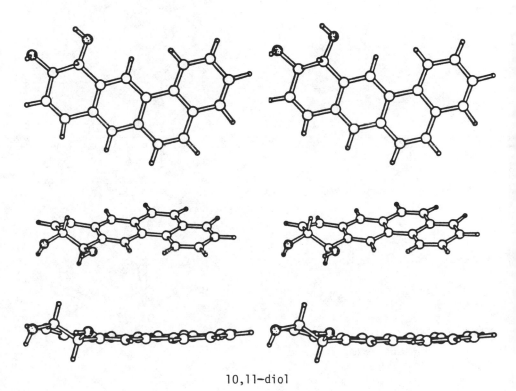

10,11-diol

Figure 12.   Continued.

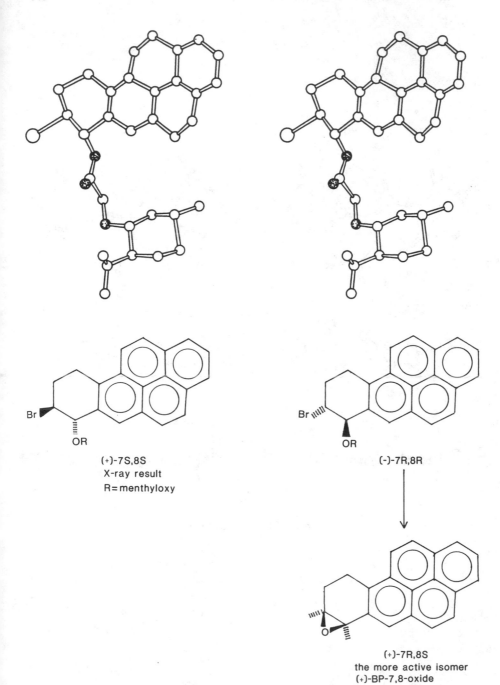

(+)-7S,8S
X-ray result
R=menthyloxy

(-)-7R,8R

(+)-7R,8S
the more active isomer
(+)-BP-7,8-oxide

Figure 13. Absolute configuration of benzo[a]pyrene metabolites
determined by anomalous dispersion (X-ray diffraction) studies of
a bromoderivative.

There are four stereoisomers of the diol epoxide of BP (XVIII - XXI). Two (XVIII and XIX) have the epoxide oxygen atom below the plane of BP and the other two (XX and XXI) have the oxygen above this plane. In addition the 7-hydroxyl group may be on the same (syn) side (XIX and XXI) as the epoxide oxygen atom or on the opposite side (anti) (XVIII and XX). Also the trans hydroxyl groups may be diaxial or diequatorial, and this may be demonstrated by crystallographic and other, such as NMR, studies (92, 93).

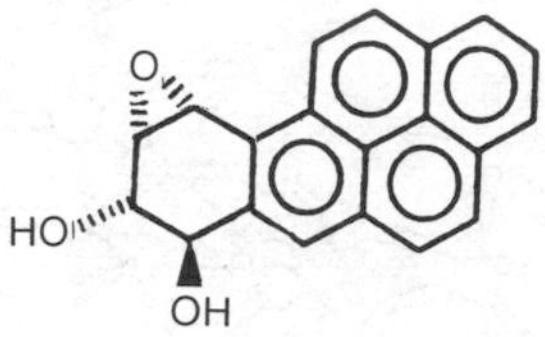

(+)-anti
XVIII

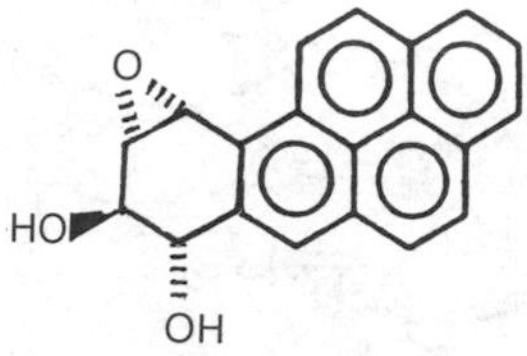

(+)-syn
XIX

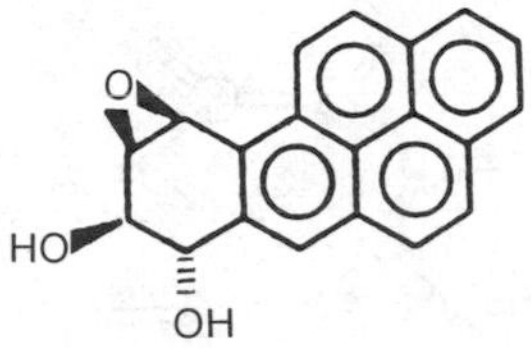

(-)-anti
XX

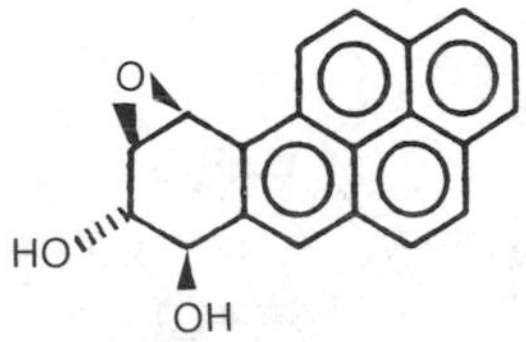

(-)-syn
XXI

The structure of the anti-diol epoxide of BP (XVIII) was determined by Neidle and co-workers (94); the structure is shown in Figure 14, together with the structure of the diol (before epoxide formation) and the tetrol (opening of the epoxide ring). The two hydroxyl groups in the diol epoxide are diequatorial (as shown to be the case in solution by NMR studies) and the epoxide ring lies in a plane nearly perpendicular to the PAH system. Other analogous epoxides with one or two hydroxyl groups in the same ring (XXII, XXIII) have similar conformations (95, 96) (Figure 15). The first structure of a syn-compound was done by X-ray diffraction on a naphthalene derivative with a methyl group in place of one of the hydroxyl group (XXIV) (97). Here the formation of an internal hydrogen bond, as predicted by Hulbert (98), is observed. However, this hydrogen bond is not the cause of the presence of axial groups, but rather the result, since if the 7-hydroxy group is replaced by a methoxy group, as in a dimethoxy epoxide (XXV), the conformation of the hydroxyl groups is also diaxial (99). These structures are illustrated in Figure 16. In the (±)-anti-tetrahydrobenzo[a]pyrene diol epoxide the ring bearing the epoxide and diol has a C8 half-chair conformation (with C8 as the atom most out of the plane). In the syn-hydroxy epoxides (XXIV and XXV) the ring has a similar conformation. A syn-diol epoxide (XIX) studied by Neidle (100), illustrated in Figure 17, was shown by X-ray studies to have equatorial

(a)

(b)

(c)

(d)

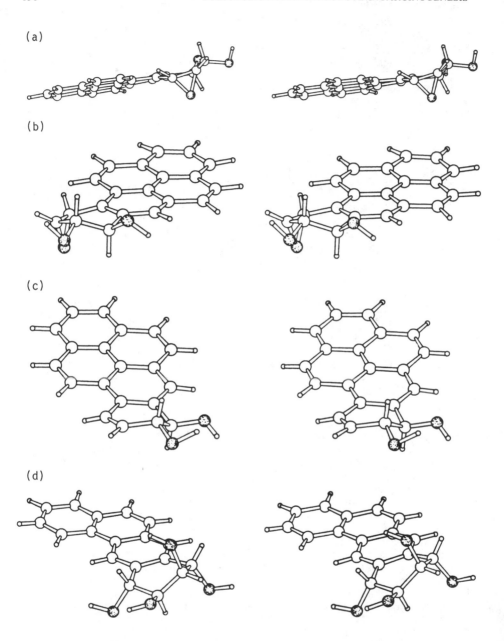

Figure 14. (a,b) Views of the structure of an anti-diol epoxide of BP (94). Note the diequatorial conformation of the hydroxyl groups. (c) A dihydrodiol of BP and (d) a tetrol of BP, all determined by Neidle and co-workers.

(a)

(b)

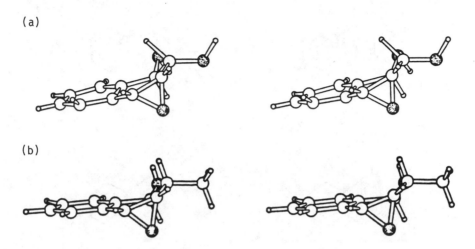

Figure 15.   Structures of analogs of the anti-diol epoxide.
(a) Klein and Stevens (95), (b) Zacharias, Glusker and Whalen
(96).

(a)

(b)

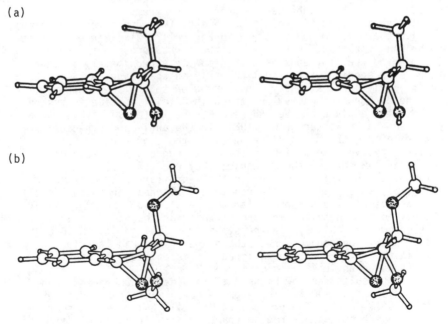

Figure 16.   Structures of analogs of the syn-diol epoxide.
(a) Glusker, Zacharias, Whalen and co-workers (97), (b) Klein
and Stevens (99).

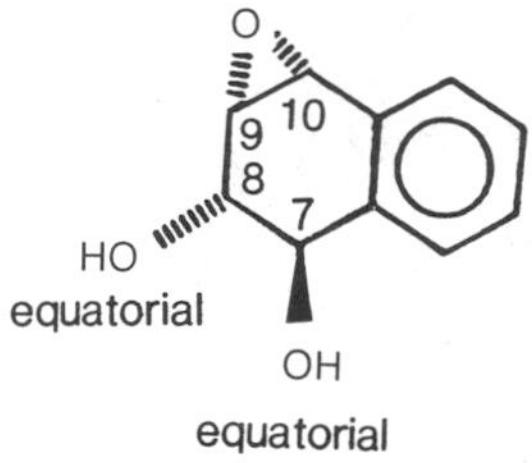

XXII

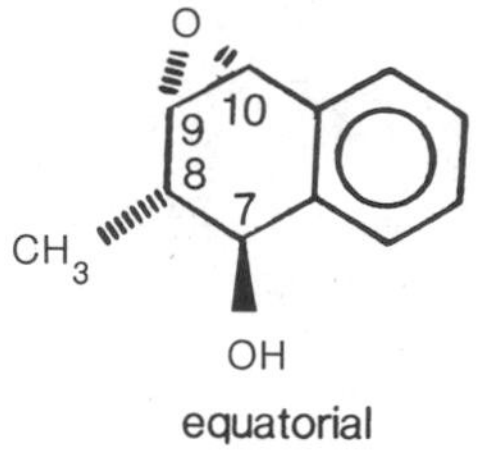

XXIII

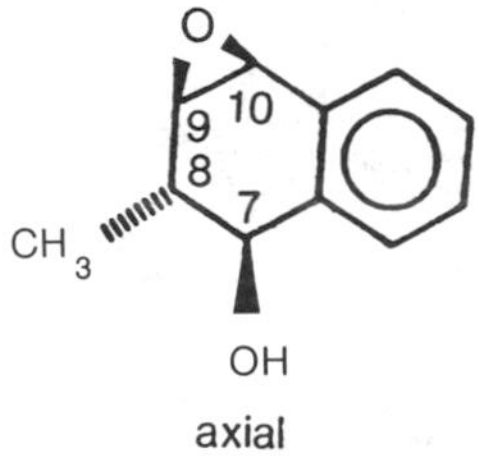

XXIV

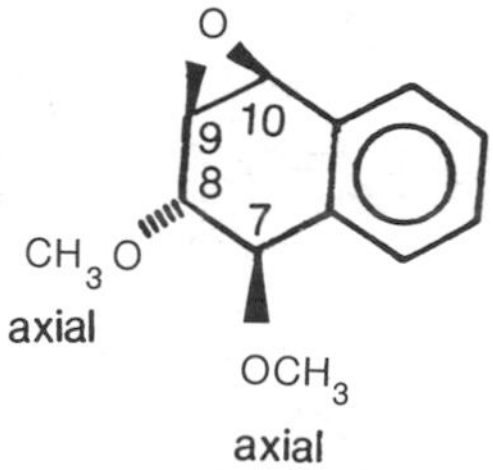

XXV

hydroxyl groups, even though potential energy calculations had
indicated that this might be a higher energy form.

The epoxides in Figures 14 to 16 show two general conformations,
of which the more common is shown in Figure 18 (a). Since the rest
of the PAH is planar, or nearly so, the first bonds in the saturated
ring bearing the diol epoxide must also lie approximately in that
plane. The epoxide will also lie nearly in the plane, but with C9
(or its equivalent) slightly above or below this plane. However,
only the carbon atom equivalent to C8 in BP diol epoxide is pushed
far out of the plane of the PAH (see Figure 18 (b)). As discussed
by Whalen and co-workers (101, 102), in the anti- and syn-diol
epoxides of BP, the major differences in the conformations of diol
epoxides involve orientation of the epoxide group with respect to
the planar PAH part of the molecule; they used the relative orientation of the benzylic C-O bond to the π-orbitals of the aromatic
rings (101) as a measure (referred to as "aligned" or "nonaligned").
These π-orbitals lie perpendicular to the plane of the PAH. X-ray
data give a more detailed measure of this for the diol epoxides of
BP, as shown in Table I. In the syn-isomer ("aligned") the O-C10
bond is nearly perpendicular to the plane of the PAH (torsion angle
98°); in the anti-isomer "nonaligned" it is at about 40° to this
plane (torsion angle 49°) (see Figure 18 (c)).

These results can now be used to consider what happens when a
diol epoxide attacks DNA. The epoxide group will open and trans
addition will occur. The product (XXVI) will have the DNA substituted adjacent to the bay region (particularly if it is hindered so
that the epoxide group is made more reactive) and will lie axial to
the PAH ring system. This means that the plane of the PAH and the
alkylated base of DNA must have a perpendicular relationship to each
other as indicated in Figure 19. In this Figure those sites in

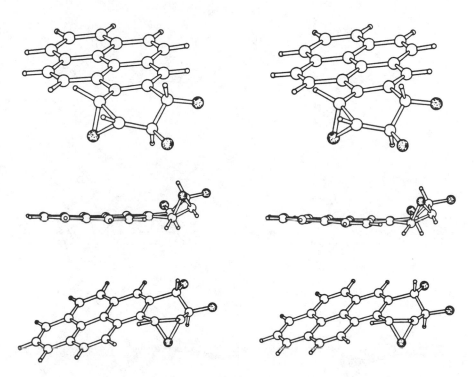

Figure 17. Views of the structure of a syn-diol epoxide of BP determined by Neidle and co-workers (100).

(a)

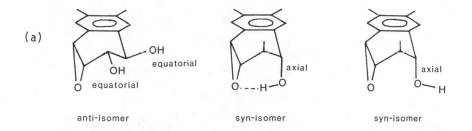

Figure 18. Conformations of diol epoxides of BP. (a) Diagrams of the anti- and syn-isomers of a diol epoxide showing the possibility for internal hydrogen-bonding in the syn-diol epoxide. (b) Conformations of the rings in a portion of the anti- and syn-diol epoxides of BP showing the different ring conformations. (c) Relative orientation of the epoxide group and the PAH system in diol epoxides of BP.

Continued on next page.

(b)

(c)

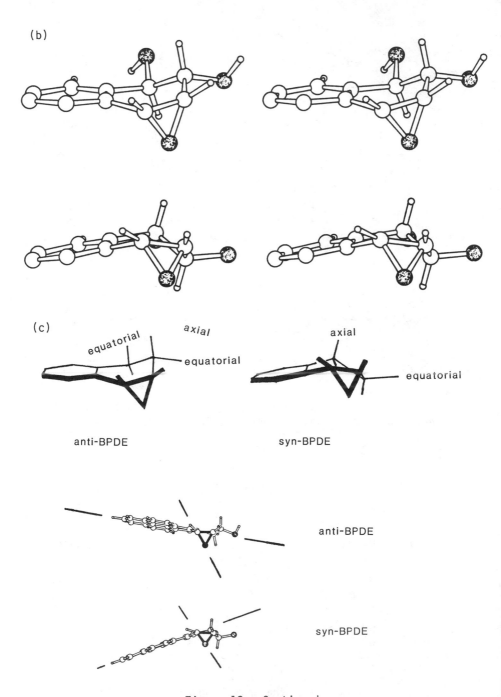

Figure 18.   Continued.

Table I.   Torsion Angles in Diol Epoxides of BP

Torsion Angle	Anti-BPDE	Syn-BPDE	
O-9-8-7	35	-111	
O-10-10A-6A	-49*	98**	
8-9-10-10A	2	-1	
9-10-10A-6A	19	29	
10-10A-6A-7	-7	-12	aromatic ring
10A-6A-7-8	-25	-33	
6A-7-8-9	44	59	dihydroxy group
7-8-9-10	-34	-45	
8-9-10-10A	2	1	epoxide

 *"aligned"
**"nonaligned"

Source: Adapted from Refs. 93 and 99.

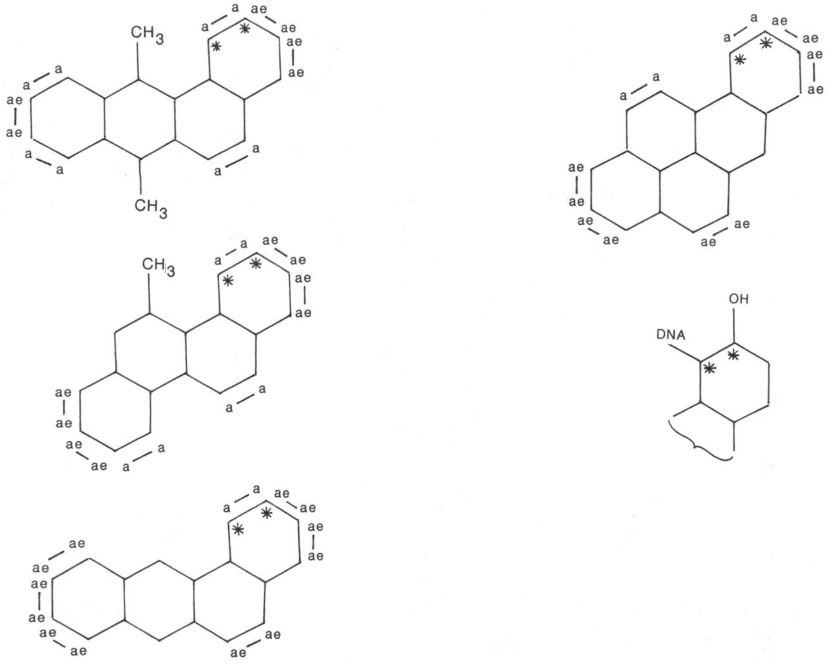

Figure 19. Perpendicularity of the PAH and the base.
Conformations of trans-diols. The steric hindrance in a bay
region will cause hydroxyl or other substituents (such as DNA) to
lie in an axial orientation. Sites where this occurs are marked
"a". Those sites where there is no steric hindrance are marked
"ae" (axial or equatorial). The conclusion is that when a diol
epoxide alkylates DNA the base on DNA will be bonded axially to
the PAH group.

carcinogenic molecules that are hindered are indicated by "a" (axial), while those that are unhindered are designated "ae" (either conformer possible).

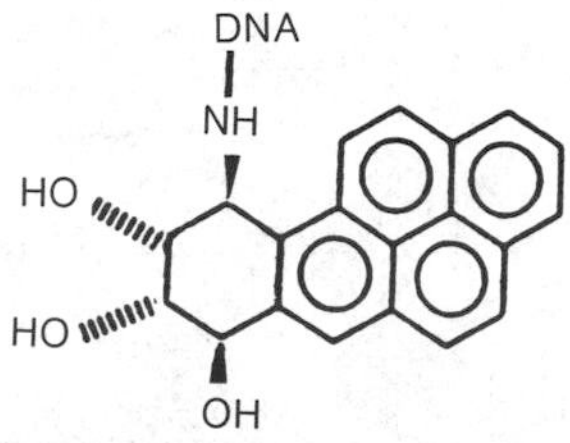

XXVI

Interactions of Activated Carcinogens with Polypeptides

Since our interest is in the stereochemical result of alkylation of a biological macromolecule, we have studied models of both proteins and nucleic acids. In proteins there are several ways in which the polypeptide chain can fold; of these the most common organized folding is the alpha-helix (103) and the beta-pleated sheet (104). Other parts of the polypeptide may fold in a random manner. The stereochemical effect of introducing a large hydrophobic group into a protein has been demonstrated (105) by the determinations of the crystal structures of the tripeptide sarcosylglycylglycine, $CH_3NH_2CH_2CONHCH_2CONHCH_2CO_2^-$ (XXVII), and its alkylated product (XXVIII). The aromatic alkylating agent (XXIX) crystallized with

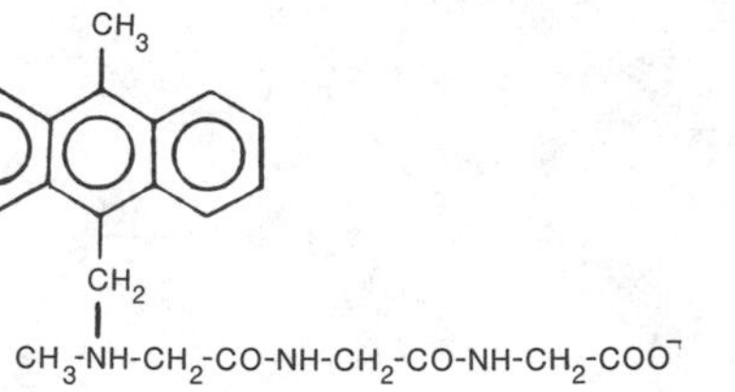

$CH_3-NH-CH_2-CO-NH-CH_2-CO-NH-CH_2-COO^-$

XXVIII

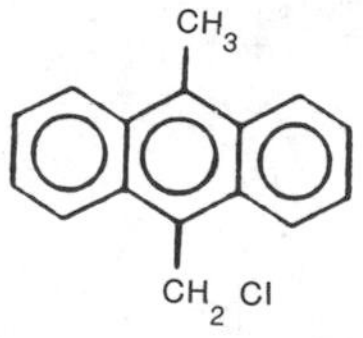

XXIX

its aromatic ring systems slightly overlapping. The tripeptide (XXVII) crystallized with an intricate three-dimensional network of hydrogen bonds, but no water of crystallization. The alkylated tripeptide (XXIX), however, crystallized in a "bilayer" type structure as shown in Figure 20. The aromatic groups packed in much the same way that they did in the simple alkylating agent (XXIX), and the peptide groups formed sheets held together by hydrogen bonds. Hydrogen bonding between these sheets of peptides involved additional water of crystallization. While the simple tripeptide had torsion angles near those expected for an alpha-helix, on alkylation the torsion angles changed to lie more nearly those expected for a beta-pleated sheet. Thus a simple model was provided of the possibly loss of alpha-helicity on alkylation by a bulky hydrophobic group.

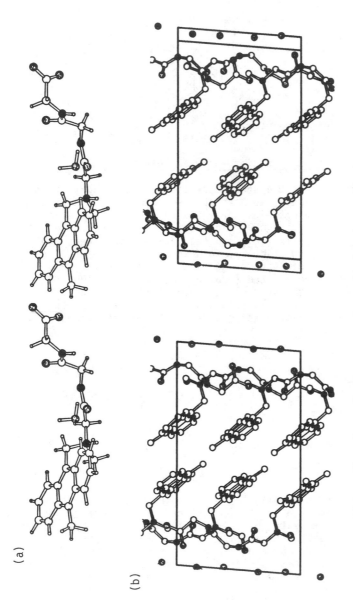

Figure 20. Crystal structure of an alkylated peptide.
(a) Alkylated peptide structure showing the non-planar peptide group (upper right of diagram). (b) Packing in the crystal structure of the alkylated peptide. This diagram shows the segregation of the various portions of the structure into areas occupied by aromatic groups, areas occupied by the more hydrophilic peptide groups, and the introduction of water of crystallization into the structure.

(a)

(b)

It might also be expected that the aromatic portion of the PAH alkylating agent will tend to lie in the interior of the protein, as far as possible from the hydrophilic surface and its aqueous environment. Thus the result of alkylation can be extensive and can lead to a loss of any function, such as catalytic activity (if the protein is an enzyme) and/or loss of ability to recognize an attachment site or another macromolecule.

Another interesting point arising from this structure determination was that the peptide became distorted on alkylation; the peptide group farthest removed from the site of alkylation had a non-planar peptide group (torsion angle of 159° instead of the expected value of 180°). This may have been caused by the crystal packing, since if that peptide group were planar a carboxyl group would project into an area of the crystal occupied by the aromatic groups. Presumably less energy was lost in distorting the peptide group than in revising the packing of the aromatic groups. A similar situation could be envisioned to occur in biological systems where there are extensive hydrophobic areas that might sequester a PAH group; such a distortion might also render the peptide group more liable to attack by other agents. However, until the nature of the relevant alkylated protein is known, it is not possible to deduce the direct effects of alkylation by an activated carcinogen.

Interactions of Activated Carcinogens with Nucleotides

In a similar way the effect of alkylation of DNA by an activated PAH may be drastic, as the hydrophobic aromatic group of the alkylating agent tries to avoid the aqueous environment of the nucleic acid. The exact nature of the lesion in DNA is unknown, and so is the type of DNA that is attacked. Recent X-ray crystallographic studies, as well as other physicochemical studies, have made it clear that DNA is not simply a polynucleotide, folded as Watson and Crick (106) proposed. There are three main conformational types of DNA; they each keep the hydrogen-bonded bases in the center of the helix, but may tilt them by a "propellor twist," may slide them from the center of the helix in the plane of the base pairs, and may vary the amount of rotation from one base pair to the next up the helical axes. These variations are listed in Table II and represent data from the X-ray structures of various polynucleotides. The three main conformational types are B-DNA (which is like Watson and Crick's right-handed helical model and was demonstrated in the structure of a dodecamer CGCGAATTCGCG, 107, 108, Figure 21 (a)), A-DNA (another right-handed helical form of DNA that has some of the characteristics of RNA, 109, 110, Figure 21 (b)) and Z-DNA (a conformation found for nucleic acids consisting of alternating purine-pyrimidine sequences and at high salt concentrations, with a left-handed helical conformation, 111, 112, Figure 21 (c)).

Other important conformational parameters of nucleic acids are the sugar pucker and the conformation about the glycosidic linkage. The puckering of the five-membered sugar ring in ribose or deoxyribose derivatives, is usually described with respect to the plane through the four most planar atoms and then defining how the fifth atom (one of O1', C1', C2', C3' or C4') lies with respect to this plane; if this fifth atom lies on the same side of the plane as does

Table II. Three Major Conformations of DNA

	B-DNA	A-DNA	Z-DNA
Handedness of helix	Right	Right	Left
Base pairs per turn	10	11	12
Axial rise per nucleotide residue	3.37 Å	2.56 Å	3.7 Å
Degrees of winding per	36°	32.7°	-30°
van der Waals diameter	19.3 Å	23 Å	18.2 Å
Sugar pucker	C2'-<u>endo</u>	C3'-<u>endo</u>	C3'-<u>endo</u> or C1'-<u>exo</u>
Glycosidic bond	anti	anti	anti alternating with syn
Major groove	deep and wide	very deep	shallow
Minor groove	shallower	very shallow	very deep

Note: See Refs. 113 and 114 for more details.

C5', then the atom is said to be <u>endo</u>, and if it is on the opposite
side it is defined as <u>exo</u>. The conformation about the sugar-base
linkage is defined as <u>anti</u> when the torsion angle (chi, O4'-Cl'-
N1'-C2 for pyrimidines and O4'-Cl'-N9-C4 for purines) lies near 180°
and <u>syn</u> when it lies near 0°. These situations are illustrated in
Figure 22. The glycosidic linkage are all <u>anti</u> in B-DNA and A-DNA,
but in Z-DNA the guanine bases are in the <u>syn</u> conformation.

In Figure 23 the sites on DNA accessible to an alkylating agent
are shown. In the major groove of B-DNA O6 and N7 of guanine, N4 of
cytosine, N6 and N7 of adenine and O4 of thymine are available for
alkylation. In the minor groove of B-DNA N2 of guanine, O2 of cytosine and of thymine and N3 of adenine are available for alkylation.
The amino groups N2 on guanine and N6 on adenine are the commonest
sites of interaction with halomethyl PAH alkylating agents (<u>75</u>, <u>76</u>);
this implies attack of guanine in the minor groove and adenine in
the major groove of B-DNA. However the groove size varies with
other conformers of nucleic acids (<u>113</u>, <u>114</u>) as shown in Table II.
In B-DNA the major groove is deep and wide and the minor groove is
narrower and shallow; in A-DNA the major groove is deeper; in Z-DNA
only the minor groove remains. So there are different possiblities
for accommodating a bulky PAH group for each conformational type of
DNA.

Very interesting information relevant to the stereochemical
results of alkylation of DNA comes from studies of nucleosides alkylated by activated PAHs. Eight such structures have been reported
(<u>115-118</u>). Three are products of the interaction of chloromethyl
PAHs with N6 of adenosine XXX-XXXII, two with deoxyadenosine
(XXXIII, XXXIV), two are para-substituted benzyl derivatives of
guanosine, alkylated at O-6 (XXXV, XXXVI) (<u>117</u>) and one is an ace-
tylaminofluorene derivative of guanosine, alkylated at C8 (XXXVII)
(<u>118</u>).

The general scheme for preparation of the alkylated adenosine
and deoxyadenosine involved the interaction of the appropriate amino-
hydrocarbon with a 6-chloropurine riboside (<u>116</u>). This allowed for
the most specific interaction without too many byproducts. The
molecular structures determined from diffraction studies are shown
in Figures 24 and 25. In Figure 24 the shape of an alkylated
nucleoside (<u>115</u>) is compared with that of the unalkylated form,

XXX   XXXI   XXXII

(a)

(b)

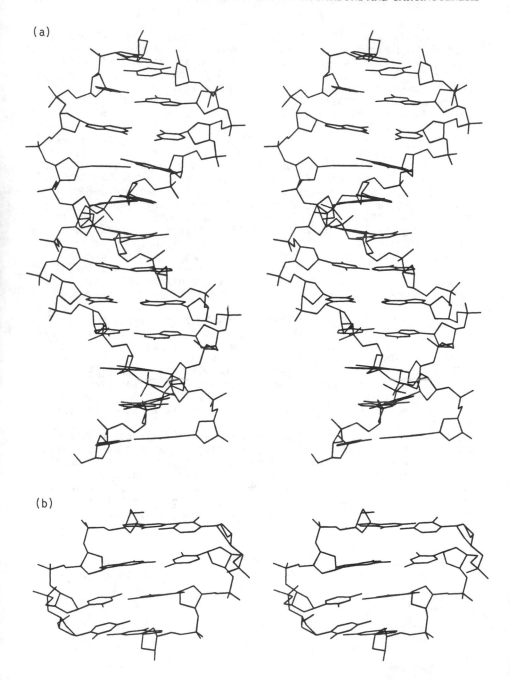

Figure 21.   Diagrams of the structures of polynucleotides,
representative of portions of (a) B-DNA, (b) A-DNA, (c) Z-DNA.

Continued on next page.

(c)

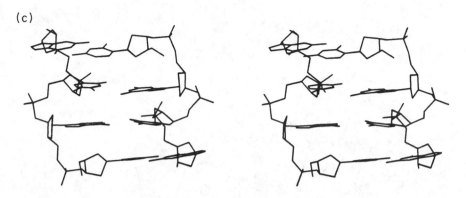

Figure 21.   Continued.

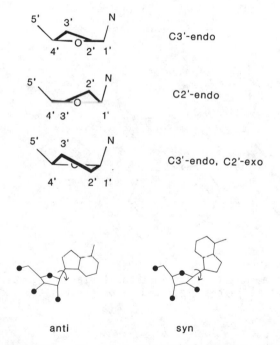

Figure 22.   (a) Ribose sugar conformations illustrating the meanings of C3'-<u>endo</u>, C2'-<u>endo</u> and C2'-<u>endo</u>, C3'-<u>exo</u>. (b) Glycosidic bond conformations of ribosides showing <u>anti</u>- and <u>syn</u>-conformations.

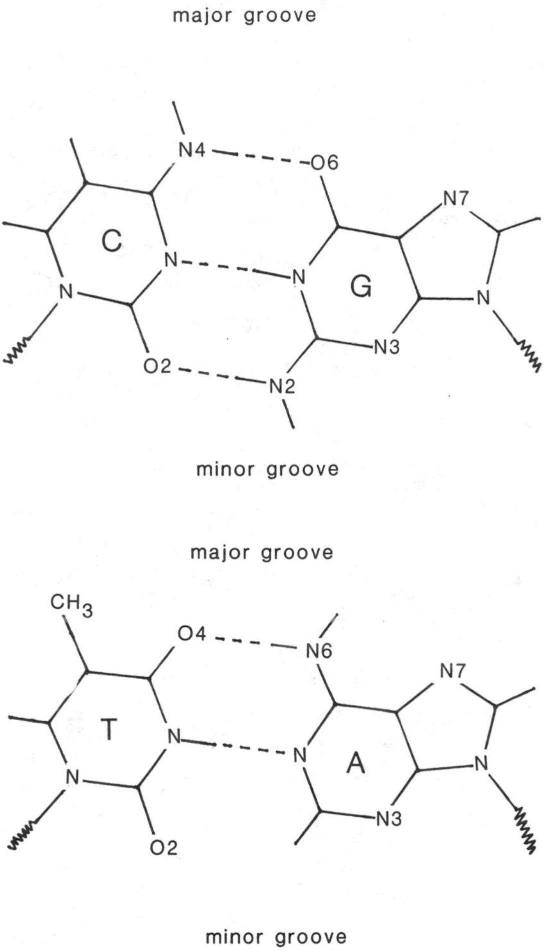

Figure 23. Sites on DNA accessible to alkylating agents. Oxygen and nitrogen atoms in the major and minor grooves are indicated.

(a)

(b)

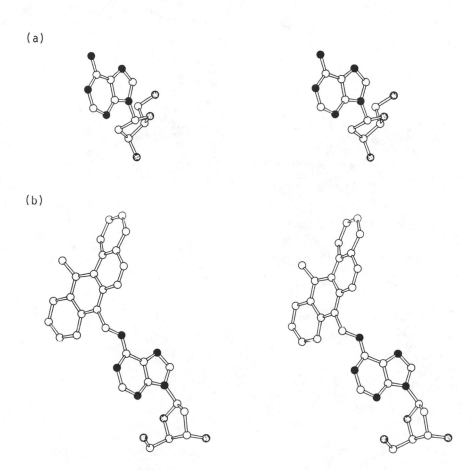

Figure 24.   Comparison of the structures of (a) deoxyadenosine and (b) an alkylated derivative.   Note the syn conformation of the product of alkylation.

(a)

(b)

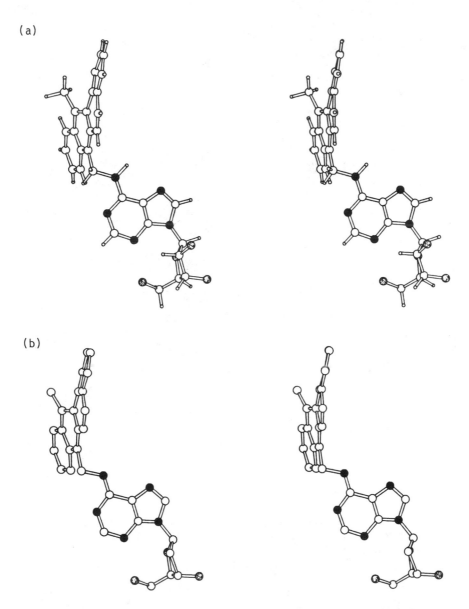

Figure 25. Shapes of five alkylated derivatives of adenosine
and deoxyadenosine. (a) Adenosine derivative of DMBA (XXXII).
(b) Deoxyadenosine derivative of DMBA (XXXIV). No hydrogen
atom positions for this diagram. (c) Adenosine derivative of
dimethylanthracene (XXXI). Note the different conformation.
(d) Deoxyadenosine derivative of dimethylanthracene (XXXIII).
(e) Adenosine derivative of methylanthracene (XXX).

                                        Continued on next page.

(c)

(d)

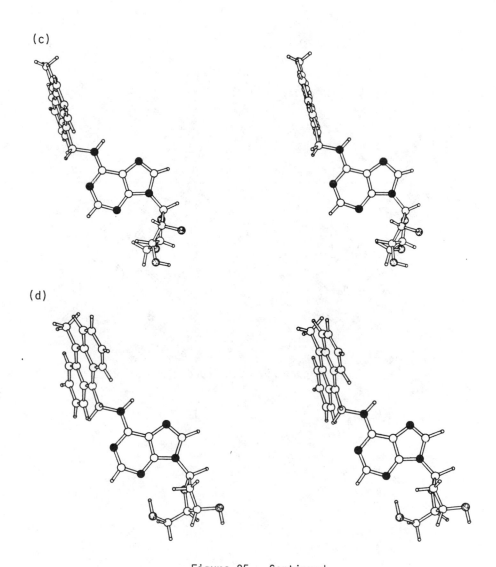

Figure 25.   Continued.

Continued on next page.

(e)

Figure 25. Continued.

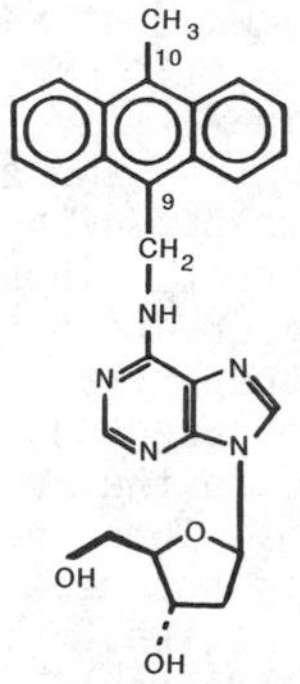

XXXIII

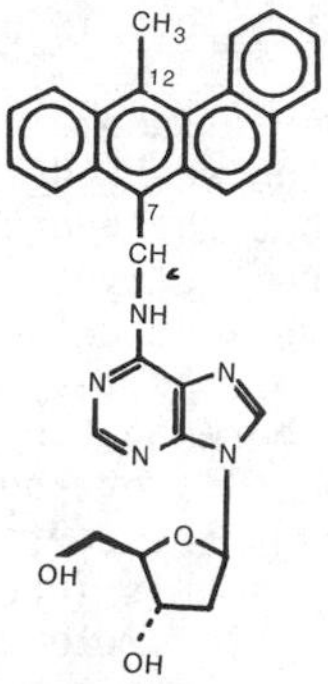

XXXIV

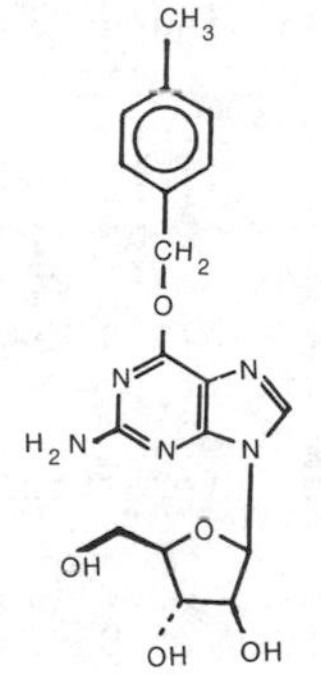

XXXV

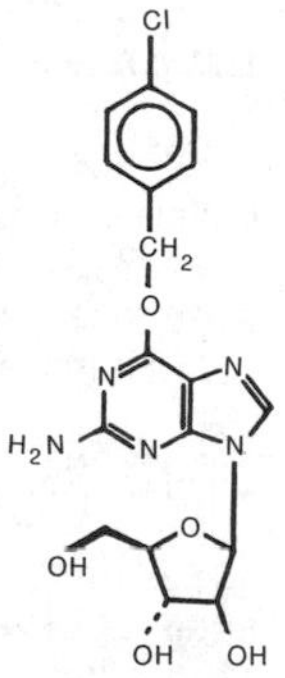

XXXVI

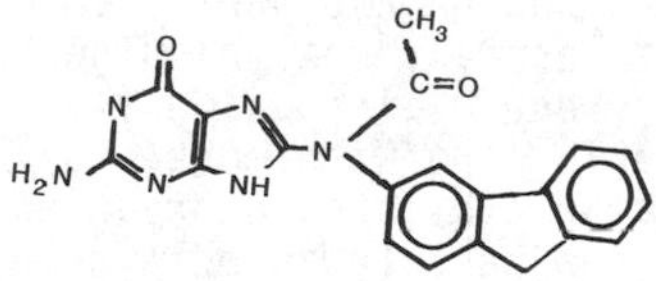

XXXVII

deoxyadenosine (119). Note that in the alkylated nucleoside the adenine and PAH residues lie nearly perpendicular to one another. It can be seen that alkylation has changed the sugar-base conforma- tion (distant from the site of alkylation) from anti (as in deoxy- adenosine and in B-DNA) to syn (as in alkylated deoxyadenosine and in some bases in Z-DNA). The sugar puckers are either C2'-endo or C3'-endo. A second structural feature of great interest in these crystals is that the more planar anthracene portion of the PAH is stacked between adenine residues of other molecules throughout the crystal. The highly buckled region of the PAH does not take part

in this stacking, but is positioned out of the way, as shown in
Figure 26.
        This general scheme of packing is found in all five adenine
derivatives studied by diffraction methods; the PAH and adenine are
interleaved, but the hydrogen bonding arrangements in the crystals
of different derivatives (XXX-XXXIV) are different. This led us to
suppose that this stacking is a significant interaction. Interestingly in four of the five compounds there is the intramolecular
hydrogen bond between the 5'-hydroxyl group and adenine, but in the
fifth compound this hydrogen bond is not found in the crystal studied, even though the conformation is still syn. This indicates that
the formation of this internal hydrogen bond is not necessary for
the adoption by the molecule of the syn-conformation. Finally there
seems to be a weak (but possibly significant) interaction between
the ribose ring oxygen atom of one molecule and the carbon atom of a
12-methyl group in the 12-methyl benz[a]anthracenyl derivative; such
interactions have been noted in other crystal structures (84, 108,
111, 120). Such an interaction, illustrated in Figure 27, may be
important in aiding in the positioning of the PAH portion of the
adduct within DNA.
        The two O^6-alkylated guanosine derivatives (117), shown in
Figure 28, do not demonstrate a perpendicularity of hydrocarbon and
base, presumably because the hydrocarbon is much smaller. However,
again the nucleoside has the syn-conformation. This perpendicularity, however, is found for the acetylamino fluorine derivative.

Computer Modelling of DNA-Activated Carcinogen Interactions

The computer has provided an excellent medium for modelling the
interactions of such activated PAHs with portions of DNA. As mentioned earlier, it is generally believed that the extracyclic base
amino groups, N2 of guanine, and more rarely N6 of adenine, are
attacked by activated PAHs. There are two general classes of models
for DNA alkylated by a diol epoxide. In the first the PAH lies in
one of the grooves of DNA but on the perimeter of the helix (121-
128); this is referred to as "external binding." The pyrene-like
chromophore of benzo[a]pyrene diol epoxide lies at about 35° to the
DNA helix axis. This is in line with model building experiments by
us in which we "docked" the alkylated deoxyadenosine compounds studied (XXX-XXIV) onto the syn-base (guanine) in Z-DNA. We used Z-DNA
because we had found a syn-conformation for the alkylated adenosines
and deoxyadenosines and Z-DNA contains guanine residues in the syn-
conformation. It meant that an adenine group was "docked" onto a
guanine residue; the result is that major groove alkylation (on O6
of guanine or N6 of adenine) is modelled. The result is shown in
Figure 29. A rotation of 180° about the C6-N6 bond (which is free
to rotate) to the PAH placed the PAH on the surface of the helix and
did not disrupt normal interbase hydrogen bonding within the helix
in any way. Some other rotation angles about this bond led to
unacceptable steric interactions. The interesting feature of the
rotated model is that the curvature of the dimethylbenz[a]anthracene
moiety conforms to some extent to the shape of the helix. Attempts
to dock the alkylated bases into B-DNA led to more steric problems.
Such model building by others (129-131) has led to the suggestion

(a)

(b)

(c)

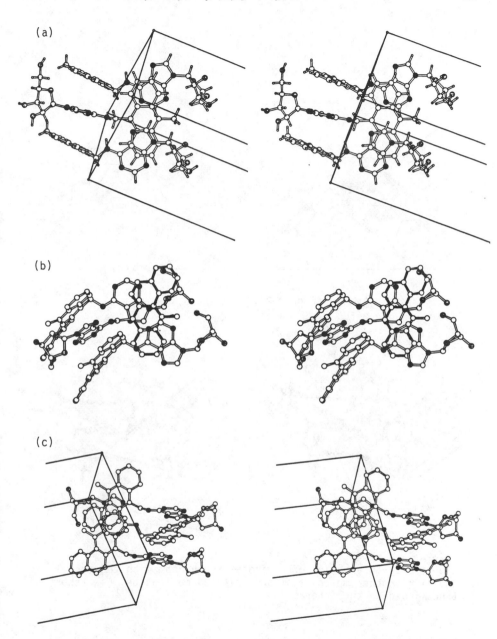

Figure 26.   Stacking of PAHs and bases in alkylated adenosine and deoxyadenosine derivatives. (a) The methylanthracenyl derivative of adenosine. (b) The DMBA derivative of deoxyadenosine viewed onto the planes of adenine groups. (c) Side view of this. (d) Diagram of overlap of a DMBA derivative in the adenosine and deoxyadenosine adducts Ar denotes the aromatic ring and pu indicates purine. Areas of overlap are black.

Continued on next page.

(d)

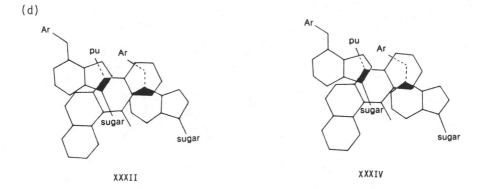

XXXII   XXXIV

Figure 26.   Continued.

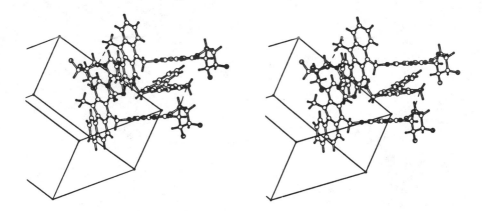

Figure 27.   Interaction of a ribose oxygen atom with the methyl
group in another molecule of an alkylated nucleoside.   Two short
lines indicate the interaction.

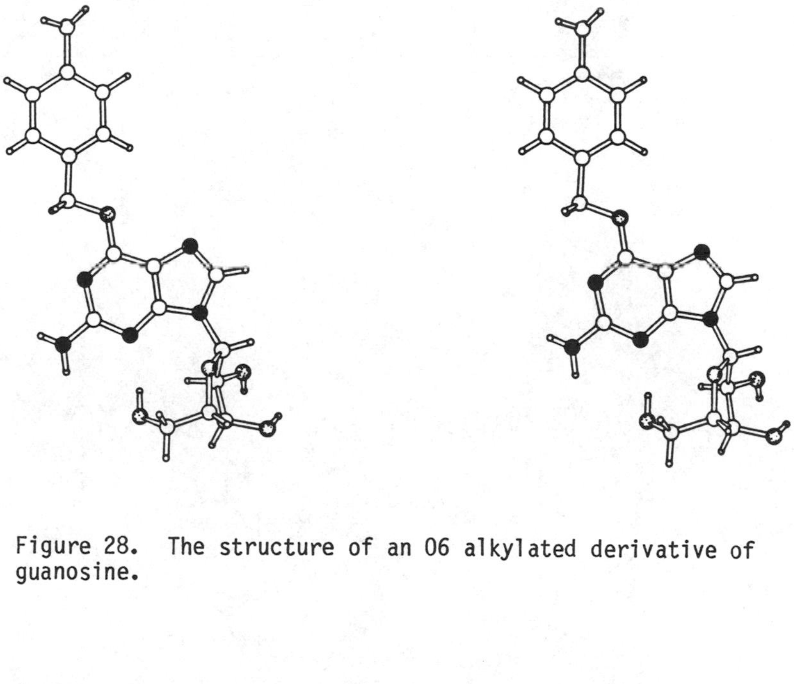

Figure 28.   The structure of an O6 alkylated derivative of guanosine.

(a)

(b)

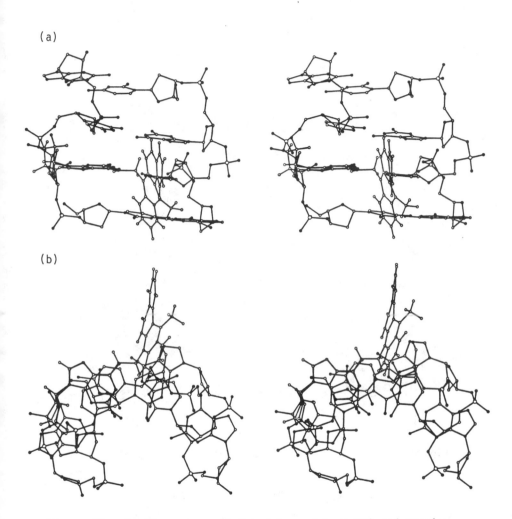

Figure 29.   Docking of an alkylated base into Z-DNA and torsion
of the PAH group. (a, b) Two views of the alkylated base (as
determined by X-ray diffraction techniques) docked onto the syn-
guanine group in Z-DNA. (c, d) Two views of the same modelling
experiment with a torsional rotation of 180 about C6-N6.   Note
how the curvature of the PAH group conforms to that of the helix.
This is a model of the carcinogen lying in a groove of DNA.
                                        Continued on next page.

(c)

(d)

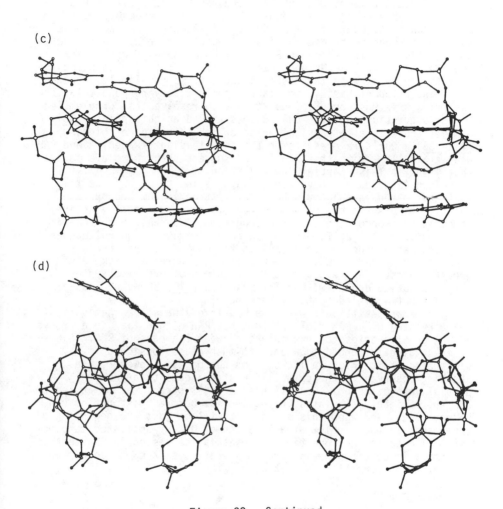

Figure 29. Continued.

that the benzo[a]pyrene adduct formation could promote a local
B-to-A DNA transition, since A-DNA has a wider but less deep minor
groove into which the aromatic portion could fit.
    An alternative model, in which the PAH is internally bound
("quasi-intercalated"), is shown in Figure 30. Presumably some
distortion must occur if a PAH is both covalently bound to a base
and lies between that base and its neighboring base up the DNA helix
axis. This is illustrated in Figure 31 which shows first a diol
epoxide docked onto B-DNA and then the effect of trying to bring the
aromatic group between bases; because it is covalently attached
some distortion of bases from planes approximately perpendicular to
the helix axis must occur. One attempt to resolve this problem led
to the suggestion that the DNA becomes kinked at this point (138).
However there is also physical evidence for this kind of binding
(132-138). Our X-ray results on the interleaving of adenine and PAH
also are in line with this model, and suggest that there are forces
that favor at least partial intercalation.
    To date we do not have enough information to be able to tell
what the lesion in DNA would be on alkylation in the appropriate
position of DNA (that is, the one that is carcinogenically active).
Attempts to overcome this are in progress. A synthetic heptameric
duplex d(GTCA*GAC):d(GTCTGAC) prepared by Stezowski and colleagues
(139) has an adenosine group (A*) in the center of one strand
alkylated by a PAH group. Attempts are in progress to make a minor
groove adduct. The report of the crystal structure of a synthetic
oligonucleotide duplex with a T:T mismatch (140) shows no loss of
helicity but modified (wobble) base-pairing.
    So we are still left with two models of the stereochemistry of
DNA alkylated by a PAH diol epoxide; the PAH either lies in a groove
of DNA or else tries to intercalate between the bass of DNA. Since
it is covalently bonded to a base it must cause considerable
distortion if it tries to lie between the bases. However, the
stacking observed in the crystalline state seems to argue for
partial intercalation. We will need crystal structures of at least
one appropriately alkylated polynucleotide before this problem can
be resolved. And when this is done it will be just the beginning of
the answer to the problem of alkylation of DNA by activated carcinogens. The subsequent question is, what is the lesion in DNA that is
important in carcinogenesis, and then what does it cause to happen
so that tumor formation is initiated?

Acknowledgments

I thank Drs. H. L. Carrell, A. Dipple, R. G. Harvey, S. Hecht,
R. Moschel, J. J. Stezowski, and D. E. Zacharias for many helpful
discussions and collaborations. Most of the diagrams were drawn
using the computer programs VIEW (141) and DOCK (142). Figure 21
was drawn with data obtained from the Protein Data Bank, Brookhaven.
The work of the author was supported by grants CA-10925, CA-22780,
CA-06927, RR-05539 from the National Institutes of Health, BC-242
from the American Cancer Society, and by an appropriation from the
Commonwealth of Pennsylvania.

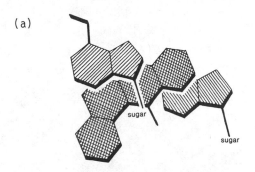

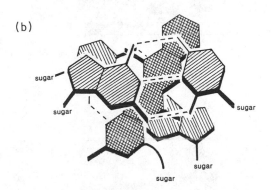

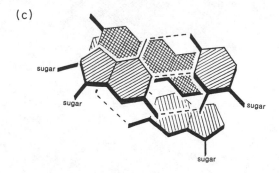

Figure 30. Schematic diagrams of overlap of ring systems in (a) the alkylated nucleoside described here, (b) daunomycin inter-calated in a hexanucleotide (143), (c) proflavine intercalated in a dinucleoside phosphate (70), (d) a model of a diol epoxide of DMBA semi-intercalated in a manner analogous to that in (a) so that N2 of guanine may be alkylated, (e) a model of a diol epoxide of DMBA semi-intercalated in a manner analogous to that in (c) so that O6 of guanine may be alkylated.

Continued on next page.

(d)

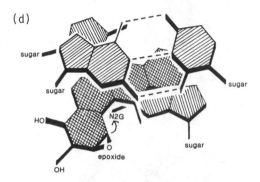

(e)

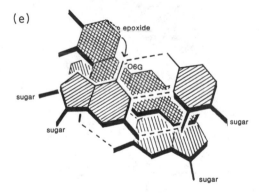

Figure 30. Continued.

(a)

(b)

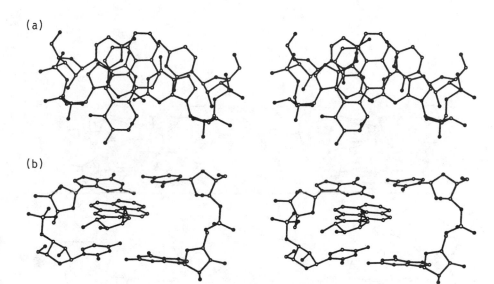

Figure 31. Schematic diagrams of alkylation of DNA by a diol epoxide by (a) semi-intercalated (viewed from above), (b) semi-intercalated (viewed from the side). The guanine in the lower right of the diagram will be alkylated when the epoxide ring opens. (c) alkylation of a guanine with the PAH lying in a groove of DNA, and (d) the problem that results when a covalent bond is formed. The base and PAH group cannot both lie in planes perpendicular to the helix axis. Therefore, if the PAH is semi-intercalated, as shown, the base must distort out of the plane of the base pair with subsequent distortions along the helix. Various models have suggested kinks or conformational changes to account for this. One way in which this might occur is illustrated in this schematic diagram. The conformation that actually results in reality on such alkylation is still unknown.

Continued on next page.

(c)

(d)

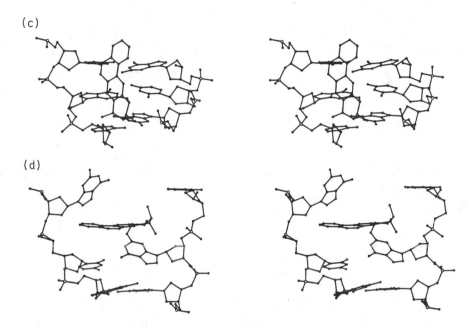

Figure 31. Continued.

Literature Cited

1. Pott, P. "Chirurgical Observations Relative to the Cataract, the Polypus of the Nose, the Cancer of the Scrotum, the Different Kinds of Ruptures and the Mortification of the Toes and Feet"; Hawes, Clarke, and Collins: London, 1775.
2. Earle, H. Medico-Chirurgical Trans., 1823, 12, part II, pp. 296-307.
3. Curling, T. B. "A Practical Treatise on the Diseases of the Testes and of the Spermatic Cord and Scrotum," Second edition; Blanchard and Lea: Philadelphia, 1856.
4. Potter, M. National Cancer Institute, Monograph No. 10; Conference: Biology of Cutaneous Cancer, Philadelphia, PA, 1962; pp. 1-4.
5. Yamagiwa, K.; Ichikawa, K. J. Cancer Res. 1918, 3, 1 (1916 Mitt. Fak. Univ. Tokyo, 15, 296).
6. Passey, R. D. Brit. Med. J. 1922, 2, 1112.
7. Kennaway, E. Brit. Med. J. 1955, 2, 749.
8. Sawicki, E. Chemist-Analyst 1964, 53, 24.
9. Doll, R.; Hill, A. B. Brit. Med. J. 1950, 2, 739.
10. Wynder, E. L.; Wright, G. Cancer 1957, 10, 255.
11. Miller, J. A. Cancer Res. 1970, 30, 559.
12. Nemoto, N.; Gelboin, H. V. Biochem. Pharmacol. 1976, 25, 1221.
13. Sims, P.; Grover, P. L. Nature (London) 1974, 252, 326.
14. Glusker, J. P.; Trueblood, K. N. "Crystal Structure Analysis: A Primer"; 1972, Oxford University Press: New York (Second edition, in press).
15. Friedrich, W.; Knipping, P.; von Laue, M. Sitzber. math.-physik. Klasse Bayer. Akad. Wiss. München, 1912, p. 303.
16. Bragg, W. L. Proc. Roy. Soc. (London) 1913, A89, 248.
17. Bragg, W. H.; Bragg, W. L. Nature 1913, 91, 557.
18. Lonsdale, K. Nature (London) 1928, 122, 810.
19. Cox, E. G.; Cruickshank, D. W. J.; Smith, J. A. S. Proc. Roy. Soc. 1958, A247, 1.
20. Bacon, G. E.; Curry, N. A.; Wilson, S. A. Proc. Roy. Soc. 1964, A279, 98.
21. Stoicheff, B. P. Can. J. Phys. 1954, 32, 339.
22. Abrahams, S. C.; Robertson, J. M.; White, J. G. Acta Crystallogr. 1949, 2, 233.
23. Banerjee, K. Nature (London) 1930, 125, 456.
24. Mathieson, A. McL.; Robertson, J. M.; Sinclair, V. C. Acta Crystallogr. 1950, 3, 245.
25. Pauling, L. "The Nature of the Chemical Bond," 1960, Third edition; Cornell Univ. Press: Ithaca, New York.
26. Pullman, A.; Pullman, B. Adv. Cancer Res. 1955, 3, 117.
27. Jerina, D. M.; Daly, J. W. In "Drug Metabolism"; Parks, D. V.; Smith, L., Eds.; Taylor and Francis: London; p. 13.
28. Pullman, B. In "Polycyclic Hydrocarbons and Cancer. 2. Molecular and Cell Biology"; Gelboin, H. V.; Ts'o, P.O.P., Eds.; Academic Press, Inc.: New York, 1978; p. 419.
29. Iball, J. Z. Kristallogr. 1936, 94, 7.
30. Iball, J. Am. J. Cancer 1939, 35, 188.
31. Burns, D. M.; Iball, J. Proc. R. Soc. 1960, A257, 491.
32. Iball, J. Z. Kristallogr. 1935, 92, 293.

33. Iball, J.; MacDonald, S. G. G. Z. Kristallogr. 1960, 114, 439.
34. Iball, J.; Scrimgeour, S. N. Acta Crystallogr. 1975, B31,
     2517.
35. Iball, J.; Scrimgeour, S. N.; Young, D. W. Acta Crystallogr.
     1976, B32, 328.
36. Iball, J.; Young, D. W. Nature (London) 1956, 177, 985.
37. Sayre, D.; Friedlander, P. H. Nature (London) 1960, 187, 139.
38. Iball, J. Nature (London) 1964, 201, 916.
39. Zacharias, D.; Glusker, J. P., unpublished observations.
40. Glusker, J. P. In "Crystallography in North America";
     McLachlan, D. Jr.; Glusker, J. P., Eds.; Section F, Chapter
     17, American Crystallographic Association: New York, 1983;
     p. 404.
41. Sayre, D.; Friedlander, P. H. Nature (London) 1956, 178, 999.
42. Iball, J.; Morgan, C. H.; Zacharias, D. E. J. Chem. Soc.
     Perkin Trans. II 1975, p. 1272.
43. Robertson, J. M.; White, J. G. J. Chem. Soc. 1947, p. 1001.
44. Robertson, J. M.; White, J. G. J. Chem. Soc. 1956, p. 925.
45. Iball, J. Nature (London) 1936, 137, 361.
46. van Soest, T. C.; Peerdeman, A. F. Acta Crystallogr. 1970,
     B26, 1940.
47. van Soest, T. C.; Peerdeman, A. F. Acta Crystallogr. 1970,
     B26, 1947.
48. van Soest, T. C.; Peerdeman, A. F. Acta Crystallogr. 1970,
     B26, 1956.
49. Swenson, D. H.; Miller, J. A.; Miller, E. C. Biochem.
     Biophys. Res. Comm. 1973, 53, 1260.
50. Haddow, A. Proc. Can. Cancer Res. Conf. 1957, 2, 361.
51. Busetta, B.; Hospital, M. Acta Cryst. 1972, B28, 560.
52. Huggins, C.; Yang, N. C. Science 1962, 137, 257.
53. Cook, J. W.; Hazelwood, G.A.D. Chem. and Ind. (London) 1933,
     11, 758.
54. Coombs, M. M.; Bhatt, T. S.; Young, S. Br. J. Cancer 1979,
     40, 1914.
55. Jones, D. W.; Sowden, J. M.; Hazell, A. C.; Hazell, R. G.
     Acta Cryst. 1978, B34, 3021.
56. Hoffmann, D.; Bondinell, W. E.; Wynder, E. L. Science 1974,
     183, 215.
57. Kashino, S.; Zacharias, D. E.; Prout, C. K.; Carrell, H. L.;
     Glusker, J. P. Acta Cryst. 1984, C40, 536.
58. Zacharias, D. E.; Kashino, S.; Glusker, J. P.; Harvey, R. G.;
     Amin, S.; Hecht, S. S. Carcinogenesis 1984, 5, 1421.
59. Slaga, T. J.; Iyer, R. P.; Lyga, W.; Secrist, A., III; Daub,
     G. M.; Harvey, R. G. In "Polynuclear Aromatic Hydrocarbons:
     Chemistry and Biological Effects"; Bjorseth, A.; Dennis, A. J.,
     Eds.; Battelle Press: Columbus, Ohio, 1980; p. 753.
60. Lerman, L. S. J. Mol. Biol. 1961, 3, 18.
61. Craig, M.; Isenberg, I. Biopolymers 1970, 9, 689.
62. Craig, M.; Isenberg, I. Proc. Natl. Acad. Sci. U.S. 1970, 67,
     1337.
63. Pauling, L. In "The Nature of the Chemical Bond." Second
     Edition. Oxford University Press, 1940; p. 189.
64. Srikrishnan, T.; Parthasarathy, R. Nature 1976, 264, 379.

65. Jain, S. C.; Tsai, C. C.; Sobell, H. M. J. Mol. Biol. 1977, 114, 301.
66. Tsai, C. C.; Jain, S. C.; Sobell, H. M. J. Mol. Biol. 1977, 114, 317.
67. Tsai, C. C.; Jain, S. C.; Sobell, H. M. Proc. Natl. Acad. Sci. U.S.A. 1975, 72, 628.
68. Sakore, T. D.; Jain, S. C.; Tsai, C. C.; Sobell, H. M. Proc. Natl. Acad. Sci. U.S.A. 1977, 74, 188.
69. Sobell, H. M.; Jain, S. C.; Sakore, T. D.; Reddy, B. S.; Bhandary, K. K.; Seshadri, T. P. Biomol. Struct. Conform. Funct., Evol., Proc. Int. Symp. 1981, 1, 441 (Pergamon: Oxford, England).
70. Berman, H. M.; Stallings, W.; Carrell, H. L.; Glusker, J. P.; Neidle, S.; Taylor, G.; Achari, A. Biopolymers 1979, 18, 2405.
71. Neidle, S.; Achari, A.; Taylor, G. L.; Berman, H. M.; Carrell, H. L.; Glusker, J. P.; Stallings, W. C. Nature (London) 1977, 269, 304.
72. Dickerson, R. E.; Drew, H. R. J. Mol. Biol. 1981, 149, 761.
73. Geis, I. J. Biomol. Struct. and Dynamics 1983, 1, 581.
74. Westhof, E.; Sundaralingam, M. Proc. Natl. Acad. Sci. U.S.A. 1980, 77, 1852.
75. Dipple, A.; Brookes, P.; Mackintosh, D. S.; Rayman, M. P. Biochemistry 1971, 10, 4323.
76. Dipple, A.; Slade, T. A. Eur. J. Cancer 1970, 6, 417.
77. Peck, R. M.; Tan, T. K.; Peck, E. B. Cancer Res. 1976, 36, 2423.
78. Glusker, J. P.; Zacharias, D. E.; Carrell, H. L. Cancer Res. 1976, 36, 2428.
79. Zacharias, D. E. Acta Crystallogr. 1977, B33, 902.
80. Carrell, H. L., unpublished observations, 1981.
81. Sims, P.; Grover, P. L. Adv. Cancer Res. 1974, 20, 165.
82. Glusker, J. P.; Carrell, H. L.; Zacharias, D. E.; Harvey, R. G. Cancer Biochem. Biophys. 1974, 1, 43.
83. Glusker, J. P.; Zacharias, D. E.; Carrell, H. L.; Fu, P. P.; Harvey, R. G. Cancer Res. 1976, 36, 3951.
84. Murray-Rust, P.; Glusker, J. P. J. Amer. Chem. Soc. 1984, 106, 1018.
85. Zacharias, D. E.; Glusker, J. P.; Harvey, R. G.; Fu, P. P. Cancer Res. 1977, 37, 775.
86. Silverton, J. V.; Dansette, P. M.; Jerina, D. M. Tetrahedron Lett. 1976, p. 1557.
87. Conney, A. H. Cancer Res. 1982, 42, 4875.
88. Harvey, R. G. Acc. Chem. Res. 1981, 14, 218.
89. Zacharias, D. E.; Glusker, J. P.; Fu, P. P.; Harvey, R. G. J. Am. Chem. Soc. 1979, 101, 4043.
90. Boyd, D. R.; Gadaginamath, G. S.; Kher, A.; Malone, J. F.; Yagi, H.; Jerina, D. M. J. Chem. Soc. Perkin I 1980, p. 2112.
91. Bijvoet, J. M.; Peerdeman, A. F.; van Bommel, A. J. Nature (London) 1951, 168, 271.
92. Yagi, H.; Hernandez, O.; Jerina, D. M. J. Am. Chem. Soc. 1975, 97, 6881.

93. Sayer, J. M.; Yagi, H.; Croisy-Delcey, M.; Jerina, D. M. J. Am. Chem. Soc. 1981, 103, 4970.
94. Neidle, S.; Subbiah, A.; Cooper, C. S.; Ribeiro, O. Carcinogenesis 1980, 1, 249.
95. Klein, C. L.; Stevens, E. D. Cancer Research 1984, 1523.
96. Zacharias, D. E.; Glusker, J. P.; Whalen, D. L. (unpublished observations).
97. Glusker, J. P.; Zacharias, D. E.; Whalen, D. L.; Friedman, S.; Pohl; T. M. Science 1982, 215, 695.
98. Hulbert, P. B. Nature (London) 1975, 256, 146.
99. Klein, C. L.; Stevens, E. D. Acta Cryst. C 1984, 315.
100. Neidle, S.; Cutbush, S. D. Carcinogenesis 1983, 4, 415.
101. Sayer, J. M.; Yagi, H.; Silverton, J. V.; Friedman, S. L.; Whalen, D. L.; Jerina, D. M. J. Am. Chem. Soc. 1982, 104, 1972.
102. Sayer, J. M.; Whalen, D. L.; Friedman, S. L.; Paik, A.; Yagi, H.; Vyas, K. P.; Jerina, D. M. J. Am. Chem. Soc. 1984, 106, 226.
103. Pauling, L.; Corey, R. B.; Branson, H. R. Proc. Natl. Acad. Sci. U.S.A. 1951, 37, 205.
104. Pauling, L.; Corey, R. B. Proc. Natl. Acad. Sci. U.S.A. 1951, 37, 261.
105. Glusker, J. P.; Carrell, H. L.; Berman, H. M.; Gallen, B.; Peck, R. M. J. Am. Chem. Soc. 1977, 99, 595.
106. Watson, J. D.; Crick, F. H. C. Nature 1953, 171, 737.
107. Wing, R.; Drew, H.; Takano, T.; Broka, C.; Tanaka, S.; Itakura, K.; Dickerson, R. E. Nature (London) 1980, 287, 755.
108. Drew, H. R.; Wing, R. M.; Takano, T.; Broka, C.; Tanaka, S.; Itakura, K.; Dickerson, R. E. Proc. Natl. Acad. Sci. U.S.A. 1980, 78, 2179.
109. Shakked, Z.; Rabinovich, D.; Cruse, W. B. T.; Egert, E.; Kennard, O.; Sala, G.; Salisbury, S. A.; Viswamitra, M. A. Proc. Roy. Soc. Lond. 1981, B213, 479.
110. Conner, B. N.; Takano, T.; Tanaka, S.; Itakura, K.; Dickerson, R. E. Nature 1982, 295, 294.
111. Wang, A. H-J.; Quigley, G. J.; Kolpak, F. J.; Crawford, J. L.; van Boom, J. H.; van der Marel, G.; Rich, A. Nature (London) 1979, 282, 680.
112. Wang, A. H.-J.; Quigley, G. J.; Kolpak, F. J.; van der Marel, G.; van Boom, J. H.; Rich, A. Science 1981, 211, 171.
113. Saenger, W. "Principles of Nucleic Acid Structure." Springer-Verlag: New York, 1984.
114. Dickerson, R. E.; Kopka, M. L.; Drew, H. R. in "Structure and Dynamics: Nucleic Acids and Proteins." ed. E. Clementi and R. H. Sarma. Adenine Press: New York, 1983, pp. 149-179.
115. Carrell, H. L.; Glusker, J. P.; Moschel, R.; Hudgins, W. R.; Dipple, A. Cancer Res. 1981, 41, 2230.
116. Stezowski, J. J.; Stigler, R-D., Joos-Guba, G.; Kahre, J., Lösch, G. R., Carrell, H. L.; Peck, R. M.; Glusker, J. P. Cancer Res. 1984, 44, 5555.
117. Zacharias, D. E.; Glusker, J. P., Moschel, R., Dipple, A. (manuscript in preparation).

118. Kuroda, R.; Neidle, S.; Evans, F. E.; Broyde, S.; Hingerty, B. E. International Union of Crystallography Meeting, Hamburg, Germany, 8-18 August 1984. Abstract 03.2-3.
119. Watson, D. G.; Sutor, D. J.; Tollin, P. Acta Cryst. 1965, 19, 111.
120. Brennan, R. G.; Privé, G. G.; Blonski, W. J. P.; Hruska, F. E.; Sundaralingam, M. J. Biomol. Struct. Dynam. 1983, 1, 939.
121. Beland, F. A. Chem.-Biol. Interactions 1978, 22, 329.
122. Geacintov, N. E.; Gagliano, A.; Ivanovic, V.; Weinstein, I. B. Biochemistry 1978, 17, 5256.
123. Geacintov, N. E.; Gagliano, A. G.; Ibanez, V.; Harvey, R. G. Carcinogenesis 1982, 3, 247.
124. Geacintov, N. E.; Gagliano, A. G.; Ibanez, V.; Lee, H.; Jacobs, S. A.; Harvey, R. G. J. Biomol. Struct. Dynam. 1983, 1, 913.
125. Lefkowitz, S. M.; Brenner, H. C. Biochem. 1982, 21, 3735.
126. Prusik, T.; Geacintov, N. E. Biochem. Biophys. Res. Commun. 1979, 88, 782.
127. Ridler, P. J.; Jennings, B. R. FEBS Letters 1982, 139, 101.
128. Undeman, O.; Lycksell, P-O.; Gräslund, A.; Astlind, T.; Ehrenberg, A.; Jernström, B.; Tjerneld, F.; Nordén, B. Cancer Res. 1983, 43, 1851.
129. Aggarwal, A. K.; Islam, S. A.; Neidle, S. J. Biomol. Struct. Dynam. 1983, 1, 873.
130. Hingerty, B.; Broyde, S. J. Biomol. Struct. Dynam. 1983, 1, 905.
131. Anderson, R. W.; Whitlow, M. D.; Teeter, M. M.; Mohr, S. C. 188th National Meeting of American Chemical Society, Philadelphia, PA, August 27, 1984.
132. Frenkel, K.; Grunberger, D.; Boublik, M.; Weinstein, I. B. Biochem. 1978, 17, 1278.
133. Pulkrabek, P.; Leffler, S.; Weinstein, I. B.; Grunberger, D. Biochem. 1977, 16, 3127.
134. Pulkrabek, P.; Leffler, S.; Grunberger, D.; Weinstein, I. B. Biochem. 1979, 18, 5128.
135. Drinkwater, N. R.; Miller, J. A.; Miller, E. C.; Yang, N.-C. Cancer Res. 1978, 38, 3247.
136. Gamper, H. B.; Straub, K.; Calvin, M.; Bartholomew, J. C. Proc. Natl. Acad. Sci. U.S.A. 1980, 77, 2000.
137. Hogan, M. E.; Dattagupta, N.; Whitlock, J. P. Jr. J. Biol. Chem. 1981, 256, 4504.
138. Taylor, E. R.; Miller, K. J.; Bleyer, A. J. J. Biomol. Struct. Dynam. 1983, 1, 883.
139. Stezowski, J. J.; Joos-Guba, J., Glusker, J. P.; Reily, M. D.; Marzilli, L. G. (manuscript in preparation).
140. Kennard, O. International Union of Biophysics Meeting, Bristol, UK, 1984.
141. Carrell, H. L. Program VIEW. Philadelphia: The Institute for Cancer Research, The Fox Chase Cancer Center, 1976.
142. Badler, N.; Stodola, R. K.; Wood, W. Program DOCK. Philadelphia: The Institute for Cancer Research, The Fox Chase Cancer Center, 1982.
143. Quigley, G. J.; Wang, A. H.-J., Ughetto, G., van der Marel, G.; van Boom, J. H.; Rich, A. Proc. Natl. Acad. Sci. U.S.A. 1980, 77, 7204.

RECEIVED May 2, 1985

Polycyclic Aromatic Hydrocarbon–DNA Adducts
Formation, Detection, and Characterization

ALAN M. JEFFREY

Cancer Center, Institute of Cancer Research and Department of Pharmacology, Columbia University, New York, NY 10032

This chapter reviews the role of the covalent binding of metabolites of polycyclic aromatic hydrocarbons to DNA. A variety of methods for the measurement of such binding are discussed, including the use of radiolabeling, and immunological and fluorescence techniques. The structures of some of the DNA adducts which have been characterized to date are also described.

Chemical carcinogenesis by polycyclic aromatic hydrocarbons (PAHs) is a multi-step process in which each of the steps must occur if a neoplasm is to develop. Thus, exposure to PAHs alone is not necessarily sufficient for the induction of a tumor. Many of these factors are summarized below and are discussed in various chapters of this volume. Considered here will be those factors influencing the reactions of the metabolically activated forms of the PAHs with DNA and the ways in which adducts may be detected and characterized.

Chemical Carcinogenesis: A Multistage Process

1) Exposure: concentration and time
2) Adsorption: presence of carcinogens/cocarcinogens
3) Metabolism: activation/inactivation
4) Reactivity/stability of metabolites: chemical and metabolic
5) Potential of reaction with DNA: steric and electronic/specific target sequences
6) Stability of DNA adducts: chemical and effects of repair enzymes
7) Potential of adduct to initiate cell
8) Presence of tumor promoters
9) Cell escapes normal control of proliferation
10) Cancer/metastasis

Direct evidence for the importance of DNA adduct formation being one

0097–6156/85/0283–0187$06.25/0
© 1985 American Chemical Society

of the critical events in chemical carcinogenesis has been difficult
to obtain, although there is considerable data implicating the
probable importance of such reactions. Good correlations have been
obtained between binding of PAHs to DNA, but not to protein or RNA,
in mouse skin and their potential as chemical carcinogens (1). Such
simple relationships do not always hold and later studies (2,3) found
some carcinogens, such as dibenz[a,c]anthracene, bound better than
did some stronger carcinogens such as dibenz[a,h]anthracene. One
problem with this type of approach is the possibility of release of
tritium from the labeled hydrocarbons used which can then become
incorporated into the DNA itself. For these reasons, a better index
of DNA binding has been obtained by only looking at those PAH-DNA
adducts which are retained by Sephadex LH20 chromatography (4).

This lack of a simple relationship between DNA binding and
carcinogenicity is perhaps not surprising if carcinogenesis is a
multistage process in which carcinogen-DNA binding is only one step.
In addition, specific adducts may differ in their ability to induce
mutations (5), and different cell lines may be equipotent in their
ability to metabolize, for example benzo[a]pyrene (B[a]P), and yet
may not be equally sensitive to transformation (6-8). Some studies
demonstrate a wide range linear dose-response curve for such binding
(9), while others do not (10); this difference has important
implications for human risk (11). Some evidence has been presented
that binding to adenine residues (12) is more important than the
major guanine sites. Other work has suggested that depurination of
DNA by PAHs is critical at least for mutagenesis. Metabolites, such
as the 7,8-dihydrodiol of B[a]P, which are on the pathway towards the
ultimate metabolite which binds to DNA (13) are generally more potent
as carcinogens (14) while those not on such pathways are less
carcinogenic (15-18). Instances have been noted where there is
reduced activity of such intermediates, but these have often been
explained in terms of the lower absorption or greater instability of
these more polar metabolites. In addition, results vary depending
upon the animal model used (14,19-22). Synthetically prepared
derivatives of weak carcinogens, which might be expected to be
further metabolized to derivatives which bind to DNA, are often
highly carcinogenic (23). Data from other classes of chemical
carcinogens show a correlation between the need for metabolic
activation to products which bind to DNA and the ability of such
precarcinogens to induce tumors, lending support to the idea that DNA
binding is critical to carcinogenesis. Although not complete, there
is a strong correlation between the potential carcinogenicity of
compounds and their ability to act as mutagens (24). The latter
process is much more clearly related to DNA damage. Although
unrelated directly to PAH carcinogenesis, additional evidence for the
critical role of DNA comes from studies on the sensitivity of cells
to transformation by light when they have been previously exposed to
bromodeoxyuridine (25). This thymidine analog specifically
misincorporates into DNA and then can be photochemically decomposed
to induce DNA strand breaks. Other data on the transformation of
cells by transfection with protooncogenes previously modified with
nitrosomethylurea (26) or, more recently, B[a]P diol epoxide (27),
provides the strongest evidence that DNA is the target for ultimate
carcinogens. These experiments raise interesting questions regarding

the mechanism of chemical carcinogenesis. Will this prove to be a
general mechanism or will different tumors arise by different
mechanisms? Although the cells used in this instance may be
considered to be partially transformed they are the first experiments
in which it can ascertain that only DNA is modified to the exclusion
of other potential targets such as RNA of proteins.

Detection of DNA Adducts

The level of DNA modification in any biological system is low,
ranging up to about one adduct per 10^5 bases in the DNA helix. Any
methods used to detect such damage must therefore either be very
sensitive or involve some method by which the effect of such damage
is amplified to measurable levels.

Indirect Methods.   These involve the assay of consequences resulting
from DNA damage but not the adducts themselves. An example of this
approach is the Ames assay (28–30) in which damage to the DNA can
induce a variety of mutational events resulting in the bacteria
which becoming capable of growth in the absence of histidine. Since
a large number of bacteria can be exposed to the active metabolites
of PAHs mutational events occurring even with very low frequencies
can be detected. Additional amplification of DNA damage can be
obtained in suitable strains by the presence of plasmids containing
error prone repair enzymes (31). Thus, damage at a level which
itself would not produce a measurable mutation frequency may, through
the repair process, become detectable. Some systems appear to be
entirely dependent upon the repair process for mutation (32). These
Processes are discussed in greater detail in Chapter 13. Similar
approaches have been developed for mammalian cells using, for
example, the HGPRT locus or oubain resistance (for a review see 33).
This type of assay, while sensitive, rapid, and relatively simple to
perform, has serious limitation when attempts are made to extrapolate
risks to the cancer in human populations. First, there is not a
simple quantitative relationship between carcinogenesis and
mutagenesis (24); secondly, although it has been possible to obtain
some correlation (e.g., 5) within defined test systems using pure
chemicals, normally humans are exposed to complex mixtures of
compounds, components of which may have significant synergistic or
antagonistic effects on each others' carcinogenicity. Exposures to
mixtures are even more difficult to evaluate by these mutagenesis
assays. Such problems may be, at least partially, overcome by more
direct quantitation of DNA adducts which represents the end point of
this complex series of interactions. Although no completely general
methods are available, several approaches to the problem are
available.

Direct Methods.   The classical approach has been to prepare tritiated
or carbon labeled analogs of the parent hydrocarbons which may then
be used in animal or in vitro experiments. Tritiated compounds are
generally easier to prepare, using exchange reactions on the parent
hydrocarbon, than their ^{14}C analogs and have higher specific
activities. However, during the metabolism of such compounds, some
of the tritium is released as tritiated water, either directly or

through the NIH shift (34), when phenols and quinones are formed. This tritiated water can become incorporated into the normal bases of DNA and may contribute significantly to the overall apparent binding levels measured in DNA (4). Such problems are circumvented in the case of ^{14}C labeled hydrocarbons. Tritium labeled compounds can be prepared at sufficiently high specific activities (often 20–80 Ci/mmole) that fentomole amounts of adduct may be detected. Depending upon the amount of DNA which can be isolated, detention of adducts down to the level of about one adduct per 10^9 bases is possible, although more typically, the range is one per 10^{6-7} bases.

DNA can be isolated from the test material by a variety of approaches depending upon the source of the material. Proteolytic enzymes are often helpful in solublizing material prior to phenol extraction, the most commonly used method. Phenol extraction has the advantage that it will not only help to purify the DNA from protein, but also extract some of the unwanted, non-covalently bound PAH. Final alcohol precipitation not only allows for removal of the phenol and any remaining non-covalently bound hydrocarbon but also concentrates the DNA. Ribonuclease treatment removes any contaminating RNA. Additional purification by cesium chloride centrifugation (35) is also often performed. This is particularly suited to small quantities of DNA. Hydroxyapatite chromatography is also effective in separating RNA, proteins, and DNA (36,37).

Post-labeling Techniques. The methods described above, while valuable in a laboratory experimental situation, are not suited to either complex mixtures or to measurement of human exposures. An alternative approach, developed by Dr. Randerath and his group (38,39), is shown in Figure 1. DNA is isolated from any appropriate source, which may be from the blood of exposed individuals or animals, tissue samples or cells in culture. The DNA is then digested enzymatically to the 3'-phosphates which are then phosphorylated in the 5' position with [γ-^{32}P] ATP. Because of the extremely high specific activities (~5000 Ci/mmole) and counting efficiencies which can be achieved with ^{32}P, less material is needed than when either tritiated or carbon labeled PAHs are used as radiochemical tracers. The major disadvantage of this approach is that, in contrast to the separation of non-covalently bound radiolabeled PAHs from DNA, which is generally quite easy, the separation of the modified and normal deoxyribonucleoside diphosphates and any radiochemical decomposition products from the high specific activity ATP is very difficult when the adducts are only present at one part per 10^{6-7}. To achieve such resolution a multi-directional tlc system on PIE cellulose was developed in which samples are applied to the center of the plate and the first two directions of development are washes intended to remove the bulk of the undesired radioactivity. The plate is cut after each run so that the unwanted radioactivity is discarded. The final two dimensions of chromatography allows for separation of the modified bases which are then finally detected by autoradiography at levels down to a few attomoles of adduct. The main restriction upon the sensitivity of this method is the higher background as more material is applied to the tlc plate and the time for autoradiography is increased. In general for PAHs, the separation of modified and unmodified

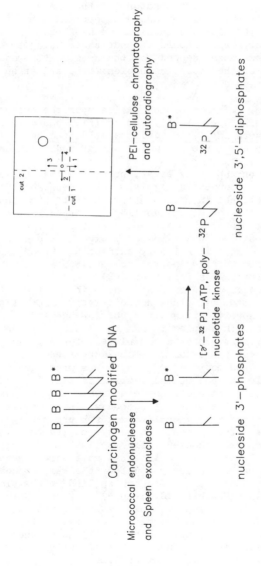

Figure 1. Postlabeling technique for detection of modified bases in DNA.

nucleotides occurs quite well and detection levels around one
adduct/10^7 bases can be obtained. Alternative methods involving
prior separation of modified and unmodified bases prior to labeling
with ^{32}P result in difficulties in quantitation of the extent of
adduct formation. Separation of the adducts on reverse phase plates
with subsequent transfer of the modified bases to the PIE cellulose
plates results in the method becoming more sensitive but even more
labor intensive and complex.

Identification of the chemical nature of these adducts is,
however, possible only by chromatographic comparison with standards
prepared by reacting the ultimate carcinogens with DNA. Proof of
identity by co-chromatography should be undertaken using several
chromatographic systems which depend for their separation upon
different properties of the adduct molecules: derivatization prior to
reanalysis may be included.

Immunological Methods.   This approach relies upon the preparation of
antisera which recognize specifically the modified bases in DNA.
Such antibodies prepared initially against methylated bases (40), are
becoming increasingly available for more complex modifications
including PAHs (41-43).

Many variations on the assay exist, but the ELISA, shown
schematically below, is currently highly favored because of its
simplicity once established in a laboratory; sensitivity, detecting
about one adduct per 10^7 bases; and ability to screen many samples
because of easy automations. The current prerequisite for the assay
is that DNA can be modified to sufficiently high levels with the
ultimate carcinogen to make it suitably antigenic. These types of
antigens have been used to raise polyclonal antibodies in rabbits
(41) and monoclonal antibodies from mice (42).

About 5 ng of modified DNA is coated onto the surface of a 96
well plate (Figure 2, Step 1). In a competitive version of this
assay, antibody with specificity against the PAH-DNA adduct of
interest is incubated with samples if DNA to be tested (Step 2).
This mixture is then transferred to the plate (Step 3). The less PAH
modification of the test sample of DNA, the more free antibody there
will be to react with the modified DNA which was precoated onto the
plate. After suitable incubation and washing of the plate (Step 4),
only the excess antibody which did not react with the test DNA in
solution will be left on the plate. A second antibody now is needed
which is directed against the general class, for example IgG, of that
adsorbed on the plate. This second antibody, coupled to suitable
marker enzymes, is generally commercially available. For example, in
the case of the anti-BPDE-DNA IgG polyclonal antibodies prepared in
rabbits, a goat anti-rabbit antibody coupled to alkaline phosphatase
was used (41). The second antibody is now applied to the plate and
it binds to any remaining rabbit IgG (Step 5). After washing (Step
6), the substrate for the enzyme, in this example p-nitrophenyl
phosphate, is added (Step 7) As hydrolysis proceeds (Step 8), a
yellow develops under the slightly basic conditions of the final
assay which is inversely proportional to the extent of modification
of the test sample of DNA. It is this amplification of detection by
the catalytic activity of the enzyme linked to the second antibody
which gives this assay such high sensitivity. Even more sensitive

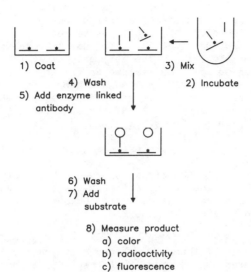

1) Coat   3) Mix
   4) Wash   2) Incubate
5) Add enzyme linked
   antibody

6) Wash
7) Add
   substrate

8) Measure product
   a) color
   b) radioactivity
   c) fluorescence

Figure 2. Schematic picture of a competitive enzyme–linked immunosorbent assay (ELISA).

methods have been developed using hydrolysis of ^{32}P labeled ATP (41),
biotin—aviden—enzyme complexes (44), or the formation of fluorescent
product as a result of enzyme catalyzed reactions (45). However, one
of the limiting factors at the moment is the availability of
antibodies with high specificity toward modified DNA which do not
react with unmodified DNA when present at 10^{7-9} times greater
concentrations. The limits of this assay are about 0.1 fmole
adduct/µg DNA or about one adduct/10^8 bases.

The simplicity of this assay makes it highly attractive for
further development. Some of the limitations include the need for
the preparation of relatively large amounts of appropriately modified
DNA for the initial immunization. This may be reduced with the
development of in vitro immunization techniques in which as little as
5ng of antigen can be effective. The antibodies must also be
characterized once prepared. For example, a variety of monoclonal
antibodies against B[a]PDE-modified DNA have been prepared (42) which
show varying specificity. At one extreme, some require the full
structure of the adduct bound to DNA, at the other, B[a]P tetraol
will effectively compete. What is not clear at the moment is how
specific such antisera are for a particular PAH. Will these antisera
recognize only B[a]P tetraol structures or those of any diol epoxide
modified DNA?

Thus, as more specific monoclonal antibodies are prepared, it
should become possible to determine not only the total extent of DNA
modification but also to gain insight into the structural and
conformational properties of such modified DNA.

Fluorescence Measurements. PAHs are often highly fluorescent and in
addition, many of those of interest as chemical carcinogens have
sufficiently conjugated ring systems that they absorb light at longer
wavelengths than the DNA to which they may have become bound.
Metabolic activation may reduce the extent of conjugation, and hence
the longest wavelengths at which they absorb, and binding to DNA
often reduces the quantum yields of these hydrocarbons. However,
fluorescence studies have played an important role in investigating
both the structure and conformation of DNA adducts.

Because of the extents of modification of DNA are low,
relatively high concentrations of DNA are used, taking advantage of
the fact that DNA does not absorb strongly beyond about 310nm. In
addition, the quantum yield of adducts may be increased by cooling to
liquid nitrogen temperatures (46). Photon counting techniques (47)
allow averaging of signal over time, increasing the signal to noise
ratios. Combinations of these approaches have been used quite widely
for various hydrocarbons (46,48-54). The method has the advantage
that it requires no prior knowledge of the structure of the adduct
nor its synthetic preparation. Success, however, depends
considerably upon the type of metabolic activation which occurs. For
example, in the case of 7,12-dimethylbenz[a]anthracene (DMBA),
several groups have reported fluorescence spectra for DNA isolated
from a variety of sources after exposure to the hydrocarbon
(50,55,56). All describe metabolic activation having occurred in the
1-4 positions. However, dimethylanthracene absorbs strongly at long
wavelengths to give a very characteristic spectrum. Had metabolism
occurred in the 8-11 positions, the residual phenanthrene chromophore

would have been much harder to detect. Only after more thorough HPLC
analysis of the bound products was it possible to be certain that the
majority of adduct resulted from activation in the bay region
(50,55,57).

Another limitation of this approach is that, because of the
relatively broad excitation and emission spectra, it is severely
limited when attempts are made to analyze exposures to complex
mixtures of PAHs. In the case of PAHs themselves, significant
improvement in the quality of spectra have been obtained by matrix
isolation techniques (58) in conjunction with working liquid helium
temperatures to remove the effects of thermal broadening of the
spectral lines. Under such conditions, it has been possible to
analyze directly quite complex mixtures of hydrocarbons. However,
the materials generally used for the matrix, such as nitrogen or
hexane, are quite unsuited to PAH-modified DNA samples. Mixtures of
water, ethylene glycol and ethanol do, however, form suitable glasses
in which DNA samples can be embedded. In such glasses, the bound
hydrocarbon moieties will be in any one of a large variety of micro-
environments and, while the individual molecules may themselves have
sharp excitation and emission spectra, the result of these
overlapping spectra is again a broad fluorescent spectrum. These
subpopulations, or isochromats, may be individually excited with very
narrow band pass light such as produced by a laser. This approach,
called 'fluorescence line narrowing', has been applied to a study of
B[a]P bound to DNA (59). The spectra are much more characteristic
than the broad band spectra of equivalent samples (46). Subtle
differences even allow distinction between 7,8,9,10-tetrahydroB[a]P
and its corresponding tetraol. At the resolution so far obtained,
distinction between the tetraol, BPDE modified DNA, and the BPDE-
deoxyguanosine derivative isolated from such modified DNA could not
be made. The emission spectra are highly characteristic, and complex
mixtures containing up to nine PAH metabolites, have recently been
resolved (unpublished results). In addition, it has been possible to
detect the B[a]PDE-deoxyguanosine derivative present in DNA isolated
from mouse embryo fibroblast 10T1/2 cells exposed for 24 hours to 1
µg of B[a]P at the level of one adduct per 10^6 bases without prior
digestion of the DNA (unpublished results). This approach provides
an opportunity to begin to investigate directly the types of DNA
adducts which are formed when cells are exposed to complex mixtures
of PAHs.

An alternative approach used to investigate the covalent binding
of B[a]P to mouse skin has been to release the hydrocarbon–DNA
adducts from the isolated DNA by acid hydrolysis (60,61). Since, in
the case of BPDE-DNA adducts, they are acid labile and the tetraols
produced are not only more fluorescent than the adducts themselves
but also more easily extracted from the large excess of unmodified
bases, this provides a convenient approach. The method does,
however, require the certainty that hydrolysis of the adducts will
occur and that the products will be stable or, if degradation occurs,
that they can still be recognized.

Determination of Structures of PAH–DNA Adducts

Since the levels of modification in vivo are so low, all structural

identification of such adducts have been based upon the preparation
of the active metabolite or ultimate carcinogenic form of the
original PAH which can then be reacted with DNA, homo-
deoxyribonucleotide polymers or monodeoxyribonucleotides. This
approach generally allows preparation of one or two hundred µg of
adduct which can be used to compare with the chromatographic
properties of adducts isolated after in vivo exposure to the parent
PAH and, if such sets of adducts are co-chromatographic by HPLC, the
subsequent chemical characterization of the synthetic adducts. (See
below).

Reactive Metabolites of PAHs. A wide variety of products have been
identified as metabolites of PAHs. These include phenols, quinones,
trans-dihydrodiols, epoxides and a variety of conjugates of these
compounds. Simple epoxides, especially those of the K-region, were
initially favored as being the active metabolites responsible for the
covalent binding of PAH to DNA. Little direct experimental support
exists for this idea (62,63,64) except in microsomal incubations
using preparation in which oxidations at the K-region are favored
(65,66). Evidence has been presented that a 9-hydroxyB[a]P 4,5-oxide
may account for some of the adducts observed in vivo (67,68) although
these products have never been fully characterized.
     Studies on the comparative abilities (13) of B[a]P metabolites
to bind to DNA in microsomal systems showed that the 7,8-dihydrodiol
was the most efficient. This led to the proposal (69) that
dihydrodiol epoxides were the ultimate carcinogenic metabolites.
Chemical synthesis of all possible isomers (70,71) has allowed
complete structural identification of the adducts (72-74).
     The profile of adducts formed may well differ if simultaneous
exposure to other compounds in addition to the carcinogen occurs.
Such exposures have a marked effect on the carcinogenicity of
compounds which may be a result of changes at any one of a number of
points in the carcinogenic process. Effect on metabolism, and
consequently DNA adduct formation, are know to occur by enzyme
induction, inhibition, or activation (75-81). Recent reports (82,83)
have shown that in mouse skin, the spectrum of B[a]P-DNA adducts
could be altered by simultaneous exposure to pyrene, fluoranthene or
B[e]P which can alter the carcinogenicity of B[a]P. These
synergistic and antagonistic effects have important implications for
human risk assessments when, even if exposure to a single agent is
being considered. Other factors such as dietary or life-style
variations and genetic differences all impact on the biologically
effective dose received by an individual.

Target Sites on DNA for Adducts Formation. Early studies on the
reaction of simple epoxides with DNA indicated that the N[7] position
of guanine was strongly favored (84). This does not appear to be the
case with the more complex diol epoxides. Predominantly, adducts
have been characterized as having reacted with the N[2] amino group of
guanine (see below). Additional evidence has shown that some
reaction may occur at phosphate residues (85) or the N[7] of guanine
(86-88). The N[7] guanine adducts appear to be particularly unstable
and they depurinate or open their imidazole rings rapidly. BPDE
reacts well with dC or dA homopolymer (89). However, these sites
appear to be very poor targets in the case of DNA itself (72-74,90).

Within the sequence of DNA, much less is known regarding the sequence requirements for binding. Experiments by Pulkrabek et al. (91), using plasmid DNA showed that in restriction enzyme fragments modification was roughly proportional to G:C content. However, these fragments were still possibly too long to see the influence of more local or nearest neighbor effects. Experiments by Strauss (92) and Haseltine (93), looking at inhibition of DNA polymerases or alkaline labile sites (presumably N^7, not the N^2 adducts) showed little sequence specificity. More recently direct photochemical approaches have been used (94) and this provides an interesting alternative tool to study these effects.

An additional question relating to sites of modification is the possible influence of DNA binding proteins and the conformational changes they induce in DNA as it forms chromatin, which in turn may influence available binding sites for adduct formation. Several studies have investigated the distribution of DNA adducts in chromatin (95–102), although no clear answer for the influence of these proteins has yet emerged.

Arene oxides and diol epoxides are generally unstable in aqueous, especially acidic media (103–105), and in addition, several groups have noted that DNA has a marked catalytic effect upon diol epoxide hydrolysis (106,107). However, in cells there appears to be sites, probably lipid in nature, in which these compounds can have much longer half-lives.

Structures of Specific PAH–DNA Adducts.   After enzymic digestion of the modified DNA, structural identification has used conventional techniques of HPLC purification of the adducts and initial characterization by analysis of their nmr spectra. These spectra, especially when measured at high field strength, can provide valuable information regarding both the sites of modification and the structures and conformations of the adducts. For example, if all the aromatic protons can be interpreted, it may be possible to distinguish between adenine and guanine adducts. In addition, if the C-8 proton of the imidazole ring can be seen, that position may be eliminated as a site of modification. Substitution at the N^7 position would cause protonation of the imidazole ring and subsequent change in the chemical shift of the C-8 proton. Inspection of the chemical shifts and coupling constants of the protons attached to the ring of the original diol epoxide usually indicate the point of attachment and whether addition occurred cis or trans. Further derivatization or decoupling experiments may be necessary for complete interpretation. Because of the importance of chirality in both the formation and reaction of these ultimate carcinogens with DNA, the CD spectra of these adducts have often yielded valuable information (108,109). The elution order of these DNA adducts from the HPLC columns may change as one octadecyl-substituted reverse phase column is compared with another. For these reasons, the elution order cannot be assumed to correlate with published values. The CD spectra of these adducts are, however, very characteristic and will often allow pairing of diastereoisomeric, and cis and trans pairs of adducts (108,109) and is often the most sensitive method for the detection of such conformational changes in structure.

One of the major problems has been to determine the site of
attachment of the PAH to the base. Some information may be obtained
directly from the nmr spectra eliminating certain points of
attachment. As mentioned above, if the C-8 proton of guanine or
adenine can be identified, then this cannot be the point of
attachment of the carcinogen. Estimation of the pKa's of the adducts
either by titration (**108**) or partition (**110**) has, however, provided
additional valuable information. Mass spectral fragmentation
patterns can be of help in determining the site of substitution as
well as in determining which bases are involved in binding (**108,111-
113**). Substantial advances have been made in recent years on the
mass spectral analysis of involatile compounds and derivatization is
not always essential (**114-118**). X-ray analysis of DNA adducts has,
to date, only been applied to model systems (**119-121**).

Examples of Specific PAHs

Benzo[a]pyrene. The best documented adduct is that shown below which
results from the trans opening of the anti-diol epoxide of B[a]P
(anti-BPDE) by the N^2-amino group of guanine (Figure 3). Of particular
interest is the fact that when racemic **anti**-BPDE is reacted with B
form DNA about 95% of the adducts result from reaction of the 7R
enantiomer (**73,74,90,111,122**). Very little of the 7S reacts.
Explanations have been proposed to explain this but further work is
needed to to fully clarify this stereoselectivity. This finding is
also relevant when any theoretical treatments of carcinogenesis
(**123,124**) which currently do not consider such differences. Clearly,
the stereoselectivity of both metabolism and the subsequent covalent
binding can play dominant roles in determining the overall
carcinogenicity of a particular PAH.   The presence of another
compound such as glutathione (**77,125,126**), butylated hydroxyanisole
(**127**), ellagic acid (**128,129**), or plant flavanoids (**76**), may also
reduce the level of binding to DNA. In addition, although diol
epoxides are relatively resistant to epoxide hydrolase, they are not
to dehydrogenases (**130**) or GSH transferases (**131,132**).
     The particular sites to which the B[a]P becomes attached in the
DNA will probably greatly influence its role. This may be especially
important if the adduct functions directly rather than through a
nonspecific, indirect process, such as induction of DNA repair.
Efforts have been made to determine the specific sites to which B[a]P
becomes bound (**92,94**) and its distribution on chromatin (**97,98**).
     In addition to this major adduct which, in some biological
systems, such as human bronchus (**90**) or mouse skin (**74**) or fibroblast
10T1/2 cells in culture (**133**), accounts for almost all of the DNA
adducts, other derivatives have been detected in many systems which
have been investigated. These may vary with respect to tissue
(**134,135,136**) or time of exposure (**135,137**).
     As described earlier in Chapter 2, B[a]P is metabolized mainly
to the 7R 7,8-transdihydrodiol. However, both enantiomers are
further epoxidized on the lower face of the molecule. Thus, while
the 7R enantiomer forms mainly the anti-diol epoxide, the 7S
enantiomer forms mainly syn-diol epoxide. In some systems, such as
human colon (**138**) or hamster embryo cells (**139**), significant
quantities of syn-diol epoxide adducts are found, including adducts

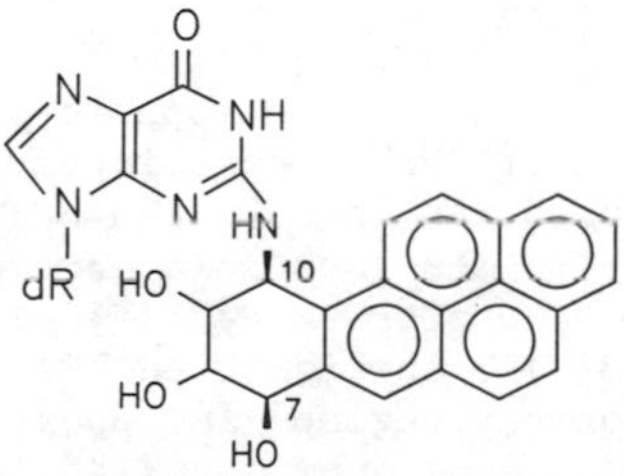

Figure 3. Structure of major DNA adduct detected in many _in vivo_ systems as a result of metabolic activation of benzo[a]pyrene or the reaction of _anti_-B[a]PDE with DNA _in vitro_. dR=deoxyribose moiety.

formed by reaction with adenine and cytosine residues (137,140).
Some of the following adducts also have been characterized, although
not all thoroughly: adenosine (108,111), the O⁶ (87) and N⁷-guanine
derivatives (86-88), phosphate (85) and adducts resulting from the
further epoxidation of the 9-hydroxy B[a]P at the 4,5-position
(67,68).

Benzo[e]pyrene. This PAH is noncarcinogenic despite the presence of
two bay regions. The reason for this, and the lack of DNA adducts,
is that each step towards adduct formation is blocked. Rather than
formation of a 9,10-dihydrodiol, metabolism is directed towards
phenols and K-region dihydrodiols (141). Even when chemically
prepared, the 9,10-dihydrodiol is poorly metabolized to diol
epoxides, probably because of the quasi axial conformation of the
hydroxyl groups. When the diol epoxide is chemically prepared, it
reacts poorly with DNA (unpublished results), probably again because
of the induced conformational change by steric hindrance of the
hydroxyl group with the aromatic ring resulting in a quasi axial
orientation. Removal of this constraint in the case of 9,10,11,12-
tetrahydroB[e]P 9,10-oxide results in a compound which reacts well
with DNA (142) and is mutagenic. The presence of B[e]P can,
however, effect the metabolism of a number of carcinogenic PAHs
including B[a]P and DMBA (143).

Benz[a]anthracenes. The parent hydrocarbon is a weak carcinogen and
the extent of modification of DNA in most in vivo systems is low.
Diol epoxides have been prepared in both the bay and non-bay regions
and reacted with DNA (144). The former is more reactive. When
either mouse skin (145) or hamster embryo (145-147) or microsomes
(148) were treated with B[a]A, the isolated DNA adducts were found to
result from mainly the non-bay region diol epoxide, although not to
the exclusion of bay-region diol epoxides (145,147,149). Thus, the
reason for the weak carcinogenicity of B[a]A appears to result from
the preferential metabolism to a diol epoxide which binds poorly to
DNA. As with B[a]P, there is a marked stereoselectivity in the
tumorigenesis of various bay region diol epoxides of B[a]A (150),
again indicating the importance of not only the site at which
metabolism occurs but also the stereochemistry of the products. Both
bay-region and non bay-region anti-diol epoxides react preferentially
with guanine (146,151) at the N²-position (144) although the non bay-
region epoxide is overall less reactive.
    Methylated derivatives of 7-methylB[a]A are particularly
carcinogenic when substitutes in the 7-, 12-, or 6- and 8-positions
(152,153). The increased carcinogenicity of these compounds may
result from the inhibition of metabolism at the 8-11 positions which
increases the amounts of bay region diol epoxides formed, the greater
reactivity of such epoxides with DNA, or an intrinsic difficulty for
cells to repair such adducts (154).
    In the case of 7-methyl-B[a]A (155) and 7,12-dimethyl-B[a]A
(DMBA) (50,55,57,156-158) most of the evidence supports bay region
diol epoxide adducts. Thus, fluorescence studies (50,55,56,159),
light sensitivity of the adducts (160), and analysis of adducts
formed by further oxidations of the 3,4-dihydrodiol all suggest a bay
region diol epoxide intermediate (161,162). More recent evidence

from Dipple's group has suggested that many of the DMBA–DNA adducts may arise from syn–diol epoxides and that adenosine adducts may be more prevalent than in the case of other PAHs investigated so far (55,163,164). Some additional evidence from studies on 2– and 5– fluoro DMBA also support the role of a bay region diol epoxide derivative of DMBA (165,166). Substantial hydroxylation of the 7– methyl group occurs and evidence for DNA adducts from this metabolite has been presented (167–170) including possible activation by a sulphate ester (170). The overall importance of this type of adduct has not been fully established.

3-Methylcholanthrene. This PAH gives rise to many metabolites and a complex pattern of DNA adducts (159,171–173). All of the evidence suggests again that the bay region diol epoxides are involved with additional oxidation possibly having occurred at the methyl, or C1 or C2 positions.

Chrysenes. Chrysene is only weakly carcinogenic despite its symmetry which results in two bay regions. Although evidence for bay-region diol epoxide adducts has been presented (51,174), a triol-epoxide pathway, which suggests an alternative mechanism of activation analogous to the 9-hydroxyB[a]P 4,5-oxide, has also been proposed to account for as much as 50% of the DNA adducts (175,176). While, as with B[a]A, the parent hydrocarbon is weakly carcinogenic, substitution with methyl groups has a pronounced effect upon the carcinogenicity of these compounds. In particular, the 5-methyl derivative (5-MeC) was found to be particularly carcinogenic (177). Evidence from both animal testing with halogenated 5-MeC derivatives and direct analysis of the PAH adducts suggest that activation occurs through a 1,2-dihydroxy-5-methyl-1,2,3,4-tetrahydrochrysene 3,4-oxide (118,178-180). Adducts were prepared by reaction of the diol epoxide with DNA and the isolated adducts characterized by their nmr, mass, and uv spectra and pKa determinations.

Fluoranthenes. With the exception of 3-methylcholanthrene, much less work has been undertaken on nonalternant PAHs. Several recent studies have reported on the major metabolites and mutagenicity of various fluoranthenes (181–185), but little is known about the DNA adduct which they form. Some studies on dibenzo[a,e]fluoranthene showed that several adducts are formed by microsomal incubations (185) and additional studies will be required to provide complete structural elucidation of the products formed.

Conclusions

Over the past few years substantial advances have been made in our understanding of the chemistry of the major adducts formed when PAH carcinogens are metabolically activated to derivatives which bind to DNA. Unlike many simple alkylating agents the N^2 of guanine is a major target for reaction. Metabolic activation through diol epoxides has proved an important pathway for many PAHs and substitution with a methyl group in the bay region adjacent to the activated benzo ring generally enhances carcinogenicity. Information is also available on factors governing the reaction of these metabolites with DNA and the conformation in which some of these adducts exist.

Areas in which additional information is still needed relates to the role and relative importance of different adducts and the mechanisms by which they initiate cells. General principles are developing which will allow better predictions to be made at each of the stages of chemical carcinogenesis outlined in Table I. The ultimate goal therefore, would be, by a combined analysis of all these steps, to predict accurately the carcinogenicity of newly discovered or untested PAH derivatives.

Acknowledgments

Part of the work described here was supported by NCI grant CA-021111. I thank Alycia Osborne for her assistance in preparing this manuscript.

Literature Cited

1. Brookes, P. Lawley, P.D. Nature London 1964, 202, 781-4.
2. Goshman Heidelberger Cancer Res. 1967, 27, 1678-88.
3. Phillips, D.H., Grover, P.L. Sims, P. Int. J. Cancer 1979, 23, 201-8.
4. Wigley, C.B., Newbold, R.F., Amos, J. Brookes, P. Int. J. Cancer 1979, 23, 691-6.
5. Brookes, P. Osborne, M.R. Carcinogenesis 1982, 3, 1223-6.
6. Lo, K.Y. Kakunaga, T. Cancer Res. 1982, 42, 2644-50.
7. Gehly, E.B. Heidelberger, C. Cancer Res. 1982, 42, 2697-2704.
8. Tejwan, R., Jeffrey, A.M. Milo, G.E. Carcinogenesis 1982, 3, 727-32.
9. Dunn, B.P. Cancer Res. 1983, 43, 2654-8.
10. Lutz, W.K., Viviani, A. Schlatter, C. Cancer Res. 1978, 38, 575 -8.
11. Ehling, U.H., Averbeck, D., Cerutti, P., Friedman, J., Greim, H., Kolbye, A.C. Mendelsohn, M.L. Mut. Res. 1983, 123, 281-341.
12. DiGiovanni, J., Romson, J.R., Linville, D., Juchau, M.R. Slaga, T.J. Cancer Lett. 1979, 7, 39-43.
13. Borgen, A., Darvey, H., Castagnoli, N., Crocker, T.T., Rasmussen, R.E. Wang, I.Y. J. Med. Chem. 1973, 16, 502-6.
14. Kapitulnik, J., Levin, W., Conney, A.H., Yagi, H. Jerina, D.M. Nature London 1977, 266, 378-80.
15. Cohen, G.M., MacLeod, M.C., Moore, C. Selkirk, J.K. Cancer Res. 1980, 40, 207-211.
16. Slaga, T.J., Bracken, W.M., Dresner, S., Levin, W., Yagi, H., Jerina, D.M. Conney, A.H. Cancer Res. 1978, 38, 678-681.
17. Levin, W., Buening, M.K., Wood, A.W., Chang, R.L., Kedzierski, B., Thakker, D.R., Boyd, D.R. Gadadinamath, G.S. J. Biol. Chem. 1980, 255, 9067-74.
18. Kapitulnik, J., Levin, W., Yagi, H., Jerina, D.M. Conney, A.H. Cancer Res. 1976, 36, 3625-3628.
19. Levin, W., Wood, A.W., Chang, R.L., Slaga, T.J., Jerina, D.M. Conney, A.K. Cancer Res. 1977, 37, 2721-2725.
20. Slaga, T.J., Bracken, W.M., Viaje, A., Levin, W., Yagi, H., Jerina, D.M. Conney, A.H. Cancer Res. 1977, 37, 4130-3.

21. Kapitulnik, J., Wislocki, P.G., Levin, W., Yagi, H., Jerina, D. M. Conney, A.H. Cancer Res. 1978, 38, 354–358.
22. Wood, A.W., Chang, R.L., Levin, W., Ryan, D.E., Thomas, P.E., Mah, H.D., Karle, J.M. Yagi, H. Cancer Res. 1979, 39, 4069–77.
23. Buening, M.K., Levin, W., Wood, A.W., Chang, R.L., Lehr, R.E., Taylor, C.W., Yagi, H. Jerina, D.M. Cancer Res. 1980, 40, 203–6.
24. Sugimura, T., Sato, S., Nagago, M., Yahagi, T., Matsushima, T., Seino, Y., Takeuchi, M. Kawachi, T. In 'Overlapping of carcinogens and mutagens': Magee, P.N., Ed. University of Tokyo Press: Tokyo, 1976: pp.191–215.
25. Barrett, J.C., Tsutsui, T. Ts'o, P.O.P. Nature London 1978, 274, 229–32.
26. Sukumaran, S., Notano, V., Martin–Zanca, D. Barbacid, M. Nature London 1983, 306, 658–61.
27. Marshall, C.J., Vousden, K.H. Phillips, D.H. Nature London 1984, 310, 586–9.
28. Yamasaki, E. Ames, B.N. Proc. Natl. Acad. Sci. USA. 1977, 74, 3555–9.
29. McCann, J., Choi, E., Yamasaki, E. Ames, B.N. Proc. Natl. Acad. Sci. USA. 1975, 72, 5135–9.
30. Ames, B.N., McCann, J. Yamasaki, C. Mut. Res. 1975, 31, 347–64.
31. McCann, J., Springarn, N.E., Jobori, J. Ames, B.N. Proc. Natl. Acad. Sci. USA. 1975, 72, 979–83.
32. Ivanovic, V. Weinstein, I.B. Cancer Res. 1980, 40, 3508–11.
33. 'Identifying and Estimating the Genetic Impact of Chemical Mutagens': National Academy Press: Washington,DC, 1983:.
34. Jerina, D.M. In 'Drug metabolism concepts':, Ed. American Chem. Society:, 1977:.
35. Harris, C.C., Frank, A.L., van Haaften, C., Kaufman, D.G., Jackson, F. Barrett, L.A. Cancer Res. 1976, 36, 1011–1018.
36. Shoyab, M. Chem. Biol. Interact. 1979, 25, 71–85.
37. Beland, F.A., Dooley, K.L. Casciano, D.A. J. Chemistry 1979, 174, 177–86.
38. Gupta, R.C., Reddy, M.V. Randerath, K. Carcinogenesis 1982, 3, 1081–92.
39. Reddy, M.V., Gupta, R.C., Randerath, E. Randerath, K. Carcinogenesis 1984, 5, 231–43.
40. Rajewsky, M.F., Muller, R., Adamiewicz, J. Drosdziok, W. In 'Immunological detection and quantification of DNA components structurally modified by alkylating carcinogens ethylnitrosourea': Pullman, P., Ts'o, P.O.P., and Gelboin, H. V., Eds. D. Reidel Press: Doudrecht,Holland, 1980: pp.207–18.
41. Hsu, I.C., Poirier, M.C., Yuspa, S.H., Grunberger, D., Weinstein, I.B., Yolken, R.H. Harris, C.C. Cancer Res. 1981, 41, 1090–5.
42. Santella, R.M., Lin, C.D., Cleveland, W.L. Weinstein, I.B. Carcinogenesis 1984, 5, 373–7.
43. Wallin, H., Borrebaeck, C.A.K., Glad, C., Mattiasson, B. Jergil, B. Cancer Lett. 1984, 22, 163–70.
44. Shamauddin, A.M. Harris, C.C. Arch. Pathol. Lab. Med. 1983, 107, 514–7.
45. Poirer, M.C., Reed, E., Zwelling, L.A., Ozols, R. Yuspa, S.H. Environ. Health Perspec. 1985, in press.
46. Ivanovic, V., Geacintov, N. Weinstein, I.B. Biochem. Biophys. Res. Comm. 1976, 70, 1172–1179.

47. Daudel, P., Duquesne, M., Vigny, P., Grover, P.L. Sims, P. FEBS Lett. 1975, 57, 250-3.
48. Rahn, R.O., Chang, S.S., Holland, J.M., Stephens, T.J. Smith, L. H. J. of Biochem. and Biophys. Methods 1980.
49. Moschel, R.C., Hudgins, W.R. Dipple, A. Chem. Biol. Interact. 1979, 27, 69-79.
50. Vigny, P., Kindts, M., Cooper, C.S., Grover, P.L. Sims, P. Carcinogenesis 1981, 2, 115-9.
51. Vigny, P., Spiro, M., Hodgson, R.M., Grover, P.L. Sims, P. Carcinogenesis 1982, 3, 1491-3.
52. Ridler, P.J. Jennings, B.R. FEBS Lett. 1982, 139, 101-4.
53. Vigny, P., Ginot, Y.M., Kindts, M., Cooper, C.S., Grover, P.L. Sims, P. Carcinogenesis 1980, 1, 945-54.
54. Jernstrom, B., Ominius, S., Undeman, O., Graslund, A. Ehrenberg, A. Cancer Res. 1978, 38, 2600-2607.
55. Moschel, R.C., Pigott, M.A., Costantino, N. Dipple, A. Carcinogenesis 1983, 4, 1201-4.
56. Ivanovic, V., Geancintov, N.E., Jeffrey, A.M., Fu, P.P., Harvey, R.G. Weinstein, I.B. Cancer Lett. 1978, 4, 131-40.
57. Cooper, C.S., Ribeiro, O., Hewer, A., Walsh, C., Grover, P.L. Sims, P. Chem. Biol. Interact. 1980, 29, 357-67.
58. Stroupe, R.C., Tokousbalides, P., Dickinson, R.B., Wehry, E.L. Mamantov, G. Anal. Chem. 1977, 49, 701-5.
59. Heisig, V., Jeffrey, A.M., McGlade, M.J. Small, G.J. Science 1984, 223, 289-91.
60. Rahn, R.O., Chang, S.S., Holland, J.M. Shugart, I.R. Biochem. Biophys. Res. Comm. 1982, 109, 262-8.
61. Shugart, L., Holland, J.M. Rahn, R.O. Carcinogenesis 1983. 4. 195-8.
62. Baer-Dubowska, W., Frayssinet, C. Alexandrov, K. Cancer Lett. 1981, 14, 125-9.
63. Baird, W.M., Harvey, R.G. Brookes, P. Cancer Res. 1975, 35, 54-57.
64. Platt, K., Bucker, M., Golan, M. Oesch, F. Mut. Res. 1982, 96, 1-13.
65. Jeffrey, A.M., Blobstein, S.H., Weinstein, I.B. Harvey, R.G. Anal. Biochem. 1976, 73, 378-85.
66. Santella, R.M., Grunberger, D. Weinstein, I.b. Mut. Res. 1979, 61, 181-9.
67. Robertson, J.A., Nordenskjold, M. Jernstrom, B. Carcinogenesis 1984, 5, 821-6.
68. Guenthner, T.M., Jernstrom, B. Orrenius, S. Carcinogenesis 1980, 1, 407-18.
69. Sims, P., Grover, P.L., Swaisland, A., Pal, K. Hewer, A. Nature London 1974, 252, 326-7.
70. Yagi, H., Thakker, D.R., Hernandez, O., Koreeda, M. Jerina, D.M. J. Amer. Chem. Soc. 1977, 99, 1604-11.
71. Fu, P.P. Harvey, R.G. Tetrahedron Lett. 1977, 2059.
72. Jeffrey, A.M., Jennette, K.W., Blobstein, S.H., Weinstein, I.B., Beland, F.A., Harvey, R.G., Kasai, H. Nakanishi, K. J. Amer. Chem. Soc. 1976, 98, 5714-5.
73. Meehan, T. Straub, K. Nature London 1979, 277, 410-2.
74. Koreeda, M., Moore, P.D., Wislocki, P.G., Levin, W., Conney, A. H., Yagi, H. Jerina, D.M. Science 1978, 199, 778-81.
75. Gelboin, H.V. Physiol. Rev. 1980, 60, 1107-66.

76. Huang, M.T., Wood, A.W., Newmark, H.L., Sayer, J.M., Yagi, H., Jerina, D.M. Conney, A.H. Carcinogenesis 1983, 4, 1631–7.
77. Jernstrom, B., Babson, J.R., Moldeus, P., Holmgren, A. Reed, D. J. Carcinogenesis 1982, 3, 861–6.
78. Sydor, W., Chou, M.W., Yang, S.K. Yang, C.S. Carcinogenesis 1983, 4, 131–6.
79. Huang, M.T., Chang, R.L., Fortner, J.G. Conney, A.H. J. Biol. Chem. 1981, 256, 6829–36.
80. Oesch, F. Guenthner, T.M. Carcinogenesis 1983, 4, 57–65.
81. Sawicki, J.T. Dipple, A. Cancer Lett. 1983, 20, 165–71.
82. Rice, J.E., Hosted, T.J. LaVoie, E.J. Cancer Lett. 1984, 24, 327–33.
83. Smolarek, T.A. Baird, W.M. Carcinogenesis 1984, 5, 1065–1070.
84. Lawley, P.D. Jarman, M. Biochem. J. 1972, 126, 893–900.
85. Koreeda, M., Moore, P.D., Wislocki, P.G., Levin, W., Conney, A. H., Yagi, H. Jerina, D.M. J. Amer. Chem. Soc. 1976, 98, 6720–2.
86. Osborne, M.R., Harvey, R.G. Brookes, P. Chem. Biol. Interact. 1978, 20, 123–30.
87. Osborne, M.R., Jacobs, S., Harvey, R.G. Brookes, P. Carcinogenesis 1981, 2, 553–8.
88. King, H.W., Osborne, M.R. Brookes, P. Chem. Biol. Interact. 1979, 24, 345–53.
89. Jennette, K.W., Jeffrey, A.M., Blobstein, S.H., Beland, F.A., Harvey, R.G. Weinstein, I.B. Biochemistry 1977, 16, 932–8.
90. Jeffrey, A.M., Weinstein, I.B., Jennette, K.W., Grzeskowiak, K., Nakanishi, K., Harvey, R.G., Autrup, H. Harris, C. Nature London 1977, 269, 348–50.
91. Pulkrabek, P., Leffler, S., Grunberger, D. Weinstein, I.B. Biochemistry 1979, 18, 5128–34.
92. Moore, P. Strauss, B.S. Nature London 1979, 278, 664–6.
93. Sage, E. Haseltine, W.A. J. Biol. Chem. 1984, 259, 11098–11102.
94. Boles, T.C. Hogan, M.E. Proc. Natl. Acad. Sci. USA. 1984, 81, 5623–7.
95. Yamasaki, H., Roush, T.W. Weinstein, I.B. Chem. Biol. Interact. 1978, 23, 201–13.
96. Mironov, N.M., Grover, P.L. Sims, P. Carcinogenesis 1983, 4, 189–93.
97. Jack, P.L. Brookes, P. Carcinogenesis 1982, 3, 341–4.
98. Jack, P.L. Brookes, P. Nucl. Acids Res. 1981, 9, 5533–52.
99. Oleson, F.B., Mitchell, B.L., Dipple, A. Lieberman, M.W. Nucl. Acids Res. 1979, 7, 1343–61.
100. Arrand, J.E. Murray, A.M. Nucl. Acids Res. 1982, 10, 1547–55.
101. Kootstra, A., Slaga, T.J. Olins, D.E. Chem. Biol. Interact. 1979, 28, 225–236.
102. Jahn, C.L. Litman, G.W. Biochem. Biophys. Res. Comm. 1977, 76 (2), 534–540.
103. Yang, S.K., McCourt, D.W. Gelboin, H.V. J. Amer. Chem. Soc. 1977. 99, 5130–5134.
104. Keller, J.W., Heidelberger, C., Beland, F.A. Harvey, R.G. J. Am. Chem. Soc. 1976, 98, 8276–7.
105. Whelan, D.L., Ross, A.M., Montemarano, J.A., Thakker, D.R., Yagi, H. Jerina, D.M. J. Am. Chem. Soc. 1979, 101, 5086–8.
106. Michaud, D.P., Gupta, S.C., Whalen, D.L., Sayer, J.M. Jerina, D. M. Chem. Biol. Interact. 1983, 44, 41–52.

107. Geacintov, N.E., Ibanez, V., Gagliano, A.G., Yoshida, H. Harvey, R.G. Biochem. Biophys. Res. Comm. 1980, 92, 1335–42.
108. Jeffrey, A.M., Grzeskowiak, K., Weinstein, I.B., Nakanishi, K., Roller, P. Harvey, R.G. Science 1979, 206, 1309–11.
109. Jeffrey, A.M., Blobstein, S.H., Weinstein, I.B., Beland, F.A., Harvey, R.G., Kasai, H. Nakanishi, K. Proc. Natl. Acad. Sci. USA. 1976, 73, 2311–15.
110. Moore, P.D. Koreeda, M. Biochem. Biophys. Res. Comm. 1976, 73, 459–64.
111. Straub, K.M., Meehan, T., Burlingame, A.L. Calvin, M. Proc. Natl. Acad. Sci. USA. 1977, 74, 5285–9.
112. Wong, L.K., Wang, C.L. Daniel, F.B. Biomed. Mass Spectrom. 1979, 6, 305–8.
113. Wiebers, J.L., Abbott, P.J., Coombs, M.M. Livingston, D.C. Carcinogenesis 1981, 2, 637–43.
114. Straub, K.M. Burlingame, A.L. Biomed. Mass Spectrom. 1981, 8, 431–5.
115. Straub, K., Meehan, T., Kambara, H., Burlingame, A.L. Evans, S. 26 Annual Conf. on Mass Spectrometry and Allied Topics 1978, 499–500.
116. Martin, C.N., Beland, F.A., Roth, R.W. Kadlubar, F.F. Cancer Res. 1982, 42, 2678–86.
117. Busch, K.L. Cooks, R.G. Science 1982, 218, 247–60.
118. Melikian, A.A., Amin, S., Hecht, S.S., Hoffmann, D., Pataki, J. Harvey, R.G. Cancer Res. 1984, 44, 2524–29.
119. Carrell, H.L., Glusker, J.P., Moschel, R.C., Hudgins, W.R. Dipple, A. Cancer Res. 1981, 41, 2230–4.
120. Carrell, H.L., Zacharias, D.E., Glusker, J.P., Moschel, R.C., Hudgins, W.R. Dipple, A. Carcinogenesis 1982, 3, 641–5.
121. Glusker, J.P., Carrell, H.L., Zacharias, D.E. Harvey, R.G. Cancer Biochem. Biophys. 1974, 1, 43–52.
122. Prusik, T., Geacintov, N.E., Tobias, C., Ivanovic, V. Weinstein, I.B. Photochem. Photobiol. 1979, 29, 223–32.
123. Silverman, B.D. Cancer Biochem. Biophys. 1981, 5, 201–12.
124. Jerina, D.M., Sayer, J.M., Thakker, D.R. Yagi, H. In 'Carcinogenicity of polycyclic aromatic hydrocarbons: The bay-region theory': Pullman, B., Ts'o, P.O.P., and Gelboin, H., Eds. D. Reidel Publishing Company: Boston, 1980: pp.1–12.
125. Jernstrom, B., Brigelius, R. Sies, H. Chem. Biol. Interact. 1983, 44, 185–93.
126. Jernstrom, B., Dock, L. Martinez, M. Carcinogenesis 1984, 5, 199–204.
127. Anderson, M.W., Boroujerdi, M. Wilson, A.E. Cancer Res. 1981, 41, 4309–15.
128. Wood, A.W., Huang, M.T., Chang, R.L., Newmark, H.L., Lehr, R.E., Yagi, H., Sayer, J.M. Jerina, D.M. Proc. Natl. Acad. Sci. USA. 1982, 79, 5513–7.
129. Mukhtar, H., Das, M., Del Tito, B.J. Bickers, D.R. Xenobiotica 1984, 14, 527–32.
130. Glatt, H., Cooper, C.S., Grover, P.L., Sims, P., Bentley, P., Merdes, M., Waechter, F. Vogel, K. Science 1982, 215, 1507–9.
131. Glatt, H., Friedberg, T., Grover, P.L., Sims, P. Oesch, F. Cancer Res. 1983, 43, 5713–7.
132. Hesse, S., Jernstrom, B., Martinez, M., Moldeus, P., Christodoulides, L. Ketterer, B. Carcinogenesis 1982, 3, 757–61.

133. Brown, H.S., Jeffrey, A.M. Weinstein, I.B. Cancer Res. 1979, 39, 1673-7.
134. Ashurst, S.W., Cohen, G.M., Nesnow, S., DiGiovanni, J. Slaga, T. J. Cancer Res. 1983, 43, 1025-9.
135. Baird, W.M., Dumaswala, R.U. Diamond, L. Basic Life Sci. 1983, 24, 565-86.
136. Daniel, F.B., Schut, H.A., Sandwitch, D.W., Schenck, K.M., Hoffmann, C.O., Patrick, J.R. Stoner, G.D. Cancer Res. 1983, 43, 4723-9.
137. Baird, W.M. Diamond, L. Biochem. Biophys. Res. Comm. 1977, 77, 162-8.
138. Autrup, H., Harris, C.C., Trump, B.F. Jeffrey, A.M. Cancer Res. 1978, 38, 3689-96.
139. Ivanovic, V., Geacintov, N.E., Yamasaki, H. Weinstein, I.B. Biochemistry 1978, 17, 1597-1603.
140. Shinohara, K. Cerutti, P.A. Proc. Natl. Acad. Sci. USA. 1977, 74, 979-83.
141. MacLeod, M.C., Cohen, G.M. Selkirk, J.K. Cancer Res. 1979, 39, 3463-70.
142. Kinoshita, T., Lee, H.M., Harvey, R.G. Jeffrey, A.M. Carcinogenesis 1982, 3, 255-60.
143. Baird, W.M., Salmon, C.P. Diamond, L. Cancer Res. 1984, 44, 1445-52.
144. Hemminki, K., Cooper, C.S., Ribeiro, O., Grover, P.L. Sims, P. Carcinogenesis 1980, 3, 277-86.
145. Cooper, C.S., Ribeiro, O., Hewer, A., Walsh, C., Pal, K., Grover, P.L. Sims, P. Carcinogenesis 1980, 1, 233-243.
146. Cary, P.D., Turner, C.H., Cooper, C.S., Ribeiro, O., Grover, P. L. Sims, P. Carcinogenesis 1980, 1, 505-12.
147. Cooper, C.S., MacNicoll, A.D., Ribeiro, O., Hewer, A., Walsh, C., Pal, K., Grover, P.L. Sims, P. Cancer Lett. 1980, 9, 53-9.
148. MacNicoll, A.D., Cooper, C.S., Ribeiro, O., Gervasi, P.G., Hewer, A., Walsh, C., Grover, P.L. Sims, P. Biochem. Biophys. Res. Comm. 1979, 91, 490-7.
149. Vigny, P., Kindts, M., Duquesne, M., Cooper, C.S., Grover, P.L. Sims, P. Carcinogenesis 1980, 1, 33-36.
150. Levin, W., Chang, R.L., Wood, A.W., Yagi, H., Thakker, D.R., Jerina, D.M. Conney, A.H. Cancer Res. 1984, 44, 929-33.
151. Cooper, C.S., Ribeiro, O., Farmer, P.B., Hewer, A., Walsh, C., Pal, K., Grover, P.L. Sims, P. Chem. Biol. Interact. 1980, 32, 209-31.
152. Huggins, C., Pataki, J. Harvey, R.G. Proc. Natl. Acad. Sci. USA. 1967, 58, 2253-2260.
153. Levin, W., Wood, A.W., Chang, R.L., Newman, M.S., Thakker, D.R., Conney, A.H. Jerina, D.M. Cancer Lett. 1983, 20, 139-46.
154. Dipple, A. Hayes, M.E. Biochem. Biophys. Res. Comm. 1979, 91, 1225-31.
155. Malaveille, C., Tierney, B., Grover, P.L., Sims, P. Bartsch, H. Biochem. Biophys. Res. Comm. 1977, 75, 427-33.
156. Bigger, C.A., Sawicki, J.T., Blake, D.M., Raymond, L.G. Dipple, A. Cancer Res. 1983, 43, 5647-51.
157. Slaga, T.J., Gleason, G.L., DiGiovanni, J., Sukumaran, K.B. Harvey, R.G. Cancer Res. 1979, 39, 1934-6.
158. Slaga, T.J., Huberman, E., DiGiovanni, J., Gleason, G. Harvey, R.G. Cancer Lett. 1979, 6, 213-20.

159. Vigny, P., Duquesne, M., Coulomb, H., Tierney, B., Grover, P.L. Sims, P. FEBS Lett. 1977, 82, 278-82.
160. Baird, W.M. Dipple, A. Int. J. Cancer 1977, 20, 427-31.
161. Tierny, B., Hewer, A., MacNicoll, A.D., Gervasi, P.G., Rattle, H., Walsh, C., Grover, P.L. Sims, P. Chem. Biol. Interact. 1978, 23, 243-57.
162. Jeffrey, A.M., Kinoshita, T., Santella, R.M. Weinstein, I.B. In 'The chemistry of polycyclic aromatic hydrocarbon-DNA adducts': Ts'o, P.P. and Gelboin, H., Eds. D. Reidel Publishing Co.: Boston, 1980: p.565.
163. Dipple, A., Pigott, M., Moschel, R.C. Costantino, N. Cancer Res. 1983, 43, 4132-5.
164. Sawicki, J.T., Moschel, R.C. Dipple, A. Cancer Res. 1983, 43, 3212-8.
165. Daniel, F.B., Cazer, F.D., D'Ambrosio, S.M., Hart, R.W., Kim, W. H. Witiak, D.T. Cancer Lett. 1979, 6, 263-72.
166. Daniel, F.B. Joyce, N.J. J. Natl. Cancer Inst. 1983, 70, 111-8.
167. DiGiovanni, J., Nebzydoski, A.P. Decina, P.C. Cancer Res. 1983, 43, 4221-6.
168. Joyce, N.J. Daniel, F.B. Carcinogenesis 1982, 3, 297-301.
169. Dipple, A., Tomaszewski, J.E., Moschel, R.C., Bigger, C.A., Nebzydoski, J.A. Egan, M. Cancer Res. 1979, 39, 1154-8.
170. Watabe, T., Ishizuka, T., Isobe, M. Ozawa, N. Science 1982, 215, 403-404.
171. Eastman, A. Bresnick, E. Cancer Res. 1979, 39, 4316-21.
172. Levin, W., Buening, M.K., Wood, A.W., Chang, R.L., Thakker, D. R., Jerina, D.M. Conney, A.H. Cancer Res. 1979, 39, 3549-53.
173. King, H.W.S., Osborne, M.R. Brookes, P. Int. J. Cancer 1977, 20, 564-571.
174. Hodgson, R.M., Cary, P.D., Grover, P.L. Sims, P. Carcinogenesis 1983, 4, 1153-8.
175. Hodgson, R.M., Pal, K., Grover, P.L. Sims, P. Carcinogenesis 1983, 4, 1639-43.
176. Hulbert, P.B. Grover, P.L. Biochem. Biophys. Res. Comm. 1983, 117, 129-34.
177. Hecht, S.S., Bondinell, W.E. Hoffmann, D. J. Natl. Cancer Inst. 1974, 53, 1121-33.
178. Melikian, A.A., LaVoie, E.J., Hecht, S.S. Hoffmann, D. Carcinogenesis 1983, 4, 843-9.
179. Melikian, A.A., LaVoie, E.J., Hecht, S.S. Hoffmann, D. Cancer Res. 1982, 42, 1239-42.
180. Hecht, S.S., Amin, S., Rivenson, A. Hoffmann, D. Cancer Lett. 1979, 8, 65-70.
181. LaVoie, E.J., Amin, S., Hecht, S.S., Furuya, K. Hoffmann, D. Carcinogenesis 1982, 3, 49-52.
182. LaVoie, E.J., Hecht, S.S., Bedenko, V. Hoffmann, D. Carcinogenesis 1982, 3, 841-6.
183. Amin, S., LaVoie, E.J. Hecht, S.S. Carcinogenesis 1982, 3, 171-4.
184. LaVoie, E.J., Hecht, S.S., Amin, S., Bedenko, V. Hoffmann, D. Cancer Res. 1980, 40, 4528-32.
185. Perin-Roussel, O., Saguam, S., Ekert, B. Zajdela, F. Carcinogenesis 1983, 4, 27-32.

RECEIVED April 1, 1985

The Intercalation of Benzo[*a*]pyrene and 7,12-Dimethylbenz[*a*]anthracene Metabolites and Metabolite Model Compounds into DNA

P. R. LeBRETON

Department of Chemistry, University of Illinois at Chicago, Chicago, IL 60680

Carcinogenic metabolites and metabolite model compounds of benzo[a]pyrene and 7,12-dimethylbenz-
[a]anthracene intercalate into DNA with physical
binding constants in the range 10^3-10^4 M^{-1}. The
association constants for hydrocarbon-base stacking are similar in magnitude to those for base-
base stacking and are much smaller than those of
intercalating drugs such as ethidium bromide which
has a value $>10^6$ M^{-1}. The intercalation of these
molecules strongly depends on DNA structure and
environment. Increases in solvent polarity enhance intercalation. The DNA stabilizers Mg^{+2} and
polyamines inhibit intercalation. Hydrocarbon-
base π interactions are much weaker in heat denatured DNA than in native double-stranded DNA. For
metabolites and metabolite models for which com-
parisons were made, binding to native single-
stranded DNA is favored over binding to circular
double-stranded DNA, and binding to poly(dA-dT) is
favored over binding to other synthetic polynucleotides. In studies of nonreactive model compounds which have steric and electronic properties
similar to those of reactive metabolites of 7,12-
dimethylbenz[a]anthracene and benzo[a]pyrene, it
is found that analogs of bay region epoxides are
better intercalating agents than those of less
carcinogenic epoxides.

Metabolites of carcinogenic polycyclic aromatic hydrocarbons (PAH)
such as (±) trans-7,8-dihydroxy-anti-9,10-epoxy-7,8,9,10-tetrahy-
drobenzo[a]pyrene (BPDE) and 7,8,9,10-tetrahydroxytetrahydroben-
zo[a]pyrene (BPT) participate in π binding interactions with nucleotide bases (1-19) which lead to the reversible formation of
physical complexes. Almost all the physically bound hydrocarbon
metabolites are intercalated between the nucleotide bases. A
great deal of information about events important to chemical car-

0097-6156/85/0283-0209$08.25/0
© 1985 American Chemical Society

cinogenesis has recently come to light as a result of the study of
reactive interactions ocurring between epoxide containing metabolites of PAH and DNA (20-34). Much less is known about the nature
and the significance of reversible physical interactions between
hydrocarbon metabolites and DNA. The potential importance of hy-
drocarbon-nucleotide physical binding to the mechanisms of PAH
carcinogenesis is shown by the data given in Table I which gives
binding constants for stacked complexes formed from the interaction of nucleotide bases with pyrene (35), BPDE (3) and ethidium
bromide (36). Table I also contains binding constants for stacked
complexes formed from the self-association of nucleosides (37).

A comparison of the DNA binding constants in Table I indicates that the π stacking interactions of hydrocarbon metabolites
such as BPDE are much weaker than those of strong intercalating
drugs such as ethidium bromide. The binding constant for BPDE intercalation into DNA is similar in magnitude to the binding constants for the hydrogen bonding of base pairs. In organic solvents ($CDCl_3$ and CCl_4) at 25° C association constants for base-
pair hydrogen bonding typically lie in the range 10^2-10^3 M^{-1} (37).

An examination of the nucleoside and nucleotide binding data
in Table I also shows that the forces responsible for the stacked
association of nucleosides are similar in magnitude to those leading to binding of pyrene to mononucleotides. In vivo the reversible physical binding of nucleotide bases to one another via base
stacking and hydrogen bonding is responsible for the storage and
transmission of genetic information and plays an important role in
determining the structure and stability of double-stranded helical
DNA. The similarity between the physical interactions of bases
with one another and the physical interactions of bases with hydrocarbons and hydrocarbon metabolites may be important to mechanisms of PAH carcinogenesis.

Recent structure-activity studies of 1-alkylbenzo[a]pyrenes
also suggest that DNA intercalation of benzo[a]pyrene (BP) metabolites plays a role in the mechanism of BP carcinogenesis (38).
The addition of bulky alkyl groups at the 1-position of BP, which
inhibit DNA intercalation of 1-alkyl-BP metabolites (19), causes a
reduction in carcinogenic activity.

In the earliest studies of the physical binding of carcinogenic hydrocarbons with DNA the effects of DNA upon the solubility
of pyrene and benzo[a]pyrene were examined. In a solution of DNA
(0.05% by weight) the solubility of these hydrocarbons is increased as much as 70 times (39, 40). The biochemical significance of
these early studies has been questioned (41), and it has been argued that mechanisms of hydrocarbon carcinogenesis depend much
more upon the interaction of parent hydrocarbons with proteins
than with DNA. This criticism is supported by current studies of
hydrocarbon carcinogenesis which point to the important role that
activation of parent hydrocarbons by microsomal enzyme systems
containing cytochrome P-450 plays in the formation of ultimate reactive carcinogens (42).

Most recent studies of hydrocarbon interactions with DNA have
focussed on the binding of hydrocarbon metabolites rather than on
the binding of the parent hydrocarbons. Much of this work deals

Table I. Comparison of Binding Constants for Stacked Complexes at
23–25 °C

Complex	$K(M^{-1})$
Nucleoside–Nucleoside[a,b]	$\sim 1.0 \times 10^1$
Pyrene–Nucleotide[c]	$\sim 3.0 \times 10^1$
BPDE–DNA[d]	6.5×10^3
Ethidium Bromide–DNA[a,e]	$> 10^6$

[a] In H_2O
[b] Taken from ref. 37.
[c] In H_2O + 5% Methanol. Taken from ref. 35.
[d] In H_2O + 2% Ethanol. Taken from ref. 3.
[e] Taken from ref. 36.

with the reaction properties of carcinogenic epoxides of BP and
7,12-dimethylbenz[a]anthracene (DMBA). There is considerable evidence suggesting that DNA lesions resulting from reactions with
those and other carcinogens are an essential step in tumor initiation (43,44).

It is found that in a significant number of cases metabolite
reactivity provides a good index for predicting PAH carcinogenic
potency (45, 46). For example, bay region epoxides and precursors
of bay region epoxides of BP and DMBA are more mutagenic and carcinogenic than less reactive epoxides, such as K-region epoxides,
derived from the same parent hydrocarbons (45-53). Figures 1 and
2 show bay region epoxides of BP and DMBA along with structures of
less active epoxides.

In vitro studies of DNA interactions with the reactive benzo[a]pyrene epoxide BPDE indicate that physical binding of BPDE
occurs rapidly on a millisecond time scale forming a complex that
then reacts much more slowly on a time scale of minutes (17). Several reactive events follow formation of the physical complex.
The most favorable reaction is the DNA catalyzed hydrolysis of
BPDE to the tetrol, BPT (3,5,6,8,17). At 25°C and pH=7.0, the hydrolysis of BPDE to BPT in DNA is as much as 80 times faster than
hydrolysis without DNA (8). Other reactions which follow formation of physical complexes include those involving the nucleotide
bases and possibly the phosphodiester backbone. These can lead to
DNA strand scission (9, 34, 54-56) and to the formation of stable
BPDE-DNA adducts. Adduct formation occurs at the exocyclic amino
groups on the nucleotide bases and at other sites (1,2,9,17,20,
28,33,34,57,58). The pathway which leads to hydrocarbon adducts
covalently bound to the 2-amino group of guanine has been the most
widely studied.

Several laboratories have examined whether BPDE covalently
bound to DNA assumes an intercalated conformation or is externally
bound. Different groups have reported different results
(5,6,8,20,34,59-69). Mobility studies using relaxed circular
pBR322 DNA indicate that reaction with BPDE gives rise to rapid
positive supercoiling which is suggestive of a conformation in
which the hydrocarbon occupies an internal site in the DNA (34).
On the other hand, from results of the most recent spectroscopic
studies it is concluded that the covalent adduct formed from the
more carcinogenic (+) enantiomer of BPDE is in an external conformation (68,69).

On the basis of PAH physical binding studies (18) it has been
previously suggested that BPDE adduct conformations are strongly
dependent upon the DNA environment and that this may be playing a
role in the varying results reported by different laboratories. A
mechanism for reaction has been proposed (2) which involves initial intercalation of BPDE followed by reaction which can lead ultimately to nonintercalated complexes. This mechanism is supported by recent kinetic flow dichroism studies (69).

Two specific suggestions concerning the role that the reversible physical binding of proximate and ultimate carcinogens derived from BP play in carcinogenesis have been made. The first is
based on recognition that DNA-BPDE complex formation precedes re-

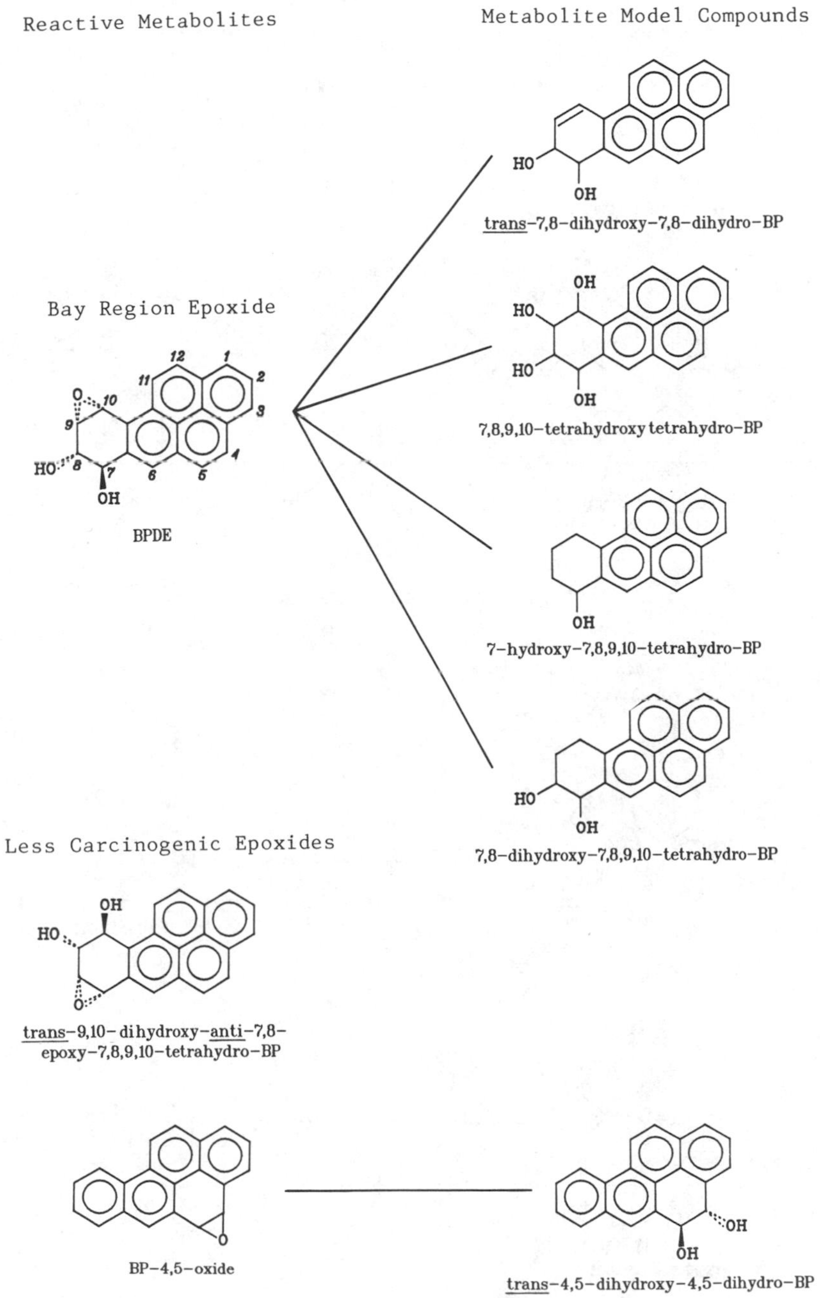

Figure 1.   Metabolites and metabolite model compounds derived from benzo[a]pyrene.

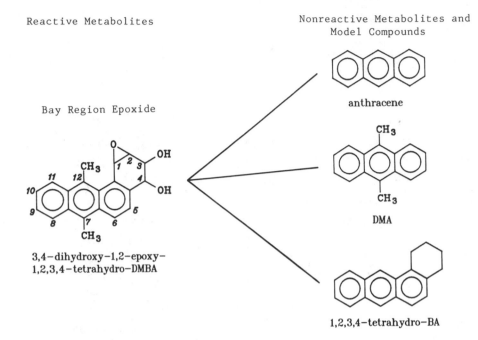

Reactive Metabolites

Nonreactive Metabolites and Model Compounds

Bay Region Epoxide

anthracene

3,4–dihydroxy–1,2–epoxy–
1,2,3,4–tetrahydro–DMBA

DMA

1,2,3,4–tetrahydro–BA

Less Carcinogenic Epoxides

DMBA–5,6–oxide

5,6–dihydro–BA

8,9–dihydroxy–10,11–epoxy–
8,9,10,11–tetrahydro–DMBA

8,9,10,11–tetrahydro–BA

Figure 2.   Reactive metabolites and metabolite model compounds derived from 7,12-dimethylbenz[a]anthracene.

action. This has led to speculation (1) that the stereochemistry associated with the formation of prereaction intercalated complexes with DNA is related to the varying reactive and carcinogenic properties of different stereoisomers of BP epoxides. It has also been suggested (9) that sites for strand scission are determined by the formation of intercalated complexes before reaction.

The second is based on the observation (18) that the nonreactive proximate carcinogen trans-7,8-dihydroxy-7,8-dihydro-BP binds favorably to DNA. This binding may lead to increased in vivo nuclear concentrations of the diol. Nuclear membrane bound cytochrome P-450 (70,71) is available for further activiation of the diol, and pooling of the proximate carcinogen in the nucleus and subsequent metabolism to BPDE may be important. This would provide an efficient mechanism by which biologically significant levels of unstable reactive metabolites reach nuclear target sites.

Several groups have investigated the DNA physical binding properties of BP (39,40,72), of nonreactive BP metabolites and of reactive BP metabolites (1-6,8,11). Nonreactive metabolites which have been studied are trans-7,8-dihydroxy-7,8-dihydro-BP (15,16,18,19), BPT (3-5,7,10,18) and trans-4,5-dihydroxy-4,5-dihydro-BP (15,18). The first of these molecules is a proximate carcinogen and a metabolic precursor of BPDE. The second and third molecules are the hydrolysis products of BPDE and of the less carcinogenic K-region epoxide, benzo[a]pyrene-4,5-oxide, respectively. Studies of the physical binding of reactive metabolites of benzo[a]pyrene have focussed on BPDE and on trans-9,10-dihydroxy-anti-7,8-epoxy-7,8,9,10-tetrahydro-BP (1-4, 6, 11). These investigations of the noncovalent binding of reactive metabolites to DNA are made difficult by the reactions, especially hydrolysis, which follow the formation of a physical complex.

In order to gain more detailed information about the physical binding of hydrocarbon metabolites to DNA, studies have also been carried out with model compounds which have many of the steric and electronic properties of carcinogenic epoxides but no reactive epoxide group. The use of nonreactive model compounds permits the clear separation of physical binding interactions from reactive interactions. Benzo[a]pyrene metabolite model compounds which have been examined include 7-hydroxy-7,8,9,10-tetrahydro-BP (4), and cis (4) and trans-7,8-dihydroxy-7,8,9,10-tetrahydro-BP (9). The DMBA metabolite model compounds which have been examined include the benz[a]anthracene (BA) derivatives, 1,2,3,4-tetrahydro-BA (12, 13), 5,6-dihydro-BA (12), and 8,9,10,11-tetrahydro-BA (12, 13, 14), as well as anthracene (12) and 9,10-dimethylanthracene (DMA) (14). Figures 1 and 2 show structures of nonreactive metabolites and metabolite model compounds derived from BP and DMBA for which DNA physical binding studies have been carried out.

The major goals of recent studies of the physical binding to DNA of BP and DMBA metabolites and metabolite models are to determine: (1) the magnitudes of the binding constants, (2) the conformations of physical complexes which are formed and the nature of DNA binding sites, (3) how DNA structure and environment influence physical binding, (4) how the structure of hydrocarbon metabolites influences physical binding properties, (5) whether the

physical binding of hydrocarbon metabolites exhibits specificity
for certain base sequences in DNA, and (6) whether different metabolites derived from the same parent hydrocarbon but with varying
carcinogenic potency exhibit different DNA physical binding properties.

Spectroscopic Probes of Physical Binding Properties

Most studies of the physical binding of hydrocarbon metabolites
and metabolite model compounds have measured the effect of DNA
binding on hydrocarbon fluorescence intensities, fluorescence
lifetimes and UV absorption spectra. Radioactive labelling has
also been used, but less frequently. Spectroscopic methods are
particularly convenient. These methods, especially fluorescence
methods, are also very sensitive. All of the hydrocarbons in
Figure 1 except the epoxides have high fluorescence quantum
yields, which permit routine detection in the 10^{-6}-10^{-7} M concentration range.

With spectroscopic methods it is possible to obtain information about the conformation of hydrocarbon–DNA complexes. The
fluorescence quantum yields of aromatic hydrocarbons are greatly
reduced when they bind to DNA in intercalated conformations.
Figure 3 shows how the intensity of the emission spectrum of DMA
decreases with increasing concentrations of DNA in 15% methanol.
(In Figure 3 and throughout this discussion DNA concentrations and
association constants have been reported in terms of PO_4^- molarity
unless otherwise indicated. The solution content of organic solvents is given in percent volume.)

The mechanism of the fluorescence quenching process which accompanies hydrocarbon intercalation is not thoroughly understood. However, in studies of 9-methylanthracene and phenanthrene, which have properties similar to the molecules considered
here, it is found that small perturbations such as those arising
from temperature variation (73) or from deuterium substitution for
hydrogen at specific positions (74) can strongly alter fluorescence quantum yields. These changes in quantum yields are due
almost exclusively to changes in the rate of intersystem crossing. It is reasonable to expect that quenching due to DNA intercalation also involves an increase in the rate of intersystem
crossing which accompanies binding. This conclusion is supported
by the observation that there is nearly a 1:1 correspondence between the disappearance of singlet excited states and the appearance of triplets in intercalated DNA complexes formed from polycyclic aromatic hydrocarbons (72,75).

In some cases it is possible to obtain a measure of the association constant for intercalation directly from fluorescence
quenching data. This method is applicable when the dynamic
quenching of the hydrocarbon fluorescence by DNA is small and when
the intercalated hydrocarbon has a negligible fluorescence quantum
yield compared to that of the free hydrocarbon. If these conditions are met, the association constant for intercalation, K_Q, is
equal to the Stern–Volmer quenching constant K_{SV} (76) and is given
by Equation 1.

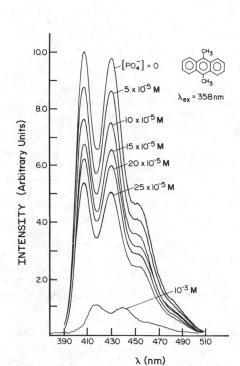

Figure 3. Uncorrected emission spectra of DMA in 15% methanol measured at varying calf thymus DNA concentrations. The spectra were measured under the conditions described in ref. 14.

$$K_Q = K_{SV} = [DNA]^{-1} [\frac{I_0}{I} - 1] \qquad (1)$$

In Equation 1, I_0 is the fluorescence intensity of the hydrocarbon measured without DNA, and I is the intensity measured with DNA.

Figure 4 shows Stern-Volmer plots measured for the DNA quenching of DMA, 1,2,3,4-tetrahydro-BA, 5,6-dihydro-BA, 8,9,10,11-tetrahydro-BA and anthracene. Figure 5 shows similar plots for trans-7,8-dihydroxy-7,8-dihydro-BP and trans-4,5-dihydoxy-4,5-dihydro-BP. Figures 4 and 5 contain data obtained both with native DNA and with heat-denatured DNA. Both figures show that the significant fluorescence quenching which is observed in native DNA is greatly diminished in denatured DNA. This strong dependence of the spectroscopic purturbation due to binding, on DNA secondary structure is indicative of an intercalative binding process (36).

Measurements of the hydrocarbon fluorescence lifetimes provide important information which is useful in interpreting the Stern-Volmer plots. In cases where Equation 1 is valid, the hydrocarbon fluorescence decay profiles must be the same with and without DNA. In some cases, BP for example, this is not the case. For BP the observed decay profile changes significantly when DNA is added (72).

However for several of the molecules shown in Figures 1 and 2, DNA has only a small effect on the observed fluorescence lifetime. These molecules include trans-7,8-dihydroxy-7,8-dihydro-BP (15,18,19), trans-4,5-dihydroxy-4,5-dihydro-BP (15,18), BPT (7,18), 1,2,3,4-tetrahydro-BA (12), 8,9,10,11-tetrahydro-BA (14), 5,6-dihydro-BA (12), anthracene (12) and DMA (14). Typical decay profiles obtained in fluorescence lifetime measurements of trans-7,8-dihydroxy-7,8-dihydro-BP and of 8,9,10,11-tetrahydro-BA are shown in Figure 6. The lifetimes extracted from the decay profiles shown here have been obtained by using a least-squares deconvolution procedure which corrects for the finite duration of the excitation lamp pulse (77).

For 8,9,10,11-tetrahydro-BA the lifetimes measured with and without DNA are the same within experimental error ($\pm$ 2 nsec). Without DNA the decay profile of trans-7,8-dihydroxy-7,8-dihydro-BP follows a single-exponential decay law. With DNA the decay profile has a small contribution from a short-lived component (τ = 5 nsec) which arises from DNA complexes. This indicates that Equation 1 is not strictly valid. However, the analysis of the decay profile with DNA also indicates that the short lifetime component contributes less than 11% to the total emission observed at $[PO_4^-]$ = 5 x 10^{-4} M. Under these conditions Equation 1 still yields a good approximate value to the association constant for intercalation.

For BP metabolites and metabolite model compounds UV absorption experiments provide an independent means by which binding constants for hydrocarbon intercalation into DNA can be measured. Intercalative binding gives rise to a red shift ($\sim$ 10 nm) in the hydrocarbon UV absorption spectrum of PAH. Figure 7 shows absorption spectra of trans-7,8-dihydroxy-7,8-dihydro-BP at varying

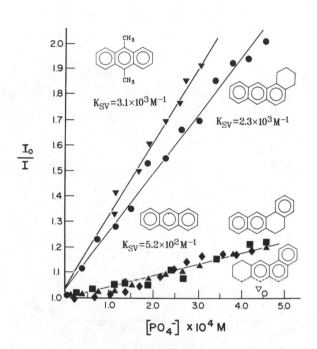

Figure 4.   Stern–Volmer plots and quenching constants derived from the fluorescence quenching of DMA (▼), 1,2,3,4-tetra-hydro-BA (●), 5,6-dihydro-BA (▲), 8,9,10,11-tetrahydro-BA (■) and anthracene (◆) by DNA in 15% methanol at 23° C.   Emission and excitation wavelengths and details concerning the experimental conditions are given in refs. 12 and 14.   The open symbols, ○ and ▽, show I_0/I for 1,2,3,4-tetrahydro-BA and DMA respectively in denatured DNA($[PO_4^-] = 4.4 \times 10^{-4}$ M).

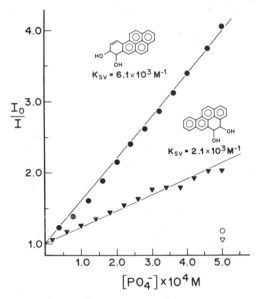

Figure 5.   Stern-Volmer plots and quenching constants derived from
            the fluorescence quenching of <u>trans</u>-7,8-dihydroxy-7,8-
            dihydro-BP and <u>trans</u>-4,5-dihydroxy-4,5-dihydro-BP in
            native DNA (closed symbols) and in denatured DNA (open
            symbols) in 15% methanol at 23° C.   Details about the
            experimental conditions are given in ref. 15.

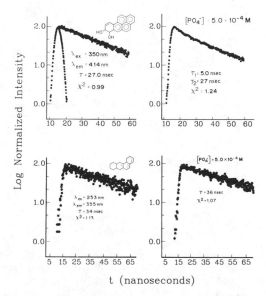

t (nanoseconds)

Figure 6.   Fluorescence decay profiles of trans-7,8-dihydroxy-7,8-
dihydro-BP and 8,9,10,11-tetrahydro-BA measured at 23
°C with and without native DNA.   Taken from refs. 14
and 15.   The upper left-hand corner contains an in-
strument response profile.   Emission and excitation
wavelengths, lifetimes, and values of χ^2 obtained from
deconvolution of the lifetime data are also given.

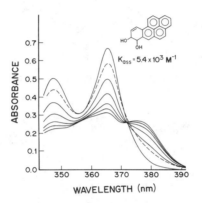

Figure 7. Absorption spectra in 15% methanol at 23°C of trans-7,8-dihydroxy-7,8-dihydrobenzo[a]pyrene in native DNA at concentrations of 0.0, 8.0 x 10^-5, 1.6 x 10^-4, 2.4 x 10^-4, 3.2 x 10^-4 and 4.0 x 10^-4 M. The broken line shows a spectrum in the presence of 3.2 x 10^-4 M DNA and 3.2 x 10^-4 M spermine. (Reproduced with permission from Ref. 15. Copyright 1985, Alan R. Liss.)

DNA concentrations. The appearance of the red-shifted UV absorption spectrum is strongly dependent upon DNA secondary structure and is greatly reduced in denatured DNA. Like the fluorescence quenching the red-shifted absorption spectrum is due to hydrocarbon intercalation (15, 18). Standard methods first derived by Benesi and Hildebrand (78) and later modified (3, 4, 9, 79) have been used to analyze the UV data. Table II contains association constants obtained from Stern-Volmer quenching data along with association constants obtained from UV binding studies of BP metabolites and metabolite models. For all three of the molecules contained in Table II, fluorescence lifetime studies (9, 15) indicate that decay profiles measured with and without DNA are very similar. In each case the fluorescence spectra and the absorption spectra were measured under identical conditions. A comparison of the results obtained from the two different methods indicates that the agreement in values for the intercalation binding constants is good.

For the DMBA metabolite models studied to date a similar comparison of results from fluorescence and absorption studies has not been carried out. In these molecules all of the more intense absorption bands occur at wavelengths below 300nm where DNA absorption interferes. This makes it difficult to determine the DNA binding constants using absorption measurements. However the good agreement between the absorption and fluorescence results for the BP derivatives supports the conclusion that when the hydrocarbon fluorescence decay profile measured without DNA is similar to that with DNA, the Stern-Volmer quenching constant provides a good measure of the association constant for PAH intercalation.

Review Of Results

Intercalation of BPDE. Several groups have studied the reversible intercalative binding of BPDE to DNA. The fluorescence quantum yield of BPDE is much lower than that of BP derivatives which do not contain an epoxide group and fluorescence techniques have not been widely used to study BPDE physical binding to DNA (4). Association constants for the DNA intercalation of BPDE have been obtained by measuring red shifts in the UV absorption spectra of BPDE which occur upon the formation of intercalated complexes (3,4,6,8) and from fluorescence studies (8) of the kinetics of DNA catalyzed hydrolysis of BPDE. The hydrolysis reaction is conveniently monitored by following the fluorescence of the hydrolysis product, BPT, which has a quantum yield many times greater than BPDE.

A summary of BPDE association constants for intercalation obtained from different studies is given in Table III. The wide variation in reported association constants can be attributed in part to differences in solvent conditions. The low binding constant obtained in ref. 4 is due to the high ionic strength and high concentration of organic solvent employed in the experiments. The difference in the values of association constants reported in refs. 3 and 8 is most likely due to the difference in the organic content of the solutions used.

Table II. Intercalation Association Constants and Stern-Volmer
Quenching Constants for Benzo[a]pyrene Metabolites and
Metabolite Model Compounds[a] with Calf Thymus DNA

	$K_Q^b(M^{-1})$	$K_{SV}^c(M^{-1})$
trans-7,8-dihydroxy- 7,8-dihydro-BP[d]	5400	6100
trans-4,5-dihydroxy- 4,5-dihydro-BP[d]	1900	2100
trans-7,8-dihydroxy- 7,8,9,10-tetrahydro- BP[e]	750-910	740

[a] The estimated uncertainty in the association constants and quenching constants is ± 10%.
[b] Association constants from UV absorption studies.
[c] Stern-Volmer Quenching Constants.
[d] Measured in 15% methanol. Taken from refs. 15 and 18.
[e] Measured in 2.5% DMSO. Taken from ref. 9.

Table III. DNA Intercalation Association Constants Reported for
trans-7,8-Dihydroxy-anti-9,10-epoxy-7,8,9,10-tetrahydrobenzo[a]pyrene

Type of DNA	Buffer (pH)	Additional Ions	Organic Solvent	T(°C)	$K(M^{-1})$
Salmon Testes[a,b] (sheared)	20 mM Tris•HCl (7.4)		2% Ethanol	21	6,580
Calf Thymus[a,c] (sonicated)	10 mM NaHCO$_3$ (9.0)	50 mM NaCl	10% Acetone	25	377
Calf Thymus[d]	5 mM Sodium Cacodylate (7.1)		0.2% Tetra- hydrofuran	25	12,000

[a] Based on UV absorption measurements.
[b] Taken from ref 3. Results with calf thymus DNA are reported to be similar to those with salmon testes DNA.
[c] Taken from ref. 4.
[d] Based on UV absorption studies and on fluorescence studies of the kinetics of DNA
catalyzed BPDE hydrolysis. Taken from ref. 8.

The derivation of an intercalation association constant from kinetic studies of BPDE hydrolysis presumes that the reaction proceeds via an intercalated complex. This mechanism is supported by the observations that the catalytic activity of denatured DNA is lower than that of native DNA (8), that catalysis is inhibited at high ionic strengths (3, 8, 17), and that mononucleotides such as GMP exhibit much greater catalytic activity than does free phosphate (80).

In addition to UV binding studies of BPDE, one other less carcinogenic epoxide, (±) trans-9,10-dihydroxy-anti-7,8-epoxy-7,8,9,10-tetrahydro-BP, was examined. Table IV which contains results obtained in this study indicates that the intercalation binding constant is 23% lower than that of BPDE (3).

Table IV also contains results of UV absorption studies of hydroxylation effects on the DNA intercalative binding of benzo[a]pyrene metabolites and metabolite model compounds. The most important feature of these results is that hydrolysis of BPDE to BPT causes a four-fold reduction in the intercalation association constant. Of all the BP derivatives studied, the tetrol has the lowest binding constant for intercalation. The small binding constant of the tetrol compared with BPDE, coupled with the DNA catalyzed hydrolysis of BPDE to the tetrol may provide a detoxification pathway for removal of a portion of unreacted intercalated BPDE.

Physical Binding of Metabolites and Metabolite Model Compounds to Secondary Sites on DNA.

For BPT a secondary DNA binding site has been reported on the basis of dialysis experiments (7). In fluorescence lifetime studies it was found that the fluorescence decay profile of BPT bound at the secondary site is very similar to that of the unbound metabolite. This suggests that the secondary site occurs on the outside of DNA. Initial results from the dialysis experiments indicated that for BPT the binding constant for external binding is 2 to 4 times lower than that for intercalation. More recent studies (75,81) indicate that the binding constant for complex formation at secondary sites is at least 2 times smaller than that originally reported. Dialysis experiments carried out on trans-7,8-dihydroxy-7,8-dihydro-BP, trans-4,5-dihydroxy-4,5-dihydro-BP and pyrene indicate that for these molecules binding at secondary sites occurs with binding constants which are about 9 times lower than those for intercalation (18). For all these molecules intercalation is by far the most important binding mode.

Base Specificity of Physical Binding.

To determine whether the physical binding of hydrocarbon metabolites to DNA exhibits base specificity, the binding of trans-7,8-dihydroxy-7,8,9,10-tetrahydro-BP was examined using fluorescence and absorption techniques (9). A comparison was also made of the varying degrees to which different synthetic polynucleotides are able to solubilize BPT (10).

Results with trans-7,8-dihydroxy-7,8,9,10-tetrahydro-BP indicate that in 100 mM NaCl and 2.5% DMSO at pH 7.0 the association constant for intercalation into poly(dA-dT) is more than 5 times

Table IV. Comparison of Results from UV Absorption Studies of the
Intercalation Binding Constants for Benzo[a]pyrene Metabolites
and Metabolite Model Compounds into DNA

Compound	$K(M^{-1})$
BPDE[a]	6580
±trans-9,10-dihydroxy-anti-7,8-epoxy-7,8,9,10-tetrahydro-BP[a]	5080
7,8,9,10-tetrahydroxytetrahydro-BP[a]	1650
ibid,[b]	44
7-hydroxy-7,8,9,10-tetrahydro-BP[b]	160
trans-9,10-dihydroxy-7,8,9,10-tetrahydro-BP[b]	177
cis-7,8-dihydroxy-7,8,9,10-tetrahydro-BP[b]	257

[a] Measured with sheared salmon testes DNA at 21°C in 2% ethanol buffered to a pH of 7.4 with 20 mM tris•HCl. Taken from ref. 3.

[b] Measured with sonicated calf thymus DNA in 200 mM NaCl, 2 mM Mg^{+2} and 10% acetone buffered to a pH of 9.0 with 10 mM $NaHCO_3$. Taken from ref. 4.

greater than the association constants for intercalation into po-
ly(dG-dC), polydG:polydC, polydA:polydT or calf thymus DNA (9).
Solubilization studies with BPT yielded similar results (10). For
BPT in 10 mM sodium phosphate, 10 mM NaCl and 1 mM EDTA at pH 7.0
the solubilizing activity of the polynucleotides studied increases
in the order polydA:polydT ≈ polydG:polydC < poly(dA-dC):poly(dG-
dT) ≈ poly(dG-dC) < poly(dA-dT). This base specificity also exhibits a pH dependence. When the polynucleotides are protonated
in 10 mM sodium citrate (pH=3.8) the solubilizing activity of po-
ly(dA-dT), poly(dG-dC) and poly(dA-dC):poly(dG-dT) are approximately equivalent.

Comparison of The Physical Binding Properties of BP and DMBA Metabolite Model Compounds. The results in Figures 4 and 5 show how
structural differences that occur in different metabolites derived
from BP and DMBA influence physical binding to DNA. The results
indicate that both trans-7,8-dihydroxy-7,8-dihydro-BP and trans-
4,5-dihydroxy-4,5-dihydro-BP are better intercalating agents than
the DMBA metabolite model compounds. While no binding studies of
the bay region diol epoxide of DMBA have yet been carried out, the
model compound studies suggest that BPDE is the better intercalating agent.
      The results in Figures 4 and 5 also show that for a given pa-
rent hydrocarbon, bay region metabolite model compounds are better
intercalating agents than model compounds of less carcinogenic metabolites. For example, the intercalation binding constants for
1,2,3,4-tetrahydro-BA and DMA are more than 4.4 times greater than
those for 5,6-dihydro-BA, 8,9,10,11-tetrahydro-BA and anthracene
(12, 14). The intercalation binding constant for trans-7,8-dihy-
droxy-7,8-dihydro-BP is 2.9 times greater than that for trans-4,5-
dihydroxy-4,5-dihydro-BP (15).

Electronic Influences on Stacking Interactions. The results of
recent studies of the intercalative binding of hydrocarbon metabolite models to DNA show how electronic factors influence π binding
interactions between nucleotides and polycyclic aromatic hydrocarbons. In some stacked complexes involving DNA and RNA bases, association constants increase as the π ionization potentials of the
bases decrease (82,83). This occurs when charge transfer or dispersion forces between the interacting partners are important
(37,82,84,). Figure 8 shows how association constants for π
stacking vary with nucleoside ionization potentials in complexes
formed from the self association of nucleosides and from the binding of riboflavin to nucleosides. In both examples the association constants increase as the nucleoside ionization potentials
decrease.
      This relationship is useful for understanding the different
intercalation binding constants of hydrocarbons with similar π
systems. This is indicated in Figure 4, by a comparison of data
for 1,2,3,4-tetrahydro-BA, DMA and anthracene. For 1,2,3,4-tetra-
hydro-BA the presence of a nonplanar alicyclic group is expected
to sterically inhibit intercalation. The nonplanar methyl groups
of DMA play a similar steric role. However for both 1,2,3,4-te-

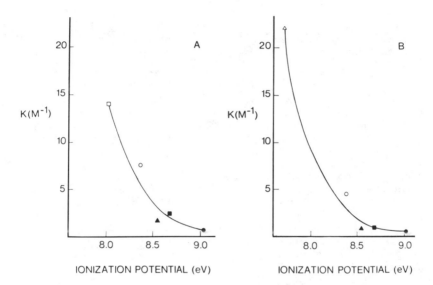

Figure 8. Dependence of π complex binding constants upon nucleo-
           side ionization potentials for (●) uridine, (■) thymi-
           dine, (▲) cytidine, (○) adenosine, (□) guanosine, and
           (△) N,N,-dimethyladenosine. Panel A shows association
           constants for the binding of nucleosides to ribofla-
           vin. Panel B shows association constants for the self
           association of nucleosides. (Reproduced from Ref. 82.
           Copyright 1981, American Chemical Society.)

trahydro–BA and DMA the association constants for intercalation
are greater than that for anthracene. For 1,2,3,4–tetrahydro–BA
the enhanced binding is accompanied by a decrease in the ionization potentials of the five highest occupied π orbitals compared
to corresponding orbitals in anthracene (85). This is shown in
Figure 9 which gives the photoelectron spectrum of 1,2,3,4–tetrahydro–BA. Previous photoelectron studies of polycyclic aromatic
hydrocarbons indicate that the methyl groups play a similar role
in destabilizing the manifold of upper occupied π orbitals in DMA
(85–87). For both 1,2,3,4–tetrahydro–BA and DMA electronic effects which accompany the addition of alicyclic groups and methyl
groups enhance intercalation. These effects are more important
than steric effects, which inhibit intercalation.

The Influence of DNA Structure and Environment on the Intercalation of Hydrocarbon Metabolites and Metabolite Model Compounds.
The physical binding of hydrocarbon metabolites to DNA is very
sensitive to DNA structure and environment. This is demonstrated
by the data in Figures 4 and 5, which show how heat denaturation
of DNA inhibits hydrocarbon quenching. These results are consistent with early studies which indicate that the ability of native
DNA to solubilize pyrene and BP is much greater than that of denatured DNA (40).
   In addition to denaturation, hydrocarbon physical binding is
sensitive to other DNA structural changes. Radioactive labelling
studies have been carried out to compare the binding of trans–7,8–
dihydroxy–7,8–dihydro–BP to double-stranded DNA versus single-
stranded DNA (16). The results show that binding to bacteriophage
single-stranded φx 174 DNA is more than 4.9 times stronger than
binding to a replicative form double-stranded circular φX DNA.
   The ability of hydrocarbon metabolites to intercalate into
DNA is strongly dependent on DNA environment as well as on DNA
structure, as demonstrated in Figure 10. The top panel of Figure
10 shows the effect of methanol on the fluorescence quenching of
DMA by DNA. The reduction of DMA binding at increasing methanol
concentration is due to the decrease in solvent polarity. The
bottom panel of Figure 10 shows how DMA quenching is greatly reduced by the addition of the DNA stabilizer Mg^{+2}. Similar results
have been obtained for BPDE (3,4,8,11,63). It has also been shown
that increasing ionic strength by adding NaCl inhibits intercalative binding of BPDE (4,63). Polyamine DNA stabilizers such as
spermine have the same effect on the intercalation of both BPDE
and on trans–7,8–dihydroxy–7,8–dihydro–BP (4,11,15). This is indicated by Figure 7, which shows how the intensity of the red-
shifted band arising from DNA complexes with trans–7,8–dihydroxy–
7,8–dihydro–BP is reduced when spermine is added. For BPDE (4,63)
the relative efficiency of DNA stabilizers for inhibiting intercalation increases in the order NaCl < $MgCl_2$ < spermine. It is important to note that in vivo polyamine and Mg^{+2} concentrations are
in the millimolar range (88, 89). It is expected that at these
levels they will efficiently protect DNA against hydrocarbon intercalation.
   The dependence of hydrocarbon intercalation on DNA conforma-

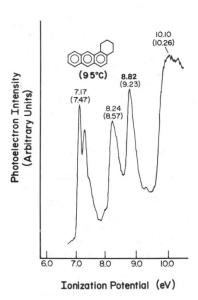

Ionization Potential (eV)

Figure 9. He(I) photoelectron spectra of 1,2,3,4-tetrahydro-BA.
Assignments are given along with probe temperatures.
Numbers in parentheses are ionization potentials for
corresponding orbitals in anthracene. (Reproduced with
permission from Ref. 12. Copyright 1983, Academic.)

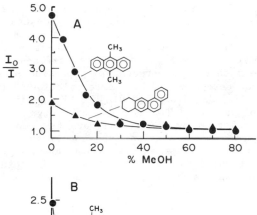

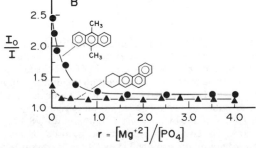

Figure 10. A.   Effects of varying methanol levels on the fluorescence quenching of 9,10-dimethylanthracene and
8,9,10,11-tetrahydro-BA by native DNA
($[PO_4]$ = 5.0 x 10^{-4} M).
B.   Effects of Mg^{+2} on hydrocarbon fluorescence quenching by native DNA ($[PO_4]$ = 5.0 x 10^{-4} M).
(Reproduced with permission from Ref. 14.   Copyright
1984, Adenine.)

tion, on the polarity of the DNA environment, and on the presence
of DNA stabilizers indicates that in vivo hydrocarbon intercalation into DNA is a process which can occur to different degrees as
DNA function changes during the course of a cell cycle.

It is interesting that trans-7,8-dihydroxy-7,8-dihydro-BP exhibits little fluorescence quenching with heat-denatured calf thymus DNA but binds strongly to single-stranded φX 174 DNA. At present there is no information concerning the nature of the binding
to single-stranded DNA. However, native single-stranded DNA is
less sterically crowded than double-stranded DNA and has secondary
structure that involves regions with a significant degree of base
stacking (90). The binding studies with trans-7,8-dihydroxy-7,8-
dihydro-BP indicate that this structure may provide a very favorable host for hydrocarbon metabolite intercalation. The greater
affinity of trans-7,8-dihydroxy-7,8-dihydro-BP to single-stranded
DNA than to double-stranded DNA and the effective inhibition of
hydrocarbon intercalation by DNA stabilizers has led to speculation that hydrocarbon intercalation into double-stranded DNA may
be most favorable when DNA is partially unwound (14, 16). Such
conformations are thought to occur when DNA is undergoing replication and when gene expression is taking place (91).

Summary

Benzo[a]pyrene   and   7,12-dimethylbenz[a]anthracene   metabolites
which are ultimate carcinogens and bay region metabolite model
compounds which simulate the π structure of ultimate carcinogens
are good intercalating agents. Bay region model compounds of BP
metabolites are better intercalating agents than comparable model
compounds of DMBA metabolites. For a given parent hydrocarbon (BP
or DMBA) bay region metabolite model compounds which have been
studied to date are better intercalating agents than model compounds of less carcinogenic metabolites examined under identical
conditions. The intercalation of aromatic hydrocarbons into DNA
is highly dependent upon the conformation and the environment of
the DNA.

There are a large number of potentially important ways in
which physical binding of hydrocarbon metabolites to DNA can influence carcinogenic activity. The physical binding of nonreactive proximate carcinogens to DNA may lead to the pooling of these
molecules at ultimate target sites. Current results indicate that
physical binding of reactive hydrocarbon metabolites to DNA precedes reaction and enhances metabolite reactivity. Following the
formation of physical complexes of reactive metabolites with DNA,
reactions occur which lead to epoxide hydrolysis and to DNA modification. The base specificity and the stereochemistry associated
with intercalated complexes may influence these reactions.

In addition to influencing hydrocarbon metabolite-DNA reactions, the physical binding properties of hydrocarbon metabolites
covalently bound to DNA may also be important to carcinogenic activity. The covalent binding of ultimate carcinogens derived from
BP and DMBA to DNA produces adducts with π binding properties similar to those of naturally occurring nucleotides. These adducts

can participate in π binding interactions with other strands of DNA, with RNA and with DNA regulating enzymes.

Acknowledgments

Support of this work by the National Institutes of Health, the American Cancer Society and the Research Board of the University of Illinois is gratefully acknowledged. The author also wishes to thank Professors Ronald G. Harvey, Nicholas Geacintov, Michael MacLeod and Nien-chu Yang for helpful discussions and Patricia Campbell and Regina Gierlowski for preparation of this manuscript.

Literature Cited

1. Meehan, T.; Straub, K. Nature 1979, 277, 410-412.

2. Lin, J.-H.; LeBreton, P. R.; Shipman, L. L. J. Phys. Chem. 1980, 84, 642-649.

3. MacLeod, M. C.; Selkirk, J. K. Carcinogenesis 1982, 3, 287-292.

4. Meehan, T.; Gamper, H.; Becker, J. F. J. Biol. Chem. 1982, 257, 10479-10485.

5. Geacintov, N. E.; Ibanez, V.; Gagliano, A. G.; Yoshida, H.; Harvey, R. G. Biochem. Biophys. Res. Commun. 1980, 92, 1335-1342.

6. Geacintov, N. E.; Yoshida, H.; Ibanez, V.; Harvey, R. G. Biochem. Biophys. Res. Commun. 1981, 100, 1569-1577.

7. Ibanez, V.; Geacintov, N. E.; Gagliano, A. G.; Brandimarte, S.; Harvey, R. G. J. Am. Chem. Soc. 1980, 102, 5661-5666.

8. Geacintov, N. E.; Yoshida, H.; Ibanez, V.; Harvey, R. G. Biochemistry 1982, 21, 1864-1869.

9. Yang, N. C.; Hrinyo, T. P.; Petrich, J. W.; Yang, D.-D. H. Biochem. Biophys. Res. Commun. 1983, 114, 8-13.

10. Chen, F.-M. Carcinogenesis 1984, 5, 753-758.

11. Meehan, T.; Becker, J. F.; Gamper, H. Proc. Amer. Assoc. Cancer Res. 1981, 22, 92.

12. Shahbaz, M.; Harvey, R. G.; Prakash, A. S.; Boal, T. R.; Zegar, I. S.; LeBreton, P. R. Biochem. Biophys. Res. Commun. 1983, 112, 1-7.

13. Prakash, A. S.; Zegar I. S.; Shahbaz, M.; LeBreton, P. R. Int. J. Quant. Chem. Quantum Biology Symposium 1983, 10, 349-356.

14. Zegar, I. S.; Prakash, A. S.; LeBreton, P. R. J. Biomol. Struct. Dynamics 1984, 2, 531–542.

15. Abramovich, M.; Zegar, I. S.; Prakash, A.S.; Harvey, R. G.; LeBreton, P. R. in "The Molecular Basis of Cancer, Part A; Macromolecular Structure Carcinogens, and Oncogenes"; Rein, R. Ed.; Alan R. Liss: New York, 1985, pp. 217–225.

16. Hsu, W.-T.; Harvey, R. G.; Weiss, S. B. Biochem. Biophys. Res. Commun. 1981, 101, 317–325.

17. Geacintov, N. E.; Hibshoosh, H.; Ibanez, V.; Benjamin, M. J.; Harvey, R. G. Biophysical Chemistry 1984, 20, 121–133.

18. Abramovich, M.; Prakash, A. S.; Harvey, R. G.; Zegar, I. S.; LeBreton, P. R. Chem.–Biol. Interactions submitted.

19. Paulius, D. E.; Prakash, A. S.; Harvey, R. G., Abramovich, M.; LeBreton, P. R. in "Proceedings of the Ninth International Symposium on Polynuclear Aromatic Hydrocarbons, Polynuclear Aromatic Hydrocarbons: Chemistry, Characterization and Carcinogenesis"; Cooke, M.; Dennis, A. G. Eds.; Battelle: Columbus, Ohio, 1985, in press.

20. Harvey, R. G. Acc. Chem. Res. 1981, 14, 218–226.

21. Remsen, J.; Jerina, D.; Yagi, H.; Cerrutti, P. Biochem. Biophys. Res. Commun. 1977, 74, 934–940.

22. Sims, P.; Grover, P. L. in "Advances in Cancer Research"; Klein, G.; Weinhouse, S., Eds.; Academic: New York, 1974 pp. 165–274.

23. Sims, P.; Grover, P. L.; Swaisland, A.; Pal, K.; Hewer, A. Nature 1974, 252, 326–328.

24. Kootstra A.; Haas, B. L.; Slaga, T. J. Biochem. Biophys. Res. Commun. 1980, 94, 1432–1438.

25. Pelling, J. C.; Slaga, T. J. Carcinogenesis 1982, 3, 1135–1141.

26. Kootstra, A. Carcinogenesis 1982, 3, 953–955.

27. Ivanovic, V.; Geacintov, N. E.; Jeffrey, A. M.; Fu, P. P.; Harvey, R. G.; Weinstein, I. B. Cancer Lett. 1978, 4, 131–140.

28. Weinstein, I. B.; Jeffrey, A. M. Jennette, K. W.; Blobstein, S. H.; Harvey, R. G.; Harris, C. Autrup, H.; Kasai, H.; Nakanishi, K. Science 1976, 193, 592–595.

29. Grover, P. L. in "Drug Metabolism–From Microbe to Man"; Parke, D. V.; Smith, R. L., Eds.; Taylor and Francis: London, 1977 pp. 105–122.

30. Leffler, S.; Pulkrabek, P.; Grunberger, D.; Weinstein, I. B. Biochemistry 1977, 16, 3133–3136.

31. Pulkrabek, P.; Leffler, S.; Grunberger, D.; Weinstein, I. B. Biochemistry 1979, 18, 5128–5134.

32. King, H. W. S.; Osborne, M. R.; Beland, F. A.; Harvey, R. G.; Brookes, P. Proc. Natl. Acad. Sci. USA 1976, 73, 2679–2681.

33. Brookes, P.; Osborne, M. R. Carcinogenesis 1982, 3, 1223–1226.

34. Agarwal, K. L.; Hrinyo, T. P.; Yang, N. C. Biochem. Biophys. Res. Commun. 1983, 114, 14–19.

35. Lianos, P.; Georghiou, S. Photochem. and Photobiol. 1979, 29, 13–21.

36. LePecq, J.-B.; Paoletti, C. J. Mol. Biol. 1967, 27, 87–106.

37. Ts'o, P. O. P. in "Basic Principles in Nucleic Acid Chemistry"; Ts'o, P. O. P., Ed.; Academic: New York, 1974; Vol. I, pp. 453–584.

38. Osborne, M. R.; Connell, J. R.; Venitt, S.; Crofton-Sleigh, C.; Brookes, P.; DiGiovanni, J.; Potaki, J.; Harvey, R. G. in "The Role of Chemicals and Radiation in the Etiology of Cancer"; Huberman, E., Ed.; Raven: New York, 1985, in press.

39. Boyland, E.; Green, B. Brit. J. Cancer 1962, 16, 507–517.

40. Boyland, E.; Green, B.; Liu, S.-L. Biochim. Biophys. Acta 1964, 87, 653–663.

41. Giovanella, B. C.; McKinney, L. E.; Heidelberger, C. J. Mol. Biol. 1964, 8, 20–27.

42. Yang, S. K.; Deutsch, J.; Gelboin, H. V. in "Polycyclic Hydrocarbons and Cancer"; Gelboin, H. V.; Ts'o, P. O. P., Eds.; Academic: New York, 1978; Vol. I, pp. 205–231.

43. Weinstein, I. B. J. Supramol. Struct. and Cell. Biochem. 1981, 17, 99–120.

44. Trosko, J. E.; Chang, C. C. in "Chemical Carcinogens and DNA"; Grover, P. L.; Ed.; CRC Press: Boca Raton, 1979; Vol. II, pp. 181–200.

45. Jerina, D. M.; Lehr, R. E. in "Microsomes and Drug Oxidations"; Ullrich, V.; Roots, I.; Hildebrandt, A.; Estabrook, R. W.; Conney, A. H. Eds.; Pergamon: Oxford, 1977; pp. 709–720.

46. Nordqvist, M. M.; Thakker, D. R.; Yagi H.; Lehr, R. E.; Wood, A. W.; Levin, W.; Conney, A. H.; Jerina, D. M. in "Molecular Basis of Environmental Toxicity"; Bhatnagar, R. S. Ed.; Ann Arbor Science Publishers: Ann Arbor, 1980 pp. 329–357.

47. Huberman, E.; Chou, M. W.; Yang, S. K. Proc. Natl. Acad. Sci.
     USA 1979, 76, 862-866.

48. Flesher, J. W.; Harvey, R. G.; Syndor, K. L. Int. J. Cancer
     1976, 18, 351-353.

49. Slaga, T. J.; Gleason, G. L.; DiGiovanni, J.; Sukumaran, K.
     B.; Harvey, R. G. Cancer Res. 1979, 39, 1934-1936.

50. Dipple, A.; Pigott, M.; Moschel; R. C.; Costantino, N. Cancer
     Res. 1983, 43, 4132-4135.

51. Vigny, P.; Kindts, M.; Cooper, C. S.; Grover P. L.; Sims, P.
     Carcinogenesis 1981, 2, 115-119.

52. Slaga, T. J.; Gleason, G. L.; DiGiovanni, J.; Berry, D. L.;
     Juchau, M. R.; Fu, P. P.; Sukumaran, K. B.; Harvey, R. G. in
     "Polynuclear Aromatic Hydrocarbons"; Jones, P. W.; Leber, P.,
     Eds.; Ann Arbor Science: Ann Arbor, 1979; pp. 753-764.

53. Slaga, T. J.; Bracken, W. M.; Viaje, A.; Levin, W.; Yagi, H.;
     Jerina, D. M.; Conney, A. H. Cancer Res. 1977, 37, 4130-4133.

54. Kakefuda, T.; Yamamoto, H. in "Polycyclic Hydrocarbons and
     Cancer"; Gelboin, H.; Ts'o, P.O.P. Eds.; Academic: New York,
     1978; Vol. II, 63-74.

55. Drinkwater, N. R.; Miller, E. C.; Miller, A. J. Biochemistry
     1980, 19, 5087-5092.

56. Gamper, H. B.; Bartholomew, J. C.; Calvin, M. Biochemisty
     1980, 19, 3948-3956.

57. Jeffrey, A. M.; Weinstein, I. B.; Jennette, K. W.; Grzesko-
     wiak, K.; Nakanishi, K.; Harvey, R. G.; Autrup, H.; Harris,
     C. Nature 1977, 269, 348-350.

58. Ivanovic, V.; Geacintov, N. E.; Yamasaki, H.; Weinstein, I.
     B. Biochemistry 1978, 17, 1597-1603.

59. Geacintov, N. E.; Gagliano, A.; Ivanovic, V.; Weinstein, I.
     B. Biochemistry 1978, 17, 5256-5262.

60. Yang, N. C.; Ng, L.-K.; Neoh, S. B.; Leonov, D. Biochem. Bio-
     phys. Res. Commun. 1978, 82, 929-934.

61. Undeman, O.; Lycksell, P.-O.; Gräslund, A.; Astlind, T.;
     Ehrenberg, A.; Jernström, B.; Tjerneld, F.; Norden, B. Cancer
     Res. 1983, 43, 1851-1860.

62. Prusik, T.; Geacintov, N. E. Biochem. Biophys. Res. Commun.
     1979, 88, 782-790.

63. Gamper, H. B.; Straub, K.; Calvin, M.; Bartholomew, J. C.
     Proc. Natl. Acad. Sci. USA 1980, 77, 2000-2004.

64. Hogan, M. E. Dattagupta, N.; Whitlock, J. P. J. Biol. Chem. 1981, 256, 4504-4513.

65. Geacintov, N. E.; Gagliano, A. G.; Ibanez, V.; Harvey, R. G. Carcinogenesis 1982, 3, 247-253.

66. MacLeod, M. C.; Mansfield, B. K.; Selkirk, J. K. Carcinogenesis 1982, 3, 1031-1037.

67. Undeman, O.; Sahlin, M.; Gräslund, A.; Ehrenberg, A. Biochem. Biophys. Res. Commun. 1980, 94, 458-465.

68. Geacintov, N. E.; Ibanez, V.; Gagliano, A. G.; Jacobs, S. A.; Harvey, R. G. J. Biomol. Struct. Dyanmics 1984, 1, 1473-1484.

69. Geacintov, N. E.; Yoshida, H.; Ibanez, V.; Jacobs, S. A.; Harvey, R. G. Biochem. Biophys. Res. Commun. 1984, 122, 33-39.

70. Pezutto, J. M.; Yang, C. S.; Yang, S. K.; McCourt, D. W.; Gelboin, H. V. Cancer Res. 1978, 38, 1241-1245.

71. Chou, M. W.; Yang, S. K.; Sydor, W.; Yang, C. S. Cancer Res. 1981, 41, 1559-1564.

72. Geacintov, N. E.; Prusik, T.; Khosrofian, J. M. J. Am. Chem. Soc. 1976, 98, 6444-6452.

73. Bennett, R. G.; McMartin, P. J. J. Chem. Phys. 1966, 44, 1969-1972.

74. O'Brien, J. J.; Henry, B. R.; Selinger, B. K. Chem. Phys. Lett. 1977, 46, 271-274.

75. Geacintov, N. E. private communication.

76. Moon, A. Y.; Poland, D. C.; Scheraga, H. A. J. Phys. Chem. 1965, 69, 2960-2966.

77. Hui, M.-H.; Ware, W. R. J. Am. Chem. Soc. 1976, 98, 4718-4727.

78. Benesi, H. A.; Hildebrand, J. H. J. Am. Chem. Soc. 1949, 71, 2703-2707.

79. Schmechel, D. E. V.; Crothers, D. M. Biopolymers 1971, 10, 465-480.

80. Gupta, S. C.; Pohl, T. M.; Friedman, S. L.; Whalen, D. L.; Yagi, H.; Jerina, D. M. J. Am. Chem. Soc. 1982, 104, 3101-3104.

81. MacLeod, M. C. private communication.

82. Yu, C.; O'Donnell, T. J.; LeBreton, P. R. J. Phys. Chem. 1981, 85, 3851-3855.

83. Hush, N. S.; Cheung, A. S. Chem. Phys. Lett. 1975, 34, 11-13.

84. Peng, S.; Padva, A.; LeBreton, P. R.; Proc. Natl. Acad. Sci. USA 1976, 73, 2966-2968.

85. Boschi, R.; Clar, E.; Schmidt, W. J. Chem. Phys. 1974, 60, 4406-4418.

86. Akiyama, I.; Harvey, R. G.; LeBreton, P. R. J. Am. Chem. Soc. 1981, 103, 6330-6332.

87. Shahbaz, M.; Akiyama, I.; LeBreton, P. R. Biochem. Biophys. Res. Commun. 1981, 103, 25-30.

88. Cohen, S. S. in "Introduction to the Polyamines"; Prentice-Hall: Englewood Cliffs, 1971; pp. 29-63.

89. Hughes, M. N. in "The Inorganic Chemistry of Biological Processes"; John Wiley: New York, 1981, p. 258.

90. Freifelder, D. in "The DNA Molecule Structure and Properties"; W. H. Freeman: San Francisco, 1978 pp. 268-270.

91. Lewin, B. in "Genes"; John Wiley: New York, 1983; pp. 503-536.

RECEIVED May 13, 1985

A Mechanism for the Stereoselectivity and Binding of Benzo[a]pyrene Diol Epoxides to DNA

KENNETH J. MILLER, ERIC R. TAYLOR[1], and JOSEF DOMMEN

Department of Chemistry, Rensselaer Polytechnic Institute, Troy, NY 12181

A molecular model is proposed for stereoselectivity during the process of covalent binding of (+)-trans-7,8- dihydroxy- anti-9,10- epoxy-7,8,9,10- tetrahydro-benzo[a]pyrene, denoted by (+)-anti-BPDE or BPDE I(+); (-)-trans-7,8-dihydroxy-syn-9,10-epoxy-7,8,9,10-tetrahydrobenzo[a]pyrene denoted by (-)-anti-BPDE or BPDE II(-), and their mirror images (-)-anti-BPDE or BPDE I(-), and (+)-syn-BPDE or BPDE II(+) to DNA. Intercalation of the BPDEs places the reactive epoxide near N2 on guanine (G), O6 on G, N6 on adenine (A), and N4 on cytosine (C). A proton catalyzed nucleophilic S_N2 reaction between C10 of the BPDEs and base atoms and thus trans addition is favored during this step. These positive H^+ atoms reside in the most negative regions of the DNA, the grooves. The DNA must kink for proper hybridization to occur during intercalative covalent binding between C10 and the base atoms. Stereoselectivity occurs during this step. The I(+) and II(+) isomers are selected by N2(G), the I(-) and II(-) isomers by O6(G), N6(A), and II(+) and I(-) by N4(C). Reorganization of the DNA places BPDE I(+) and II(+) outside the helix (most likely the minor groove for the I(+)-N2(G) adduct), whereas the I(-) and II(-) remain intercalatively covalently bound. A less favored proton addition via an S_N1 reaction forms the carbonium ion of the BPDEs which allows both trans and cis base adducts. Cis addition accounts for minor products which are the I(-) and II(-) isomers with N2(G), the I(+) and II(+) isomers with N6(A) and O6(G) and II(-) and I(+) by N4(C). A combined analysis of experimental results and theoretical modeling calculations is presented with detailed stereographic molecular structures.

[1]Current address: Department of Chemistry, University of Southwestern Louisiana, Lafayette, LA 70504

0097–6156/85/0283–0239$13.00/0
© 1985 American Chemical Society

A combined experimental and theoretical approach has been undertaken
in many research laboratories to understand the relationship between
the carcinogenicity, mutagenicity, and structure of the polynuclear
aromatic hydrocarbons (PAHs). Although any relationship may be more
complicated than the emerging models, it appears that the studies of
one particular well-known carcinogen, benzo[a]pyrene, BP, are extensive enough to permit a somewhat comprehensive proposal for the
physical binding of its diol epoxides, BPDEs, to DNA. One to four
of the diastereoisomers have been involved in extensive research.
They are denoted by BPDE I(+) for the (+)-trans-7,8-dihydroxy-anti-
9,10-epoxy-7,8,9,10-tetrahydrobenzo[a]pyrene or (+)-anti-BPDE and by
BPDE II(-) for the (-)-trans-7,8-dihydroxy-syn-9,10-epoxy-7,8,9,10-
tetrahydrobenzo[a]pyrene or (-)-syn-BPDE in Figure 1. The BPDE
I(-) and BPDE II(+) are mirror images of the I(+) and II(-) isomers,
respectively. The C10 position is involved in adduct formation with
N2(G) and it is assumed to be involved with N6(A), O6(G) and N4(C).

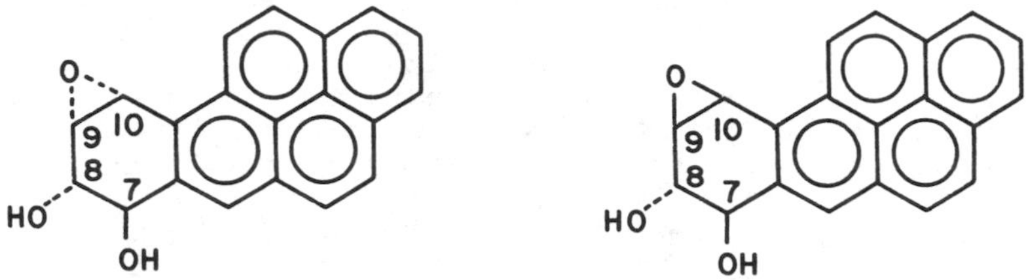

BPDE I(+)   BPDE II(-)

(+)-anti-BPDE   (-)-syn-BPDE

Figure 1. BPDE isomers.

This paper presents a unification of theoretical calculations
and experimental results in a proposed model for the stereoselectivity of the BPDEs by DNA. It is divided into five major sections: The Experimental Background presents a literature review of
the current status of the structural and conformational interpretation of the chemical and physical results for BPDE-DNA adduct formation. The Approach to the Theoretical Modeling outlines the overall considerations of the problem. The energetics of forming receptor sites, binding of the BPDEs to DNA and definitions of the energy
expressions used in the calculations are discussed in Steps in the
Binding Process. The Mechanism for BPDE-DNA Adduct Formation contains an analysis of each step in the proposed model. It constitutes the major part of the paper. Results from the literature, as
well as from our laboratory, are used to support the mechanism and
to interpret the experimental results at a molecular level. This

major section, which contains the study of all steps in the mechanism, is subdivided into six subsections: benzo-ring conformations
and hybridization of BPDEs and adducts report the dominant local
conformations which must be considered.   The possibility   of
rehybridization of amino groups is considered to insure that the
proper hybridization about C10 of the BPDE and the amino nitrogen
atoms of the bases are used.   Quantum mechanical calculations are
reported in these two subsections and they are compared with experimental results.   Then the modifications of the DNA are examined in
receptor sites in which classical intercalation, kinked DNA for
stereoselectivity and rearranged DNA to the final states of
externally and internally bound adducts are considered.   Molecular
mechanics methods are used to obtain the results reported in these
four subsections.   A summary of the entire mechanism, the stereoselectivity of BPDEs for trans and cis addition products and predictions and suggestions for experimentalists are presented in the
final major section Discussion and Conclusion.   Stereographic projections of adducts involving all four BPDE isomers are presented.
They illustrate the process of stereoselectivity by kinked DNA and
the final orientation of the pyrene moiety which depends on the
particular BPDE-isomer involved.

Experimental Background

The carcinogenicity of PAHs involves their conversion to active
metabolites (1), and their subsequent reaction with nucleic acids
(2-22).   An organism or organ may produce a unique set of
metabolites, each of which may be carcinogenic to the host or to
other systems.   These metabolites may be classified as follows (23-
25):   Primary or ultimate carcinogens, i.e., compounds that are
chemically or biologically active by virtue of their specific structure and conformation. Secondary or procarcinogens, i.e., compounds
which are chemically or biologically inert, but undergo conversions
which transform them to ultimate carcinogens.   In addition, the PAHs
or their metabolites may be promoting-agents or co-carcinogens,
i.e., non-carcinogens which serve to potentiate the effect of either
a primary or secondary carcinogen.   Research in this area has proceeded along several paths: (i) The identification of metabolites
of PAHs in biological systems, expecially the arene oxides, dihydrodiols and tetrahydrodiol-epoxides.   (ii) The determination of
the carcinogenicity and mutagenicity of these compounds.   (iii) The
identification of adducts formed with DNA and the details of the
structures and conformations of the complexes.   It appears that covalent bond formation between an active compound and cellular macromolecules may be an important step in the carcinogenic event (2,23-
26).
        The importance of the role of DNA in the genetic code has fo-
cused research on the structure of the active compound and the overall conformation of DNA as a receptor.   In addition the identification of benzo[a]pyrene as an important carcinogen (27) has stimulated extensive research to determine the origin and its activity.

This procarcinogen undergoes metabolic conversion to benzo[a]pyrene diol epoxides, BPDEs (5,28-31), which have been the focus of structural and conformational studies by theoretical and experimental methods. These chemically reactive BPDEs are involved in covalent binding to DNA (13-22).

In addition, covalent bond formation to the N7 ring nitrogen in guanine with ring opening and strand scission, and phosphorylation, which is a minor reaction pathway, are two other reaction processes which have been identified (8-11,32,33). However, the analysis of adducts to DNA with HPLC techniques entails hydrolysis with a subsequent loss of all but the BPDE bonded to the amino nitrogen of the bases, thereby focusing attention on these amino groups as active sites (13-22). Binding to N2 on guanine is most prevalent, and this has received much attention (13-20). Of particular interest is the structure of BPDE I(+) bound to N2 on guanine. To date, no crystal structure has been reported. The products resulting from the BPDE I(+)-N2(G) reaction with nucleic acids are mostly trans with a minor component of cis addition (13-17). In reactions of racemic BPDE II(±) with poly(G) the trans adduct constitutes the major component (22). It is believed that the structure and conformation of DNA and the adduct may explain the trans addition and stereoselectivity of the BPDE I(+) to the DNA. Meehan and Straub (34) and Pulkrabek et al. (35) have examined the stereoselectivity of BPDE I(+) and I(-) by DNA. BPDE I(+) binds most extensively to guanine with minor components involving adenine and cytosine. BPDE I(-) binds most extensively to adenine with minor components involving guanine. They (34) find that the level of covalent binding to N2 on guanine is of the same order as the mutagenic activity between the (+) and (-) anti diol epoxides in mammalian cells (36). This correlates with the stereoselectivity of BPDE I(+) for N2(G) (37).

In another study Brookes and Osborne (38) find that the induction of 8-azaguanine resistant mutants in Chinese hamster V79 cells by BPDE I(+) and I(-) depends on the extent of reaction with DNA and to the reaction products. For equal extents of reaction both were equally cytotoxic but BPDE I(+) was more mutagenic. The extent of binding of the I(+) isomer is 94% at N2(G), 4% to unidentified positions on G and 2% to N6(A), whereas the extent of binding of the I(-) isomer is 59% to N2(G), 21% to O6(G), 18% to N6(A) and 2% which may be O6(G) (8). However, the relative binding to DNA of I(+) and I(-) is 10:1 (34). They postulate that the difference between the I(+) and I(-) isomers results from the differences in spatial orientation. We show that stereoselective binding to N2(G) (37), N6(A) and O6(G) can occur for trans addition during the intercalative covalent binding mode, and that a minor component from cis addition yields a complementary stereoselective component. We postulate that external binding results through relaxation (denaturation and renaturation) of the DNA which allows the pyrene moiety to adjust itself to one of the grooves, but a combination of these processes may occur.

Greater insight into the conformational structure of the receptor site(s) is gained by a series of physical measurements. Lefkowitz et al. (39) performed an optically detected magnetic resonance study to show that the BPDE I(±) covalent adduct to DNA lies outside the helix, i.e., that it is not intercalated. Prusik and coworkers (40,41) find with fluorescence measurements that the BPDEs are bound externally. Drinkwater et al. (42) demonstrate the unwinding of supercoiled SV40 DNA by covalent binding of a series of hydrocarbon epoxides. Although not necessary for their work, they implied use of a classical intercalation hypothesis by comparison with the effect by ethidium bromide. Namely, parallel base pairs separated by 6.8 Å with the BPDE parallel to the base pairs and stacked at 3.4 Å. Consequently, reaction with the exocyclic amino groups (13-22) of G, A, and C presents a problem with the interpretation of a reasonable C10-nitrogen bond length and bond angles for proper hybridization about C10 and the amino group (43,44). Subsequent measurements of the orientation of the pyrene moiety and of the base pairs demonstrated that a classical intercalation site is not involved. In fact, for the covalently bound adduct Drinkwater et al. (42) and Gamper et al. (45) obtain unwinding angles of 22°±3° and 30°-35°, respectively, for relaxed DNA and Meehan et al. (46) interpret their unwinding angle of 13° for the physically bound BPDE I(+) (reversibly bound in a classical intercalation site). Thus BPDE unwinds DNA in two ways: by classical intercalation and by intercalative covalent binding.

In order to determine the orientation of the pyrene moiety and the base pairs relative to the average helix axis several approaches have been taken. Geacintov and coworkers (47) have found with electric linear dichroism spectra an upper limit of 35° for the orientation between the pyrene chromophore and the electric field vector. The angles obtained reflect average orientations of the long axis of the pyrene moiety relative to average orientations of all of the DNA bases. The DNA tends to align itself along the electric field. In order to interpret the experimental results with a physical model, it has been assumed that, on the average, the helical axis is aligned along the electric field vector. With an upper limit of 35° they proposed a model in which the pyrene chromophore is oriented on the outside of the DNA. To explain this orientation several theoretical models were proposed initially. Beland (48) and later Jeffrey et al. (49) adjusted the BPDE I(+) adduct to N2(G) to fit in the minor groove of B-DNA. Although the long axis of the pyrene moiety yields an orientation relative to the helix axis in qualitative agreement (ca. 50°-60°) with the experimental value of 35°, the steric contacts are extremely poor. To resolve this problem, either the DNA must be adjusted, i.e., unwound (43) or bent (44) to permit a better fit, or a new receptor site must be proposed.

In another recent study on the orientation of base pairs in calf thymus DNA modified by covalently bound BPDE I(+) Hogan, et al. (50) have shown that the helical axes of DNA segments are not oriented along the electric field. The orientation of the helix axis of DNA segments in the electric field yields γ(DNA) = 29° at 0.04 adducts per base pair, and the long axis of the pyrene in the BPDE adduct is oriented at α(BPDE) = 43° from the helix axis of modified DNA regions. This value agrees with that of 35° measured by Geacintov et al. (47) for an "average" helix axis of DNA aligned

with the electric field. The problem of interpretation is discussed
by Taylor et al. (44). Thus, Hogan et al. (50) postulated a kinked
receptor site with an intercalative covalently bound BPDE I(+) in
contrast to Geacintov and coworkers (47) who postulate external
binding. Alternate interpretations of this data may include local
denaturation of the duplex or some form of DNA damage. However, if
the DNA remains double stranded at the site of covalent adduct formation, then a model for the stereoselectivity of the BPDE can be
presented. Theoretically determined receptor sites for kinked DNA
were generated by Taylor et al. (44) to model this type of binding.
The kinked DNA accommodates the covalent bonding region with tetrahedral and trigonal hybrid configurations about the C10 of BPDE I(+)
and N2(G), respectively. However, the long axis of the pyrene is
nearly perpendicular to the helical axis, i.e., α(BPDE) > 80° in all
structures examined. An alternative model presented in this paper
shows that the data of Hogan et al. (50) and Geacintov et al. (47)
are consistent with an externally bound adduct to a deformed DNA
kinked to accommodate the pyrene in the minor groove. None the less
the receptor site kinked to accommodate an intercalative covalent
bound adduct yields stereoselectivity for the I(+) diastereoisomer
(of all four studied) (37) in an intermediate step which has not
been identified experimentally. Because the linear dichroism
results for the orientation of the pyrene moiety in the BPDE I(+)-
DNA adduct are very similar in experiments performed by the electric
field and flow techniques (51) it is tempting to assume that the
angles reflect measurements relative to the "average" DNA axis.
Thus, there are two interpretations to the orientation of 35°-43°
for the pyrene moiety: a purely externally bound adduct and an
intercalative covalently bound adduct. In our proposed mechanism
for I(+)-N2(G) adducts, the intercalative covalently bound form in a
kinked receptor site has an orientation α(BPDE) > 80°, whereas the
externally bound forms in a relaxed B-DNA type structure have values
of α(BPDE) $\approx$ 15° for the G base in the anti conformation and
α(BPDE) $\approx$ 50° for G in the syn conformation.

   The non-covalently bound BPDEs to DNA formed initially appear
to be intercalation complexes (46,52-55). Meehan et al. (46) report
that the BPDE intercalates into DNA on a millisecond time scale
while the BPDE alkylates DNA on a time scale of minutes. Most of
the BPDE is hydrolyzed to tetrols (53-56). Geacintov et al. (54)
have shown with linear dichroism spectral measurements that the disappearance of intercalated BPDE I(+) is directly proportional to the
rate of appearance of covalent adducts. These results suggest that
either there may be a competition between the physically non-
covalently bound BPDE I(+) and an externally bound adduct or as suggested by the mechanism in the present paper, an intercalative covalent step followed by a relaxation of the DNA to yield an externally bound adduct. Their results for the BPDE I(-) exhibit both
intercalative and externally bound adducts. The linear dichroism
measurements do not distinguish between physically bound and covalent bound forms which are intercalative in nature. Hence the assumption that a superposition of internal and external sites occurs
for this isomer.

The binding of I(-) to base atoms associated with the minor groove, specifically N6(A) (38), O6(G) and N4(C), have been observed (13-22,34,38). We propose that the DNA can be kinked to allow intercalative covalent binding to these atoms, but that the final orientation may be in such a site or externally bound to the DNA and that this final site may also be kinked.

Geacintov and coworkers (51,57) have studied the orientation of the pyrene moiety in BPDE I(+) and I(-) adducts with DNA and Undeman et al. (58) have studied the orientation of BPDE I(±) and II(±) to DNA. They interpret their electric linear dichorism measurements in terms of a superposition of two types of binding sites which they label site I and site II. To avoid confusion with the notation for our intercalation sites, their sites will be referred to as IQ and IIX. In site IQ the complex has the pyrene moiety oriented within 30° of the DNA bases, i.e., 60°-90° relative to the helix axis. They call this quasi intercalation. We refer to it as intercalative covalent binding. It results after intercalation and kinking of the DNA and covalent bond formation. In site IIX the complex has the long axis of the the pyrene moiety oriented 15°-27° for BPDE I(+) and 37°-45° for BPDE I(-) relative to the helix axis. They interpret this as external binding. They attribute site IIX binding to N2(G) and Site IQ binding to O6(G) and N6(A) in agreement with the interpretation of Brookes et al. (38), that the difference in mutagenicity between the I(+) and I(-) isomers results from differences in spatial orientation. Specifically, the I(+) isomer exhibits site IIX binding with 88-94% bound. The I(-) isomer exhibits site IIX binding with 50-60% bound, the remaining being attributed to site IQ.

The interpretation of experimental data and presentation of a detailed molecular model can be accomplished only after certain assumptions have been made about the alignment of the DNA by an electric field and in flow techniques. (1) The helix axis of the DNA is oriented along the electric field and flow axis in exactly the same manner. For DNA with superturns this assumption implies only the "average" helix axis. (2) The conformational change through the receptor site is smooth. That is, there are no sharp bends which may orient the pyrene moiety along the electric field or flow axis while it is quasi intercalated, implying that it is externally bound. (3) The average helical axis through the receptor site lies along the electric field or flow axis so that the bending is symmetrical about the kink. In Figure 2, an idealized orientation of the BPDEs in both bound forms is illustrated with one specific example. For trigonal hybridization about N2(G) and tetrahedral hybridization about C10(BPDE), values of the kink, α_x, γ(DNA) and α(BPDE) are indicated. However, in practice the pyrene moiety is not parallel to the adjacent base pair, and its long axis does not lie in the plane of the local helical axes, $\vec{l}$ and $\vec{l}'$, and the average helical axis, $\vec{E}$. Thus, α(BPDE) > 65°, the idealized value. Similarly for the outside form, α(BPDE) >20°. In this paper, a molecular model is presented in which α(BPDE) $\approx$ 80° for the intercalative covalently bound form, and α(BPDE) $\approx$ 15° and γ(DNA) $\approx$ 30° for the externally bound form. Both forms are bound to DNA with different kink conformations. In summary, a kink in the DNA need not imply intercalative

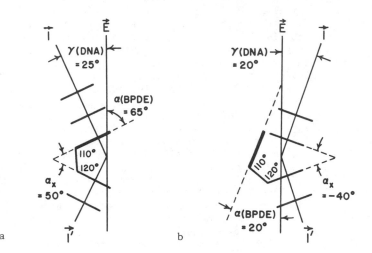

Figure 2. Geometry for idealized orientations of the pyrene moiety
(---) in (a) an intercalative covalently bound form and in (b) an
externally bound form.   The average helix axis is assumed to lie
along $\vec{E}$. The local segments of DNA lie along $\vec{l}$ and $\vec{l}'$. They are
oriented by γ(DNA) relative to $\vec{E}$ and the long axis of the
pyrene moiety is oriented by $α_x$(BPDE) relative to $\vec{E}$.

covalent binding.   Outside binding can also kink the DNA.   The interesting feature of the intercalative covalent intermediate is that
stereoselectivity can be shown to arise during this step.

Approach to the Theoretical Modeling

A theoretical analysis is presented for the binding of the four diastereoisomers of benzo[a]pyrene diol epoxides (BPDEs) to N2(G),
N6(A), O6(G) and N4(C).   Molecular models for binding and stereoselectivity involving intercalation, intercalative covalently and
externally bound forms are presented.   Molecular mechanics calculations provide the energetics which suggest possible structures for
the formation of each of the principal DNA-BPDE complexes.   Stereographic projections are used to illustrate the molecular structures
and steric fits.   The results of previous calculations on intercalation and adduct formation of BPDE I(+) in kinked DNA (37) are
summarized and extended to include the four diastereoisomers I(±)
and II(±).   The theoretical model is consistent with the observed
experimental data.
        A mechanism for the stereoselectivity of the diastereoisomers
of BPDE by duplex DNA is postulated.   It entails a series of processes: The diol epoxide intercalates with the epoxide oriented

either in the major or minor groove and adjacent to the reactive atoms of the bases. For a negatively charged DNA the most negative electrostatic potentials are in the grooves. Counterions lie in the grooves, and therefore, hydrogen ions are available for the conversion of the BPDEs to reactive carbonium ions. The DNA is shown to be flexible enough to kink about the receptor site. Adjacent base pairs undergo a change from a parallel orientation to a bend or kink by α_x = +39° and -30° about the x-axis, or kink axis, which is perpendicular to the average helix and the dyad axes through the receptor site. This permits adduct formation between C10 of BPDE and one of the base atoms with a bond length of ca. 1.5 Å and proper hybridization about the reacting atoms. The boat (B) and chair (C) forms of the benzo ring with diaxial (da) and diequatorial (de) 7,8-dihydroxy groups of each of the four diastereoisomers are adjusted in the appropriate receptor site with a minimization of the intra- and intermolecular energy. The general trends in adduct formation favor binding on the 5´ side of purines (pu) and on the 3´ side of pyrimidines (py), because the py(p)pu sequences form intercalation and kink sites most readily. For <u>trans</u> addition, the benzo[a]pyrene moiety can be oriented very favorably within the receptor site from one side (top or bottom) only. That is, either the BPDE I(+) and II(+) or the I(-) and II(-) isomers can fit in the kinked receptor site during adduct formation. This trend becomes apparent when the structures are viewed with stereographic projections. The type of addition, <u>trans</u> or <u>cis</u>, and orientation of the pyrene moiety play a major role in the selection of one of the enantiomers of the diastereomers, I and II. Optimum adduct formation occurs for BPDE I(+) and II(+)-N2(G) adducts for both the (3´) and (5´) orientation about the bases; for BPDE I(-) and II(-)-N6(A) and O6(G) adducts for the (5´) orientation; and for BPDE I(-) (5´-orientations) and BPDE II(+) (3´-orientations) adducts to N4(C). The process involving intercalation followed by intercalative covalent binding is postulated to be the step in which stereoselectivity of the BPDEs by duplex DNA occurs. For <u>cis</u> addition, stereoselectivity for the complementary isomers occurs, i.e., the benzo[a]pyrene moiety can be oriented favorably in BPDE I(-) and II(-) adducts with N2(G), and in BPDE I(+) and II(+) adducts to N6(A) and O6(G). The <u>trans</u> addition is favored over <u>cis</u> via an S_N2 mechanism, and consequently, the first case is observed.

There are two possibilities for the final conformation in which the adduct is externally bound. Both entail denaturation of the DNA, and adjustment of the adduct and a renaturization of the DNA. In the first the DNA resumes a B-type conformation after the base involved in adduct formation undergoes an <u>anti</u> → <u>syn</u> rotation about the glycosidic bond of the affected base to place the pyrene in the major groove. In the second the DNA kinks to allow the adduct to lie in the minor groove. In the latter case the orientation of the pyrene moiety relative to the helix axis agrees better with available experimental results. This change converts the complex from one in which the adduct is in an intercalative covalently bound form in a kinked site to one in which the adduct is externally bound. The distinguishing feature between these two possibilities is that the pyrene moiety lies in opposite grooves for binding to N2(G). This <u>anti</u> → <u>syn</u> modification produces a structure

with very favorable non-bonded contacts for adduct formation between
BPDE I(+) on G, somewhat poorer non-bonded contacts for BPDE II(-)
and N6(A) and O6(G) and a very unfavorable structure with BPDE II(-)
and N4(C) unless the DNA is allowed to undergo a kink from the overall B-DNA conformation.   The theoretical model presented   explains
the physical and chemical data reported on the binding of the diastereoisomers of the BPDEs to DNA.

Steps in the Binding Process

The binding of the BPDEs for adduct formation to DNA may be viewed
in terms of the ease in which the DNA can undergo a change to the
proposed receptor site, the ability of the individual diastereoisomers to fit into the receptor site and the formation of a covalent bond.   A stepwise procedure is used in the analysis, and the
energy required to accomplish the first two steps is calculated.
The process which is used to define the change begins with the alteration of B-DNA to an intercalation site (ν) followed by a bending
of the DNA to produce a kinked site (κ)

$$\uparrow \text{BP}_2 \downarrow \quad \xrightarrow{\Delta E_{DNA}^{\nu}} \quad \uparrow \text{BP}_2 \downarrow \quad \xrightarrow{\Delta E_{DNA}^{\kappa} - \Delta E_{DNA}^{\nu}} \quad \uparrow \text{BP}_2 \downarrow$$
$$\text{BP}_1 \qquad\qquad\qquad \text{BP}_1 \qquad\qquad\qquad\qquad \text{BP}_1$$

where

$$\Delta E_{DNA} = E_{DNA}^{R} - E_{DNA}^{B\text{-}DNA} \tag{1}$$

is the energy required to form a receptor site R (ν and κ) relative
to B-DNA.   It should be emphasized that the calculations   performed
to determine detailed molecular structures and conformations deal
with the thermodynamic process.   Only the steps of importance to
binding, stereoselectivity and final orientation are considered.
Detailed analyses of the pathways and specifically, the continuous
adjustment of the DNA to achieve each of the receptor sites require
costly computations.   They should be performed after the importance
of each receptor site has been established.
    Several mathematical techniques have been used to obtain atomic
coordinates for nucleic acid structures.   They incorporate several
different approaches.   (a) Systematic rotations about all backbone
torsional angles are performed and those conformations which form
helicies and have adjacent bases parallel or in a given orientation
are selected (59,60).   (b) Least squares techniques are used to relax a DNA fragment until constraints for the geometrical conditions
of proper bond lengths, bond angles with a specific placement of
bases are satisfied (61,62).   (c) Systematic adjustment of bases

through parallel, kinked or any other orientation of interest to
search for backbone completion (63-66). A large number of duplex
DNA receptor sites have been obtained for the intercalation process,
several dinucleoside monophosphate receptor sites have been used for
modeling BPDE-DNA interactions, and only one set of kinked DNA sites
are available (36,66). The question arises: which receptor sites
should be used in the modeling of the binding of molecules with DNA?
In principal, it cannot be answered without trying all of the them.
In practice, calculations with theoretically determined intercalation sites (67-71), calculations using experimental data (72,73),
dinucleoside triphosphate units (74,75), tetramer duplexes which fit
into B-DNA (61,62,66-68), and tetramer-duplexes (76) with the same
and with mixed sugar puckers yield results which demonstrate for
simple systems that the optimum binding orientations are in agreement with experimental data and that the trends in binding energies
for all possible sequences remain approximately the same. The
tetramer duplex receptor sites used in this study have the following
characteristics: (1) they fit into B-DNA, (2) they represent conformations which possess specific features, i.e., intercalation,
kinks or DNA distorted from the B-DNA form. Therefore, we proceed
with the assumption that the intercalation process will reveal the
manner in which the reactive atoms approach each other to permit intercalative covalent binding as a first step followed by an adjustment of the base pairs to non-planar orientations (kink) to accommodate the proper hybrid configurations about C10 on BPDE and
N2(G), N6(A), O6(G) and N4(C) for covalent bond formation. For
binding, two orientations along the DNA must be considered: 3´ and
5´ binding. They are illustrated with the BPDEs bound to a purine
(pu) and to a pyrimidine (py).

$$3´\uparrow \overset{BP_2}{\diagdown} \downarrow \qquad \uparrow \overset{BP_2}{\diagup}\text{-}5´ \qquad \uparrow \overset{BP_2}{\diagup}\text{-}5´ \qquad 3´\uparrow \overset{BP_2}{\diagdown} \downarrow$$
$$pu\cdot py \qquad\qquad py\cdot pu \qquad\qquad pu\cdot py \qquad\qquad py\cdot pu$$

The arrow denotes the 5´ → 3´ direction along the backbone, i.e.,
5´(sugar)3´ or equivalently 3´(p)5´. The diagram corresponds to
viewing the DNA into the minor groove as seen in most figures in
this paper. Therefore, the energy of opening the DNA to sites
favoring 3´ or 5´ binding provides the contribution of the DNA to
the process.
     Once the site is created, the ability of each BPDE to fit into
the receptor is examined. Two processes are considered: intercalation and covalent intercalative binding represented by

$$\uparrow \overset{BP_2}{\underset{BP_1}{}} \downarrow + BPDE \xrightarrow{\Delta E^R_{BPDE}} \uparrow \overset{BP_2}{\underset{BP_1}{BPDE}} \downarrow$$

The energy

$$\Delta E^R_{BPDE} = I^R_{BPDE} + \Delta C^R_{BPDE} \tag{2}$$

represents the energy for insertion of BPDE into site R. It is calculated with a Coulomb potential

$$Q = \sum_{i,j} q_i q_j / r_{ij} \tag{3}$$

a 6-14 potential

$$U = \sum_{i,j} \{-A_{ij}/\rho^6_{ij}\}\{s^{-6}_{ij}-(3/7)s^{-14}_{ij}\} \tag{4}$$

and a torsional potential

$$T = \sum_b (V_b/2)[1 + \cos \ell_b \phi_b] \tag{5}$$

for interacting atoms i and j and for rotations about each bond b. Associated with the atoms are the net atomic charges q_i and q_j, and the van der Waals radii ρ_i and ρ_j. The distance between atoms i and j is r_{ij}, the sum of van der Waals radii is $\rho_{ij} = \rho_i + \rho_j$, the reduced distance is $s_{ij} = r_{ij}/\rho_{ij}$, $A_{ij} = 1.5\alpha_i\alpha_j I_i I_j/(I_i+I_j)$ where α_i, α_j and I_i, I_j are the atomic polarizabilities and ionization potentials respectively, V_b is the barrier to internal rotation and ℓ_b is the periodicity for a complete rotation about bond b.
     The intermolecular energy I^R_{BPDE} consists of a sum Q+U. The summation in the two-body interactions proceeds over all atoms i in the BPDE and j in the DNA. The intramolecular interactions between non-bonded atoms in BPDE as well as the torsional energy are given by the change in conformational energy, ΔC^R_{BPDE}, measured relative to the global minimum for the free molecule. It consists of a sum of terms Q + U + T. In this case the summation proceeds over non-bonded atoms i and j in different fragments of BPDE, i.e., fragments separated by rotatable bonds. The conformational energy of the benzo ring is also included. For this contribution the restrictions are made that i>j and that atoms i and j are non-bonded. The definitions and the details of the energy terms and the parameters are presented elsewhere (69). The total energy change of the complex is given by

$$\Delta \varepsilon^R_{BPDE} = \Delta E^R_{DNA} + \Delta E^R_{BPDE} \tag{6}$$

For convenience, the energy will be reported relative to the global minimum as

$$\delta \varepsilon^{R}_{BPDE} = \Delta \varepsilon^{R}_{BPDE} - (\Delta \varepsilon^{R}_{BPDE})_{min} \qquad (7)$$

For the covalently bound adduct the binding energy must be added to to ΔE^{R}_{BPDE}. For the DNA, only a conformational change occurs and $\Delta E^{R}_{DNA} = \Delta C^{R}_{DNA}$ where the global minimum is assumed to be for B-DNA.

By calculating the energy changes for the intercalation process, the preference for orientation of the epoxide in the major or minor groove can be determined. This step provides a rationale for the first step in covalent binding by examining whether the reacting atoms are in proximity to each other. The important step in stereoselectivity is the ability of the BPDEs to fit into the site once a covalent bond has formed. The conformations of the benzo ring are important in the fit of each of the diastereoisomers to the kink site during adduct formation. They modify the orientation of the double bond of the benzo ring and hence the entire pyrene moiety.

Mechanism for BPDE-DNA Adduct Formation

Formation of the BPDE-DNA adduct requires a study of (1) Benzo ring conformations of BPDEs and adducts; (2) The rehybridization of amino groups on the benzo for the C10-N bond formation; (3) The receptor sites resulting from a conformational adjustment of DNA to accommodate an intercalated and finally an intercalative covalently bound BPDE, and the base sequence specificity in the formation of the receptor site; (4) Classical intercalation and the orientation of C10 of the BPDEs toward the reactive N2(G), N6(A), O6(G) and N4(C) base atoms; (5) The stereoselectivity of the BPDEs during intercalative covalent binding in kinked DNA; and (6) The possible reorientation of the complex to yield an externally bound adduct. The energetics for each of these processes will be presented to identify the important steps that influence the binding of specific isomers. It will be shown that the orientation of each diastereoisomer of BPDE about specific base atoms in kinked receptor sites in the duplex DNA during covalent bond formation is the determining factor in stereoselectivity.

The parameters which define the orientation of the BPDE adduct to N2 on guanine are given in Figure 3 in terms of the reaction coordinates R, α, β, γ, δ and ε. For binding to N6(A), O6(G) and N4(C), β is redefined by replacing N3(G) by N1(A), N1(G) and N3(C), respectively. For adduct formation, coordinates internal and external to the ring, must be considered. The ring pucker is determined by the torsional angles τ_n. These are internal coordinates. The reaction coordinates orient the entire BPDE. They are external coordinates and may be classified into three groups: The idealized reaction coordinates R = 1.5 Å, α =120° and δ = 109.5° are assumed to define a stable bond with proper hybridization about N2(G),

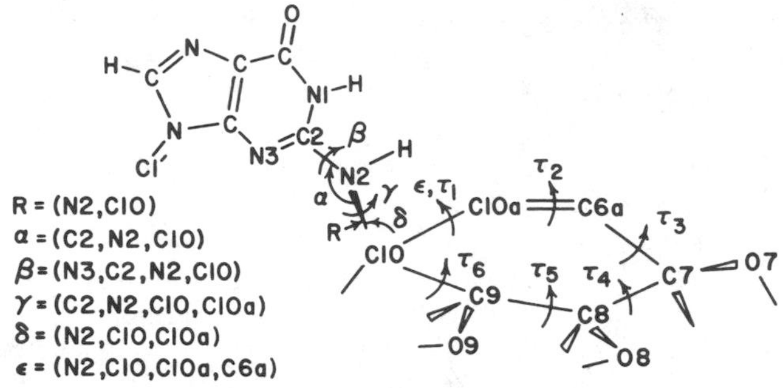

R = (N2,C10)
α = (C2,N2,C10)
β = (N3,C2,N2,C10)
γ = (C2,N2,C10,C10a)
δ = (N2,C10,C10a)
ε = (N2,C10,C10a,C6a)

Figure 3.   Reaction coordinates for trans addition of BPDE I(+) to
N2 on guanine to yield a triol adduct, BPT-N2(G).   The torsional
angles τ_1, τ_2, τ_3,... define the pucker of the benzo ring.   For
binding to the other bases, N3(G) is replaced by N1(A), N1(G) and
N3(C) for binding to N6(A), O6(G) and N4(C), respectively.

N6(A), O6(G) and N4(C) and C10 after adduct formation in our molecular mechanics calculations. The reaction coordinates β and γ orient
the pyrene moiety independently of the particular ring conformation
or isomer.   These angles reflect an overall adjustment of the BPDE
relative to the DNA.   The critical parameter is ε which is the only
parameter that depends on the ring conformation. It orients the pyrene moiety to the base through the sequence (N2,C10,C10a,C6a) for
BPDE I(+) bound to N2(G) and similarly for the other isomers.   Its
dependence on the internal torsional angles for trans addition is
$\varepsilon = \pm(120°+\tau_1)$ for the (±) isomers respectively where τ_n is defined
for the (+) isomer. For cis addition $\varepsilon = \pm(-120°+\tau_1)$.   In this pa-
per 109.3° < |ε| < 130.7°.   When an interaction with DNA occurs,
there is an interplay between the minimum energy of the benzo ring
and the ability of the pyrene moiety to fit into the receptor site.
However, the change in conformational energy is small so that the
steric fit in the receptor site becomes the determining factor in
binding. For a given ring conformation with τ_n and ε, the optimization proceeds with the remaining reaction coordinates β and γ.

Benzo-ring conformations and hybridization of BPDEs and adducts. In
order to understand the geometry, possible conformations of the diol
epoxides and BPDE-N2(G) adducts and relative energies of the various
conformations and isomers, quantum mechanical calculations were performed with the MINDO/3 approximation (77). A double precision version adapted to the IBM 3081D and modified with a damping procedure
to insure convergence was used (78). These calculations are used as
a guide to the selection of a set of idealized bond angles for
hybridization about C10 of BPDEs and the base atoms for all of the
calculations. Two model systems studied denoted by structures I and
II are shown in Figure 4.   Structure I with this hydroxyl group at
C7 trans to the epoxide is termed anti, while the structure II with
the hydroxyl group on the same side as the epoxide with respect to
the plane of the ring is termed syn.   These two diol-epoxides of
benzene (BDE) are chosen as models for the anti and syn isomers of

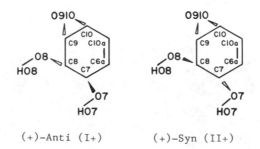

(+)-Anti (I+)   (+)-Syn (II+)

Figure 4.   Anti and syn diastereoisomers of the benzo diol-epoxide
(BDE).   For the 7,8-diol-9,10-epoxide of benzo[a]pyrene (BPDE),
these would be the (+)-anti or I(+) and (+)-syn or II(+) isomers as
indicated.   Atom numbers for BDE are those for the diol epoxide of
benzo[a]pyrene.

the 7,8-diol-9,10-epoxide of benzo[a]pyrene to reduce the cost of
the computations.   For both structures, an optimization of the
geometry for all four possible conformers was carried out.   By main-
taining (C7,C6a,C10a,C10) planar because of the double bond, the
rigidity of the pyrene moiety is modeled.   The distances, bond
angles, and the dihedral angles were allowed to vary, except for
those held rigid by the planar sequence, and fixed C-H, O-H, and C-O
(in C-O-H only) bond lengths.   Sample calculations for optimized
non-planar configurations about this double bond, and for changes in
the C-H and C-O-H bond lengths had very little effect on the energy
and no effect on the energy changes.   The hybrid configuration about
the epoxide reaction site and the ring conformation has been taken
into account properly.
    For the isolated benzo ring, the conformational angles, bond
angles and relative energies for the sixteen possibilities are reported in Table I.   Thus, eight conformations of chair (C) and boat
(B) with O7 and O8 in the diaxial (da) and diequatorial (de) orientations for the I(+) and II(+) isomers yield relative minima for the
benzo ring in this study.   They are illustrated in Figure 5.   The
I(-) and II(-) isomers yield an additional eight by replacing τ_n by
$-\tau_n$, i.e., the mirror images.
    To assign the terms boat, chair or half-chair, diaxial or di-
equatorial to the various conformations of the benzo ring of the
diol-epoxide, the following definitions based on values of the conformational angles were used.   Four different conformers were defined as follows: the same signs of the dihedral angles τ_3 =
(C8,C7,C6a,C10a) and τ_1 = (C6a,C10a,C10,C9) denote a half-chair (C)
form, and the opposite signs yield a boat (B) form.   Because of the

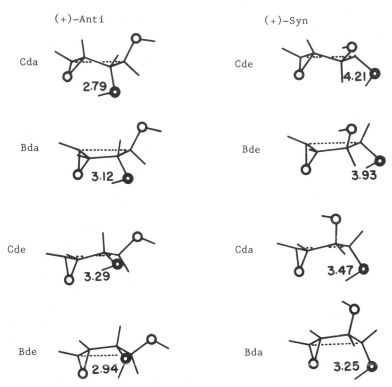

Figure 5.   Optimum conformations of diaxial (da) and diequatorial
(de) forms of the chair (C) and boat (B) structures of the (+)-anti,
I(+), and (+)-syn, II(+), benzo-ring diol-epoxide. They are ordered
with increasing energy (bottom to top).   The dotted line represents
the planar sequence (C10,C10a,C6a,C7). The epoxide oxygen and O8 or
O7 distances are given for the anti and syn isomers, respectively.

double bond and the rigid epoxide ring, the hydroxyl groups attached
to C7 and C8 can assume only a somewhat diaxial or diequatorial
position to each other.   A dihedral angle |(O7,C7,C8,O8)| > 111° is
designated as a diaxial, and less than 111° as diequatorial.   These
were obtained by minimizing the energy of conformations chosen
initially in one of the four possible extremes.   Thus, each conformation represents a relative minimum.   The anti isomer prefers a
Bde form, whereas, the syn isomer is in the Bda form.   The result
for the anti isomer is in agreement with experimental findings (79,
80) of the anti diol-epoxide of benzo[a]pyrene as seen in Table
I, with force field calculations by Silverman (81) and in results
reported by Lavery and Pullman (82) and Kikuchi et al. (83).   The
anti isomer has three conformations, Bde, Cde, Bda which lie within
2.1 kcal of the minimum, while the Cda conformation seems to be in a
relative minimum which is very unfavorable.   The conformer classified as Bda has a conformational angle, (O7,C7,C8,O8) = -116.4°,
in contrast to -89.1° for the Bde conformation.   The syn isomer has

two diaxial conformers, Bda and Cda within 1.4 kcal of the minimum. The diequatorial Bde and Cde conformations lie 3.5 and 7.8 kcal above the Bda conformer. Lavery and Pullman (82) give the energy difference between the lowest diequatorial and diaxial conformations as 2.9 and 4.1 kcal/mol for the anti and syn isomers, respectively. Kikuchi et al. (83) obtain the differences of 3.3 and 4.6 kcal/mole,

Table I. Conformational Energies and Average Benzo Ring Conformations of BPDEs Calculated with the MINDO/3 Approximation[a]

Isomer	Ring[b]	ΔC[c]	τ_1	τ_2	τ_3	τ_4	τ_5	τ_6	τ_d
I(+)	Cda	11.1	10.7	0.0	11.5	-30.7	39.8	-32.0	-136.3
II(+)	Cde	7.8							83.7
I(+)	Cde	2.3	-7.6	0.0	-6.5	19.4	-25.2	20.4	-81.7
II(+)	Cda	1.4							136.8
I(+)	Bda	3.0	-21.1	0.0	13.8	-7.3	-11.9	26.3	-116.4
II(+)	Bde	3.5							110.2
I(+)	Bde	0.9	15.6	0.0	-21.3	27.1	-12.1	- 8.6	-89.1
II(+)	Bda	0.0							133.4
I(+)	Bde	exp[d]	19.4	-6.6	-24.8	44.2	-34.0	1.9	-67.9
I(+)	Bde	exp[e]	10.6	-6.4	-35.1	47.1	-30.0	2.0	-
I(+)	Bde	FFC[f]	27.5	-3.1	-40.0	59.9	-37.5	-5.2	-56.2

a   The angles are in degrees and defined in Figure 3. The isomers I(-) and II(-) are obtained by replacing τ_n by $-\tau_n$, n = 1, 2, 3,...6, and d.
b   The Chair (C) and Boat (B) forms occur when τ_1 and τ_3 have the same and opposite signs, respectively. They may be in diaxial (da) or diequatorial (de) arrangements. τ_d = (O7,C7,C8,O8), the torsional angle which defines da ($|\tau_d|$ > 111°) and de ($|\tau_d|$ < 111°).
c   ΔC (kcal/mole) is measured relative to Bda for the II(+) isomer.
d   Experimental crystal structure (79).
e   Deduced from the experimental crystal structures for diols and epoxides of BP (80).
f   Force field calculation on the epoxide (81).

respectively. Our values of 2.1 and 3.5 kcal/mol are in good agreement. On an absolute scale the syn isomer is slightly more stable by 0.9 kcal/mol than the anti isomer. An ab initio SCF (Gaussian 70, STO-5G) calculation by Politzer et al. (84), MNDO calculations by Klopman et al. (85), and CNDO/2 calculations by Kikuchi et al. (83) give values of 2.0-3.3 kcal/mol. These results suggest from a thermodynamic point of view that there is virtually no preference for trans or cis epoxidation of the 7,8-diols to form the anti or syn diol-epoxide. Nevertheless, differences in the ratio of isomers formed may occur in the metabolic pathway because of steric interactions with the enzymatic system. Therefore, it appears that the enzymatic receptor site eventually must be factored into the calculation.

The conformational stabilization by hydrogen bonding of diol epoxides has been studied. The MINDO/3 approximation to molecular orbital theory is incapable of describing hydrogen bonding (86,87). Water dimers, for example, exhibit no minimum. To study the possibility of hydrogen bonding, we used the net charges of BDE with a Coulomb potential and a 6-14 potential in a semi-empirical procedure that is described elsewhere (69). The proximity of O910 to O8 in the anti isomer and to O7 in the syn isomer as seen in Figure 4 and 5 indicates that both isomers are potentially capable of intramolecular hydrogen bonding. The Bda and Cda conformations of the syn isomer are excellent candidates for hydrogen bonding between the epoxide O910 and HO7, and the Bde and Bda conformations of the anti isomer have the O910 and HO8 atoms in position for possible hydrogen bonding. Because the sum of van der Waals radii of the oxygen atoms is approximately 3.0 Å, these conformations were chosen for study.

As the hydrogen atom is rotated toward the epoxide in the syn isomer an improvement of -0.8 kcal occurs for the Bda conformation. The distance O910···HO7 is 2.61 Å. The Cda conformer which has a O910···HO7 distance of 2.92 Å still yields an energy change of -0.5 kcal. Lavery and Pullman (82) and Politzer, et al. (84) report hydrogen bond energy of approximately 1.7 kcal, for O910···HO7 bonds lengths of 1.98 Å and 2.25 Å, respectively. They find no hydrogen bonding in the anti isomer between the epoxide and the HO8. We applied our semi-empirical approach to the Bde and Bda conformers. In both cases the energy decreased only by -1.1 and -0.5 kcal, and the distances between HO8 and O910 are 2.46 Å and 2.84 Å. The improvement in energy in all conformations is less than expected for hydrogen bonding. The reason that the syn isomer didn't show such a high hydrogen bond energy as was found by others (82,84) may be the larger distance between the epoxide and O910 in our structures. In the case of the anti isomer the ab initio calculations of Politzer et al. (84) show a slight repulsion, but at a distance of 2.96 Å which corresponds approximately to the sum of van der Waals radii of the atoms. In all the conformations of anti and syn isomers, the O910···HO8 and the O910···HO7 sequences are not linear, and an analysis suggests that at most, the possibility for hydrogen bonding exists in the anti isomer. The improvement in energy obtained in all conformations with the use of semi-empirical techniques is not only less than expected for hydrogen bonding, but less than the energy differences between the various conformations of the anti and syn forms given in Table I. Therefore, the minimum energy conformers of the BDEs are determined by the ring system and by the

steric repulsions of the oxygen atoms and not by hydrogen bonding.
To model the adduct between BPDE and base atoms, the geometry
of the BDE adducts to N2(G) for the corresponding triol was optimized by varying the same parameters as in the case of BDE, as well
as the set of reaction coordinates and benzo ring conformations R,
α, β, γ, δ, ϵ, and τ_n defined in Figure 3. The reaction between C10
of the anti or syn isomer of BPDE and base atoms can take place
either on the same side (cis) or on the opposite side (trans) of the
epoxide to yield four different products. The notation syn and anti
is used in this investigation to designate the isomer of BPDE involved in the reaction and not that of the adducts. Each of the
isomers of the adduct may again assume four different conformations
of the 7,8-diol: Cde, Cda, Bde, and Bda. Relative minimum energy
conformations still occur when the BDE-N2(G) adduct locks into each
of the four different conformations of each isomer as found for BDE
alone. The relative energies change because of the mutual interaction of the BDE with guanine. Although the six reaction coordinates were varied (along with parameters of the BDE), the complica-
tion of convergence to a minimum was reduced by the fact that steric
constraints and hybridization limited the range of values of α
(~120° for trigonal N2), δ (~109° for tetrahedral C10), ϵ (~±120°
for tetrahedral C10) and R (~1.5 Å for a C2'-N2 bond). Finally, γ
assumes several possibilities which depend on the steric interactions between the BDE and guanine moieties. These starting values
very quickly give the relative minimum energies reported in Table II
and conformations listed in Table III.
The most constrained conformation, Cda, of anti BDE becomes the
preferred conformation for its cis and trans adducts, whereas, the
minimum energy conformation, Bda, of the syn isomer of BDE is the
most stable form in the reactants and products. A large decrease in
energy occurs for the Cde conformer relative to the Bda conformer
when the syn isomer of BDE reacts to form the BDE-guanine adduct,
while the other energies change less dramatically. No theoretical
calculations are available for comparison. The relative stability
of cis and trans adducts of the free BDE-N2(G) adduct and the observation that trans addition occurs in DNA suggests that the DNA plays
a role in the type of addition product. To model the mechanism of
adduct formation of BPDEs to N2(G), Politzer et al. (89) performed
ab initio Gaussian 70 calculations on the addition of water to
ethylene oxide. They find that protonation of the epoxide facili-
tates ring opening, and that cis addition is preferred over trans by
12 kcal/mol because of internal hydrogen bonding between the water
molecule and epoxide. The flexibility of water in this reaction may
not be available to the bases in DNA. Another mechanism has been
proposed by Loew et al. (90) who studied the attack of water on a
protonated benzene diol epoxide with the MINDO/3 molecular orbital
approximation. Their potential energy surfaces show that the activation energy for a proton assisted S_N2 reaction with water is
favored by about 6 kcal over carbonium ion formation followed by
addition of water. If one assumes that the same mechanism applies
to adduct formation to N2(G) and other base atoms then the trans
rather than cis addition products should be slightly favored with
S_N2 addition. However, if the DNA influences the protonation of the
diol epoxide to form a carbonium ion then some cis adduct should
form in an S_N1 reaction. However, no detailed theoretical calculation explaining trans over cis addition is available.

Table II. Relative Energies of Various Conformations of the BDE-N2(G) Adduct Calculated with the MINDO/3 Approximation

Conf.	ΔC kcal	ΔE kcal	R Å	α deg	β deg	γ deg	δ deg	ε deg
\multicolumn{9}{c}{(+)-Anti Isomer, I(+), cis Addition}
Cda	0.0	0.7	1.45	130.0	-35.1	-122.9	106.9	-118.4
Bde	3.2	3.9	1.46	128.9	-26.1	-125.2	104.6	-127.8
Cde	4.4	5.1	1.46	130.4	-19.7	-128.5	106.5	-135.4
Bda	5.1	5.8	1.45	135.6	-24.9	-150.3	107.6	-143.2
\multicolumn{9}{c}{(+)-Anti Isomer, I(+), trans Addition}
Cda	0.0	0.4	1.46	134.0	-10.7	127.8	105.6	134.0
Cde	3.0	3.4	1.45	133.6	-3.9	130.9	107.1	125.8
Bda	3.9	4.3	1.45	133.7	-1.7	130.1	106.4	123.7
Bde	4.2	4.6	1.46	132.8	-2.3	137.7	106.0	132.5
\multicolumn{9}{c}{(+)-Syn Isomer, II(+), cis addition}
Bda	0.0	1.6	1.46	129.2	-23.1	-124.7	104.7	-128.6
Bde	0.9	2.5	1.46	131.1	-32.6	-116.1	108.9	-142.3
Cda	1.8	3.5	1.46	130.4	-13.1	-121.8	106.3	-136.5
Cde	2.8	4.4	1.46	129.6	-33.5	-120.9	107.4	-118.0
\multicolumn{9}{c}{(+)-Syn Isomer, II(+), trans Addition}
Bda	0.0	0.0	1.45	132.3	-11.8	136.5	106.3	133.0
Bde	2.1	2.1	1.46	134.1	-6.6	132.4	107.8	127.1
Cda	2.8	2.8	1.45	133.1	-3.0	137.4	106.5	128.4
Cde	2.8	2.8	1.46	133.7	-16.1	129.9	107.6	137.4

Note: Refer to the text for definitions of conformations. ΔC is the relative energy for conformations of a given isomer, ΔE is the energy measured relative to the Bda conformation of the adduct formed by trans addition of the syn isomer. The reaction coordinates R, α, β, γ, δ, and ε are defined in Figure 3.

Table III. Benzo Ring Conformations and Conformational Energies (ΔC) of BDE–N2(G) Adducts[a]

Isomer[b]	Ring	ΔC[c]	τ_1	τ_2	τ_3	τ_4	τ_5	τ_6	τ_d
I(+)	Cda	0.0	10.7	3.1	8.7	-31.9	42.7	-35.2	-144.9
II(+)	Cde	2.8							82.4
I(+)	Cde	3.0	-5.3	-4.6	-14.2	40.3	-45.4	31.0	-73.8
II(+)	Cda	2.8							163.0
I(+)	Bda	3.9	-7.6	-8.7	13.3	-0.9	-14.0	19.2	-106.2
II(+)	Bde	2.1							113.2
I(+)	Bde	4.2	2.8	-4.3	-13.5	31.5	-30.2	15.4	-72.8
II(+)	Bda	0.0							147.5
I(+)	Cde	exp[d]	-14.7	3.7	-22.8	53.7	-65.9	44.7	-62.5

a   Refer to footnotes a and b of Table I. Angles are for <u>trans</u> addition.
b   BDE isomer involved in adduct formation.
c   ΔC is measured relative to Cda for the I(+) isomer and Bda for the II(+) isomer.
d   Experimental crystal structure for BPTOH, the hydrolysis product of the <u>anti</u>-BPDE isomer. The signs of all angles are changed from those reported in ref. 88 to correspond to the I(+) isomer.

The results of a crystal structure formed by a <u>trans</u> opening of the BPDE I(+) to yield 7,8,9,10-tetrahydroxy-7,8,9,10-tetrahydrobenzo[a]pyrene (BPTOH) shows a Cde (<u>90</u>) conformation of the ring. The O7-HO7 and O8-H8 groups are de, and the O9-H9 and O10-H10 are also de. The torsion angles of the benzo ring are in best agreement with our second most stable structure, Cde, of the anti BDE-N2(G) <u>trans</u> adduct as is seen from Table III. In adduct formation to N2(G) the <u>trans</u> adduct is the major product (<u>13-22</u>).

The quantum mechanical calculations reported in this section show that the hybridization about N2(G) is trigonal with somewhat large bond angle ($\alpha \approx 130°$). The benzo-ring conformations of the N2(G) adducts reported in Table III are within 2° of the averages of the two isomers. These calculations were performed on the BDE-N2(G) adduct without any steric considerations from the DNA. To approach the modeling of BPDE-DNA adducts for all base atoms, the idealized values of α, δ and ε will be used along with the benzo-ring conformations from Table III.

<u>Rehybridization of amino groups.</u> Adduct formation with N2(G), N6(A) and N4(C) during the intercalative covalent binding step requires a rehybridization of the amino nitrogen atoms. The loss in energy resulting from the alteration of a planar NH_2 or $NH(CH_3)$ group to a conformation of the type required for covalent binding is

given in Table IV.   The methyl substituted amine is used to   model
alkylation after adduct formation. Rotating the planar trigonal NHR
(R=H or CH$_3$) group 90° to break conjugation with the ring or rotating only the NR bond out of plane by ϕ(NR) = 120° requires ca. 5-17
kcal.   A tetrahedral amino group with ϕ(NR) = 120° or with both of
the NH or NR bonds out of plane requires approximately 7-23 kcal.
Therefore, during adduct formation we assume that C10 of the BPDEs
can form a covalent bond to the amino nitrogens even with a loss in
resonance energy which is assumed to be far less than the energy recovered during covalent bond formation.   The trigonal configuration

Table IV.   Energy Change for Loss of Conjugation of the Electron
Pair on NHR in R=H and CH$_3$ Substituted Bases[a]

| N | −NHR | | guanine | | adenine | | cytosine | |
| hybrid | orientation | | | | | | | |
	ϕ(NH)	ϕ(NR)	NH$_2$	NHCH$_3$	NH$_2$	NHCH$_3$	NH$_2$	NHCH$_3$
tr	0°	180°	0.0	0.0	0.0	0.0	0.0	0.0
tr	90°	270°	6.9	5.6	9.1	13.2	17.1	11.7
(tr)	0°	120°	6.9	6.9	6.4	3.6	10.1	7.6
(te)	0°	120°	8.5	11.4	7.8	9.0	15.9	17.6
te	30°	150°	7.1	12.4	7.0	16.4	14.0	23.2

[a] The torsion angle ϕ orienting H and R is defined relative
to atom X in the ring system.

$$\phi(NH) = (X,C,N,H)$$
$$\phi(NR) = (X,C,N,R)$$

about N2 with ϕ(NH) = 0° and ϕ(NCH$_3$) = 120° is used in this paper.
The entire NHR (R=BPDE) is rotated about the C10-N bond by β and γ
for each ϵ.
   In studies reported in this paper using molecular mechanics in
which α and δ were allowed to vary, we found that these values re-
mained close to 123° and 112°, respectively, for the adduct formation with the favored adducts and quite far from these values for

the unfavored isomers. The values $\alpha = 120°$ and $\varepsilon = 120°+\tau_1$ ranging from $112.4°$ to $130.7°$ and $\delta = 109.5°$ are consistent with the calculated values favoring trigonal and tetrahedral hybridization about N2(G) and C10 reported with quantum mechanical calculations in Table II. This suggests that the BPDEs which form adducts with DNA in receptor sites do not require a significant distortion of the bond angles to achieve a fit.

Receptor sites. The intercalation and kinked sites in DNA used in this study are listed in Table V. Three theoretically determined intercalation sites (I, II and III) permit the study to be conducted with the DNA unwound by $7°-12°$, $14-18°$ and $25°-32°$ with parallel base pairs separated by 6.76 Å and with alternating (a) sugar puckers (67,68). Attention will be confined to site I because it was found to be the most favorable in the present studies. Several kink (K) sites have been identified (66). The constraint that proper hybridization exists about N2(G) and C10 of BPDEs stimulated an investigation in kinked DNA. In an idealized structure the pyrene moiety is approximately parallel to one of the base pairs as shown

Table V. Definition of Binding Sites for Intercalation and Intercalative Covalent Binding[a]

Binding Site[b]	Sugar Puckers	Comment	Δz	α_z	Δx	α_x
I	alternating (a)	Intercalation Site I	6.76	29.2	0.0	0.0
KMs	C(2´)-endo (s)	DNA opens in M groove	7.56	33.0	-1.2	39.0
KMa	alternating (a)	DNA opens in M groove	8.24	17.3	0.0	39.0
Kma	alternating (a)	DNA opens in m groove	6.80	14.0	0.0	-30.0

[a] The binding sites with all the same (s), C(2´)-endo, and with alternating (a), C(2´)-endo to C(3´)-endo sugar puckers are kinked (K) in the major (M) or minor (m) grooves.

[b] The conformational angles are reported in Table I (66) for sites KMs and KMa. For site Kma, the backbone torsion angles are $\xi = -175.8°$, $\theta = -157.8°$, $\psi = 163.1°$, $\phi = -123.0°$ and $\omega = -97.7°$.

in Figure 2. In practice, the pyrene moiety is not oriented in this idealized manner. Consequently, representative values of the kink were selected with favorable energies of formation for this study.

The energy required to adjust the DNA to these receptor sites is given in Table VI. The DNA can kink equally well in both grooves with base pairs held at a distance sufficient for intercalation ($\Delta z = 6.76$ Å, $\alpha_x = 0°$) and for kinks ($\Delta z > 6.76$ Å, $\alpha_x \neq 0°$). These receptor sites are constructed by operations on a pair of initially coincident base pairs. Each is rotated by $+\alpha_x/2$ and $-\alpha_x/2$ about a kink axis. This axis is perpendicular to the helix and dyad axes of the base, and parallel to the C1'(py)–C1'(pu) axis. It lies approximately along the C6(py)–C8(pu) axis. Then each base pair is rotated about the helix axis by $+\alpha_z/2$ and $-\alpha_z/2$ and separated by Δz. The combinations of α_x, α_z, and Δz which permit the construction of a phosphate backbone defines families of receptor sites. With this approach, the base pairs adjacent to the BPDE are symmetrically

Table VI. Energy, ΔE_{DNA}^R in kcal/Duplex, Required to Alter B–DNA to Receptor Sites[a]

Sequence BP3 ↑BP2↓	Intercalation Site I	Kink s $\alpha_x=39°$	Kink a $\alpha_x=39°$	Kink a $\alpha_x=-30°$	Binding Orientation G	C	A
A·T ↑T·A↓	19.1	13.2	9.0	−14.9	–	–	5'
A·T ↑C·G↓	20.0	15.5	10.8	−12.4	5'	3'	5'
G·C ↑C·G↓	20.5	17.4	12.1	−10.6	5'	3'	–
A·T ↑G·C↓	23.8	17.8	15.2	− 7.2	3'	5'	5'
G·C ↑G·C↓	24.7	20.2	17.0	− 4.6	3',5'	5',3'	–
A·T ↑A·T↓	25.3	21.3	17.7	− 4.3	–	–	3',5
T·A ↑C·G↓	26.2	23.8	19.6	− 1.8	5'	3'	3'
C·G ↑G·C↓	30.7	24.5	23.6	2.7	3'	5'	–
T·A ↑G·C↓	29.6	25.6	23.5	3.0	3'	5'	3'
T·A ↑A·T↓	29.1	27.2	24.0	3.7	–	–	3'

[a]Energies reported are for intercalation site IB (67,68) and kinked sites for C(2')-endo (s) and C(2')-endo to C(3')-endo (a) (44,66). The sequence reported in ↑GC,BP2,BP3,CG↓.

displaced about the origin. The space-fixed Z-coordinate becomes
the average helical axis through the receptor site. The details of
the algorithm to generate nucleic acid structures are given elsewhere (65-67).

For kink sites there are two possibilities for the sugar
puckers, C(2´)-endo to (C2´)-endo (same, s) and C(2´)-endo to C(3´)-
endo (alternating, a) through the receptor site along the 3´ sugar
5´ direction. These alternating sugar puckers occur in intercalation sites. In the case of intercalation, parallel base pairs can
be separated to 4.58 Å (91) before sequence changes from s to a. In
the case of a kink, Taylor et al. (66) report conformations with
base pairs separated sufficiently for intercalative covalent binding. For a sugar puckers α_x ranges from ca. -40° to +64° and for s
sugar puckers α_x is confined to 0° to 70°. For external covalent
binding adjacent base pairs have minimal separation. Kinks of -40°
to +30° are found for s sugar puckers and -80° to -30° for a sugar
puckers. For covalent intercalative binding to N2(G) a kink of α_x =
+39° is used to open the DNA in the major groove (M), whereas, for
binding to N6(A), O6(A) and N4(C) an opening in the minor groove (m)
with α_x = -30° is used. Receptor sites with these kink values were
chosen because the energy required to kink the DNA increased dramatically for larger kinks and because these receptor sites yielded
favorable structures after adduct formation. These sites and the
model proposed permits an analysis on the molecular level of the
stereoselectivity of BPDEs by DNA during intercalative covalent
binding.

The most stable base sequences (67,74) are the pyrimidine(p)
purine ones: TpA, TpG and CpG. For binding to N2(G), N6(A), O6(G)
and N4(C) these lead to 5´, 5´, 5´ and 3´ type binding for receptor
sites. The 3´ and 5´ type binding for (pu) and (py) base atoms,
respectively, require the use of receptor sites which are less
stable. These binding orientations are less likely to occur if the
intercalation process precedes covalent bond formation.

Classical intercalation. The orientation of the BPDEs in the intercalation site provides an important step in the type of binding
(3´ or 5´) as well as a placement of the epoxide adjacent to the reactive base atom for trans addition. Intercalation with the epoxide
directed in both grooves has been studied (43,92). In Figure 6, two
optimum binding orientations of the BPDE I(+) Cde isomer is shown
intercalated into site I of a ↑CG,GC↓ dimer duplex unit, i.e., a
(py)p(pu) sequence. For the case of minor groove binding the
epoxide is adjacent to N2(G), whereas, for major groove binding it
is adjacent to O6(G) and N4(C). For the case of ↑TA,AT↓ sequence,
the epoxide is adjacent to N6(A) in the major groove. Intercalation
of BPDEs with all benzo ring conformations was studied. The optimum
binding orientations and relative enegies for the four isomers are
given in Table VII for intercalation site I. Sites II and III,
yielded qualitatively similar binding orientations but less favorable binding energies by 1 and 2 kcal, respectively. Preference for
intercalation site I with small unwinding angles (7°-12°) with small
contributions from sites II and III with unwinding angles of 14°-18°
and 25°-32° yields an average weighted by a Boltzmann distribution
of ca. 12°. This agrees with the experimental result of 13° (46).
The general trend shows the epoxide directed toward one of the

grooves.   This occurs because the positioning of the   pyrene moiety
is due to the steric fit, δU.   Preference for   the major groove
arises   from the electrostatic contribution, δQ.   The electrostatic
potential of the DNA is most negative in the   minor groove (93).
Hence the preference of the relatively negative epoxide   oxygen for
the major groove.   These calculations   were performed   with a   dielectric of 1.0.   In a calculation by Subbiah et al. (92) in which
the dielectric was varied from 1 to 4 with increasing   distance, the
relative energy favored minor groove binding for the BPDE   I(+) isomer and neither groove for the BPDE I(-) isomer.   Their values are
reported in Table VII.   The distant dependent dielectric reduces the
effect of the electrostatic contribution.   In their calculations the
reactive C10 of BPDE I(+) is favorably oriented for reaction with
N2(G).   To examine the effect of the counterions on the DNA, we introduced them into the most negative regions of intercalation sites
and of B-DNA, i.e., the minor groove (93,94,95).   In test calculations, +1 charges were placed along a local dyad axis of each base
pair. As the charge approached the DNA, the difference in the relative binding energies for major and minor groove orientation de-

Table III.   Optimum Binding Orientations and Relative Energies for
the Intercalation of the BPDEs into a ↑C·G,G·C↓ Dimer Duplex[a]

Isomer	Ring	Δx	Δy	θy	g	δU^I_{BPDE}	δQ^I_{BPDE}	δE^I_{BPDE}	δE^I_{BPDE} [b]
I(+)	Cde	3.0	-3.2	162	m	3.5	5.1	8.6	0.0
I(+)	Cde	0.8	2.7	-78	M	0.0	0.0	0.0	5.3
I(-)	Cda	2.6	-3.0	162	m	3.7	8.8	12.5	6.7
I(-)	Cda	-1.1	3.6	-50	M	6.2	4.0	10.2	6.6
II(+)	Cde	3.2	-3.0	159	m	2.4	9.7	12.1	
II(+)	Cde	-0.2	4.0	-62	M	6.0	8.1	14.1	
II(-)	Cda	2.8	-2.7	162	m	2.6	5.7	8.3	
II(-)	Cda	-0.2	3.0	-70	M	0.2	3.8	4.0	

[a]The optimum orientation was obtained for the epoxide directed
toward the major (M) and minor (m) grooves.   Intercalation site I
(68) with α_z = 29° was used.   The base pair sequence used is
↑GC,CG,BPDE,GC,CG↓.   The relative total energy, $\delta E = \delta U + \delta Q$, is
partitioned into relative steric, δU, and relative electrostatic
contributions.

[b]Relative energies reported by Subbiah et al. (92).

creased, and when the +1 charge approached within 4 Å of the epoxide, minor groove binding became more favorable.

The following theoretical and experimental data is presented to support the idea for activation of the BPDEs in an acid catalyzed S_N2 reaction for trans addition (90) with some contribution to carbonium ion formation in an S_N1 reaction for both trans and cis addition. The non-covalently bound BPDE adducts to DNA formed initially are intercalation complexes (46,51-55). Meehan et al. (46) and Geacintov et al. (54) reported that the BPDE intercalates into DNA on a millisecond time scale while the BPDE alkylates DNA on a time scale of minutes. Because the electrostatic potential is most negative in the grooves, positive ions are attracted to these regions (93,94,95). Consequently, protonation of the BPDE I(+) to form the activated state can occur in the ideal position for subsequent covalent bond formation to base atoms. We postulate that this activation occurs during the intercalation step. A space filling model with steric contours on which electrostatic contours are superimposed illustrate these points in Figure 6 for the BPDE I(+) isomer (37). One may interpret the presence of protons in the grooves of DNA as equivalent to a local increase in pH over the bulk value. Whalen et al. (96) observed that the rate of hydrolysis of the BPDE I(±) and II(±) isomers increases linearly with acid concentration. They postulate that a dihydrogen phosphate transfers a proton to the epoxide in a rate limiting step. They suggest that an intermediate carbonium ion is formed. Geacintov et al. (53,56,97,98), Michard et al. (99) and Kootstra et al. (100) observed that the rate of hydrolysis of the BPDEs depends on the concentration of DNA. Thus, the DNA plays the important role in the activation of the BPDEs in the receptor site and at the site of the reactive base atoms. However, in addition to adduct formation, tetrols are formed (53,55).

Lavery and Pullman (101) have calculated the relative energies for the binding of the model system; the allyl carbonium ion of the base pairs followed by deprotonation. They find that the relative energies of formation for trans addition with N2(G), N6(A) and N3(C) are 0.0, 7.6, and 7.1 kcal/mole, respectively. Their calculations are in agreement with the observed relative yields for reaction with DNA (18,19,34,38), namely that binding occurs in the order N2(G), O6(G), N3(C) and N6(A) and that BPDE I(+) prefers N2(G). Lavery et al. (102) have performed a model study of the allyl carbonium ion with guanidine. They obtain a tetrahedral intermediate. Lin et al. (103) studied geometries of intermediates involving stacked guanine and BPDE I(+) as postulated in classical intercalation, and after postulating acid catalyzed carbonium ion formation, they calculate with ab initio SCF calculations a very stable (-67 kcal/mole) bond for the model system, $CH_3^+-NH_3$ (planar). In these calculations, the proper hybrid configuration about C10-N2(G) was not studied within the total DNA-BPDE I(+) complex. Nakata et al. (104) obtain a transition state alkylation geometry with parameters corresponding to r = 2.0 Å, α = 110°, β = 100°, γ = variable, δ = 80° and ε = 0° or 180°. When the BPDE I(+), I(-), II(+) and II(-) isomers are inserted into classical intercalation sites they find that the I(+) isomer is oriented most favorably for adduct formation to N2(G) consistent with this geometry. A systematic study of cis addition on to these four base atoms has not been made.

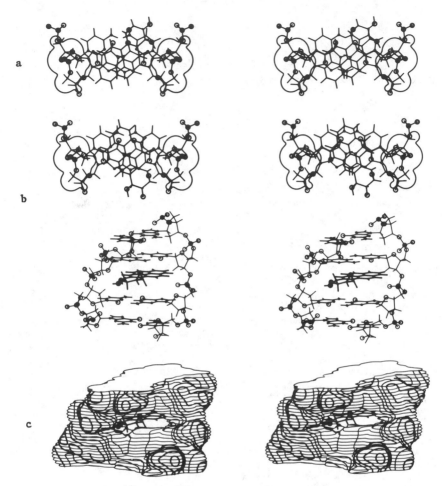

Figure 6.   Optimum binding orientations of BPDE I(+) intercalated
with the   epoxide in the major (a)   and minor (b and c) grooves in
↑C•G,G•C↓.   (c) Relative values of the electrostatic potential   are
denoted by (▬) positive, (—) intermediate and (―) negative on
steric contours of DNA viewed into the minor groove.

<u>Stereoselectivity.</u>   We find that stereoselectivity of the in-
dividual isomers   occurs during intercalative covalent binding.   To
complete the formation of a chemical bond with proper hybrid configurations, the DNA must undergo a further change to a kinked site.
For each of the kink sites the DNA was bent by α_x = +39° and –30°.
Each isomer was adjusted   to obtain an optimum fit.   For each benzo
ring conformation of each isomer, β and γ were varied until a minimum energy was achieved.

    An example of the variation of energy of adduct   formation with
ring conformation and its influence on ε and the entire pyrene
moiety is given in Figure 7.   For <u>trans</u> addition of BPDE I(+) to N2
on guanine in duplex DNA, the minimum occurs with the Cde (ε=114.7°)

conformation of the benzo ring. This assessment of the optimum conformation was made for adduct formation to all sites studied in this investigation. This example demonstrates the role of the benzo ring conformation in the orientation of the pyrene moiety through the internal angle, τ_1, and the external reaction coordinate $\epsilon = (120° + \tau_1)$.

The optimum results are illustrated for binding to N2(G), N6(A) and N4(C) in Figures 8-10. Binding to O6(G) yields the same orientations as shown for N6(A) in Figure 9. The minimum binding energies for 3´ and 5´ binding are listed in Tables VIII-XI along with the orientation of the long axis of the pyrene moiety to the average local helix axis through the kinked site. These calculations were performed for only two types of sequences: py(p)pu and pu(p)py. The DNA adjusts to the receptor sites favoring 5´ binding for N2(G), N6(A) and O6(G) and 3´ binding for N4(C) as seen from results in Table VI. The BPDE isomers favor 3´ binding for N2(G) and 5´ binding for N6(A), O6(G) and both for N4(C) as seen from the total complex energies reported in Tables VIII-XI. Intercalation into py(p)pu sequences is the favored first step. Consequently, one may assume that this step favors 5´ binding to N2(G), N6(A), O6(G) and binding to N4(C). For this reason these complexes are illustrated in Figures 8-10. However, the trends show that the I(+) and II(+) isomers fit much better than the I(-) and II(-) isomers for both 3´ and 5´ binding to N2(G) in Table VIII. Thus, stereoselectivity occurs to select this BPDE I(+)-N2(G) adduct regardless of the binding orientation, but intercalation site production favors 5´ binding.

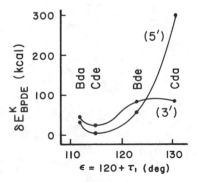

Figure 7. Intermolecular energy dependence for <u>trans</u> addition of the BPDE I(+) on ring conformation and ϵ for the $\overline{(3´)}$ and (5´) bound adducts to N2(G). δE_{BPDE}^{K} is measured relative to that for that for the Cde conformation.

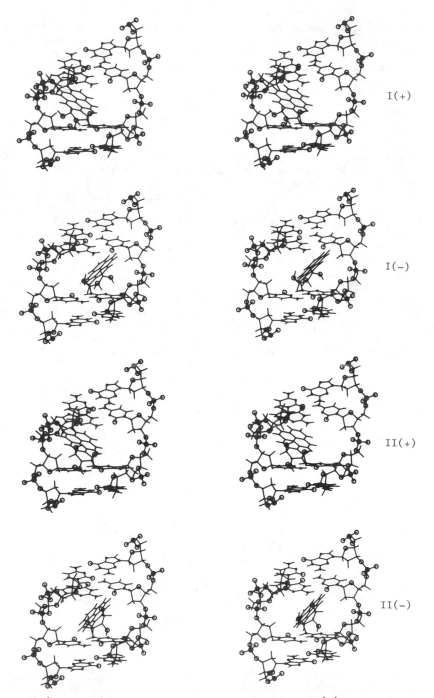

Figure 8.   Intercalative covalent binding to N2(G).   Trans BPDE-
adducts are bound to the   5′ side of G in   the ↑G•C,C•G,BPT,C•G,C•G↓
complex in DNA kinked to site KMa.

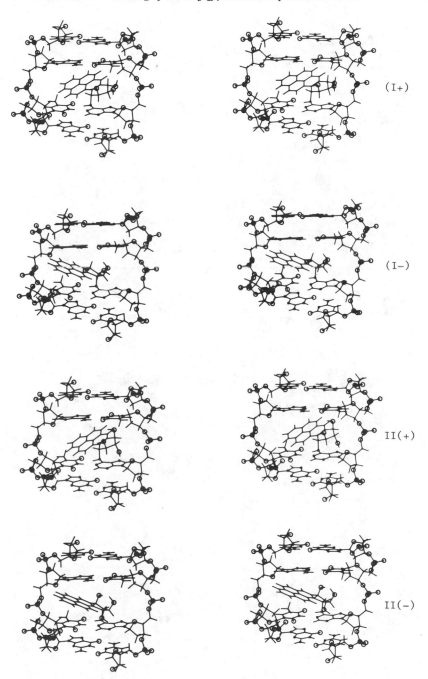

(I+)

(I−)

II(+)

II(−)

Figure 9. Intercalative covalent binding to N6(A). <u>Trans</u> BPDE-adducts are bound to the 5´ side of A in the ↑G•C,T•A,BPT,A•T,C•G↓ complex in DNA kinked to site KMa.

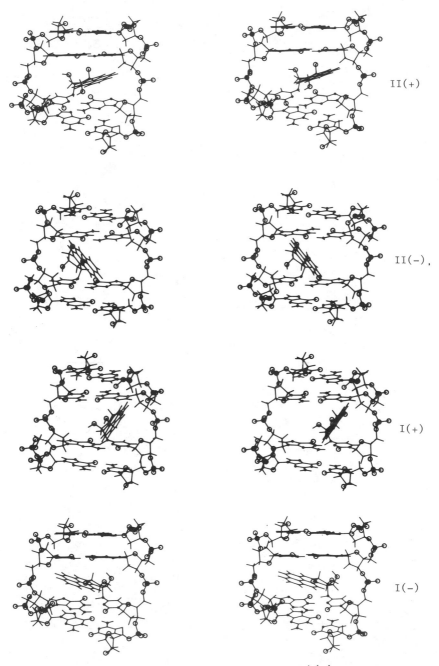

Figure 10. Intercalative covalent binding to N4(C) for four representative cases. $\underline{\text{Trans}}$ BPDE II(+) and II(−) adducts are bound to the 3′ and BPDE I(+) and I(−) adducts are bound to the 5′ side of C in the ↑G•C,C•G,BPT,G•C,C•G↓ and ↑G•C,G•C,BPT,C•G,C•G↓ complexes, respectively.

Table VIII. Binding to N2(G) via <u>trans</u> Addition of BPDEs in the 3'- and 5'-Orientations in a Kinked Site[a]

BPDE Isomer	Benzo Ring	3´-orientation[b]				5´-orientation[c]			
		β	γ	$\frac{KMa}{\delta\epsilon_{BPDE}}$	$\alpha(BPDE)$	β	γ	$\frac{KMa}{\delta\epsilon_{BPDE}}$	$\alpha(BPDE)$
I(+)	Cde:	−66	−18	0.0	83	63	99	8.6	80
II(+)	Cda:	−67	−16	3.7	84	64	99	11.4	80
I(−)	Cde:	−69	−77	67.7	84	66	6	478.5	83
II(−)	Cda:	−69	−78	71.9	84	75	18	64.4	86

[a] Kink site KMa is defined in Table III. See Figure 8. Energies are measured relative to the 3´ BPDE I(+)-N2(G) adduct.
[b] The base pair sequence is ↑G·C,G·C,BPT,C·G,C·G↓.
[c] The base pair sequence is ↑G·C,C·G,BPT·G·C,C·G↓.

Table IX. Binding to N6(A) via <u>trans</u> Addition of BPDEs in the 3'- and 5'-Orientations in a Kinked Site (Kma)[a]

BPDE Isomer	Benzo Ring	3´-orientation[b]				5´-orientation[c]			
		β	γ	$\frac{Kma}{\delta\epsilon_{BPDE}}$	$\alpha(BPDE)$	β	γ	$\frac{Kma}{\delta\epsilon_{BPDE}}$	$\alpha(BPDE)$
I(−)	Cde:	−101	18	1255.2	87	108	−112	0.0	81
II(−)	Cda:	−101	16	1405.4	87	108	−111	3.8	81
I(+)	Cde:	−96	92	101.8	88	106	−18	159.4	84
II(+)	Cda:	−96	92	104.1	88	100	−16	1636.8	85

[a] Kink site Kma is defined in Table III. See Figure 9. Energies are measured relative to the 5´ BPDE I(−)-N6(A) adduct.
[b] The base pair sequence is ↑G·C,A·T,BPT,G·C,C·G↓.
[c] The base pair sequence is ↑G·C,T·A,BPT,A·T,C·G↓.

Table X. Binding to O6(G) via <u>trans</u> Addition of BPDEs in the 3'-
and 5'-Orientations in a Kinked Site (Kma)[a]

BPDE Isomer	Benzo Ring	3'-orientation[b]				5'-orientation[c]			
		β	γ	$\delta\varepsilon_{BPDE}^{Kma}$	$\alpha(BPDE)$	β	γ	$\delta\varepsilon_{BPDE}^{Kma}$	$\alpha(BPDE)$
I(−)	Cde:	−105	25	1223.3	90	110	−114	0.0	80
II(−)	Cda:	−103	21	1723.5	90	110	−114	3.6	80
I(+)	Cde:	−96	86	131.4	81	105	−9	42.4	85
II(+)	Cda:	−96	82	134.5	86	102	−6	79.2	86

[a] Kink site Kma is defined in Table III. Energies are measured
relative to the 5' BPDE I(−)-O6(G) adduct.
b The base pair sequence is ↑G·C,C·G,BPT,G·C,C·G↓.
c The base pair sequence is ↑G·C,G·C,BPT,C·G,C·G↓.

Table XI.   Binding to N4(C) via <u>trans</u> Addition of BPDEs in the 3'-
and 5'-Orientations in a Kinked Site (Kma)[a]

BPDE Isomer	Benzo Ring	3'-orientation[b]				5'-orientation[c]			
		β	γ	Kma $\delta\varepsilon_{BPDE}$	α(BPDE)	β	γ	Kma $\delta\varepsilon_{BPDE}$	α(BPDE)
I(−)	Cde:	−98	19	3×10^4	84	109	−116	6.4	82
II(−)	Cda:	−98	19	3×10^4	84	109	−117	10.6	83
I(+)	Cde:	−102	106	32.8	80	103	−21	1245.9	89
II(+)	Cda:	−102	106	0.0	83	102	−20	1375.1	88

[a] Kink site Kma is defined in Table III. See Figure 10. Energies
are measured relative to the 3´ BPDE II-N4(C) adduct.
[b] The base pair sequence is ↑G·C,C·G,BPT,G·C,C·G↓.
[c] The base pair sequence is ↑G·C,G·C,BPT,C·G,C·G↓.

In contrast, adduct formation to N6(A) favors 5´ binding through intercalation with the base(p)pu sequences to select I(−) and II(−)
isomers through the steric fit. The base(p)py sequences form less
stable intercalation sites and the resulting 3´ binding favors the
I(+) and II(+) isomers over the I(−) and II(−) isomers, but the
overall steric fit is much poorer than for 5´ binding as seen in
Table IX.   The argument for adduct formation with O6(G) parallels
that for N6(A) and the overall results are the same as seen in Table
X.   In the case of N4(C), the I(−) and II(−) isomers fit best in 5´
binding, whereas the I(+) and II(+) isomers fit best for 3´ binding.   Intercalation sites form most easily to favor 3´ binding, and
consequently one may expect the I(+) and II(+) isomers to be selected by N4(C). Because the reaction surface is quite complex, the
activation energies were not computed. Therefore, the contribution
of this step to the relative yields cannot be assessed. However,

the dramatic difference in the energies required to fit each isomer of BPDE in the receptor sites during the intercalative covalent binding step demonstrates the stereoselectivity.

The trends can be summarized with a visual analysis of the stereographic projections. Base atoms directed toward the minor groove select the (+) isomer, whereas base atoms directed toward the major groove select the (-) isomer with the exception that N4(C) binds both. This selection is most dramatic. It is the basis of the stereoselectivity. Of the possible diastereomers, a further selection of the I(+) and II(+) or I(-) and II(-) isomers occurs. This selection is less dramatic. It arises from the slightly different fit of the I and II isomers. The orientation of the pyrene moiety relative to the base for $\underline{trans}$ addition of the (+) isomers is nearly independent of the orientation of the 7,8-dihydroxy groups. Similarly for the (-) isomers. Therefore, the orientations of the I(+) and II(+) isomers are nearly the same as are the I(-) and II(-) isomers. The further selection of I or II depends on the position in the ring of the dihydroxy groups as well as the ring conformation. The reaction coordinate $\varepsilon = \pm(120°+\tau_1)$ which orients the pyrene relative to the base for $\underline{trans}$ addition depends on minor adjustments of the benzo ring through the internal ring torsion angle τ_1. The trends obtained in the minimum conformations of the ring is Cde for isomer I and Cda for isomer II with the I(+) isomer for binding to N2(G), I(-) for binding to N6(A), O6(G), and N4(C). The pyrene moiety is adjusted in the receptor site to be as parallel as possible to the kinked base pairs. Thus, small values of $|\varepsilon|$ are favored as seen in Figure 7 and as can be deduced visually from the stereographic projections. Due to the small conformational energies of the benzo-ring, the steric fit adjusts the ring to yield conformations with $|\varepsilon| = 114.7°$, the appropriate τ_1 and the Cde and Cda forms for isomers I and II.

The possibility that both $\underline{trans}$ and $\underline{cis}$ addition can occur allows an interpretation in which adducts resulting in the major and minor components can be explained. S_N2 reactions (90) predominate to yield $\underline{trans}$ addition products. But the less favored S_N1 reaction may occur and this permits both. By visual inspection of adducts in Figures 8-10, the steric fits for $\underline{cis}$ addition can be ascertained. This is accomplished if all atoms of the BPDEs are reflected through the plane of the pyrene. Then the C10-N2(G), etc. bonds, which remain fixed relative to the base, assume an orientation for $\underline{cis}$ addition as the (+) isomers transform to (-) isomers. Beginning with Figure 8, $\underline{trans}$ I(+)-N2(G) and $\underline{trans}$-II(+)-N2(G) adducts become $\underline{cis}$ I(-)- and $\overline{II(-)}$-N2(G) adducts with approximately the same favorable steric fit. The 7- and 8-hydroxy groups do not interfere with the DNA. This becomes apparent without further calculation because (+) isomers fit nearly equally well even though the 7 and 8 hydroxy groups are reversed. Therefore, the reflection to convert $\underline{trans}$ I(+) to $\underline{cis}$ I(-) adduct can accommodate this change without steric interference by the 7- and 8-OH groups. However, the 9-OH group also undergoes a reflection. Visual inspection of the stereographic projections shows that there is room in the receptor site for this to occur. Similarily, the unfavored steric fits for the $\underline{trans}$ I(-)-N2(G) and $\underline{trans}$-II(-)-N2(G) become equally unfavored $\underline{cis}$ $\overline{I(+)}$- N2(G) and $\underline{cis}$ $\overline{II(+)}$-N2(G) adducts. In other words, these should not be observed. We attribute the observed stereoselectivity of the I(+)-N2(G) adduct to trans addition resulting from a predominantly S_N2 reaction and the fit in the kinked receptor site. The less favored S_N1 reaction which permits trans and cis addition may account for the small yield of the I(-)-N2(G) adduct (34,38) if cis addition is assumed. A similar argument applies to N6(A) and O6(G). The favored steric fit shown in Figure 9 for the trans I(-) and II(-)-N6(A) adducts become favored cis I(+) and II(+)-N6(A) adducts. Conversely, the unfavorable complexes involving trans I(+) and II(+) are equally poor candidates after a reflection through the pyrene moiety to yield cis I(-) and II(-)-N6(A) adducts. The stereo-selectivity exhibited during this step involving kinked receptor sites is consistent with the observed reactions of the isomers. The types of adducts obtained theoretically and relative experimental yields are summarized in Table XII. The % yields reported are only for adduct formation to the bases without detailed knowledge of the extent of trans and cis addition. This is our interpretation.

To place this argument in perspective assume that both trans and cis adducts were equally likely. Then stereoselectivity would not be observed. All of the adducts to base atoms listed in Table XII for trans addition correspond to the sterically forbidden possibilities for cis addition and vice versa. Therefore, I(+) and I(-) isomers would bind equally well to N2(G). This does not occur. Rather the favored S_N2 reaction over the S_N1 reaction weights trans addition heavily. Therefore, in our interpretation of experimental yields, the smaller yield of I(-) with N2(G) is predicted to be a cis addition product. The argument can be extended to demonstrate that trans I(-)-N6(A) and O6(G) and cis I(+)-N6(A) and O6(G) adducts should be expected.

Although we postulate that this receptor site results in stereoselectivity, it may not be the final state. The orientation of the long axis of the pyrene moiety is approximately 80°-90° and this implies quasi-intercalation of site IQ (56-58). The kinked site proposed by Hogan et al. (50) and studied by Taylor and Miller (44) for I(+)-N2(G) binding in retrospect appears to represent different binding sites. The orientation of the pyrene moiety of α(BPDE) = 43° and the local DNA axis in the kink of γ(DNA) = 29° (50) will be explained by external binding in the next section. The intercalative covalent binding in a kinked site is an intermediate step between intercalation and the final structure for the externally bound BPDE-N2(G) adduct, but it becomes the final structure for the quasi intercalated BPDE-N6(G), O6(G) and N4(C) adducts.

Externally and internally bound adducts. The internally bound BPDEs have been presented in a step in which the favored proton assisted S_N2 reaction results in the trans addition and the stereoselectivity of isomers toward base atoms. The DNA is dynamic. We propose a relaxation which yields both internally and externally bound adducts; the former favoring binding to N6(A), O6(G) and N4(C), and the latter favoring N2(G).

From a structural point of view, there are two possible orientations for the externally bound adduct in duplex DNA. They are easiest to discuss for BPDE-N2(G) adduct formation. The first results from an anti → syn rotation of G about its glycosidic bond. The BPDE-N2(G) adduct is transferred to the major groove and the DNA

Table XII. Stereoselectivity of BPDEs for Base Atoms by the
Intercalative Covalent Binding Step in Kinked DNA and
Predicted Final Binding Site [a]

Isomer	trans adducts ($S_{N}2, S_{N}1$)				cis adducts ($S_{N}1$ only)			
	N2(G)	N6(A)	O6(G)	N4(C)	N2(G)	N6(A)	O6(G)	N4(C)
I(+)	5´,3´ (86%,86%) (IIX)	-	-	-	-	5´ (5%,2%) (IQ)	5´ (-,4%) (IQ)	5´
I(-)	-	5´ (5%,2%) (IQ)	5´ (-,2%) (IQ)	5´	5´,3´ (4%,5%) (IIX)	-	-	-
II(+)	5´,3´ (IIX)	-	-	3´	-	5´ (IQ)	5´ (IQ)	-
II(-)	-	5´ (IQ)	5´ (IQ)	-	5´,3´ (IIX)	-	-	3´

[a] Adducts for trans addition are shown in Figure 8-10 and those assumed to be possible for cis addition are determined by symmetry
by inspection of Figure 8-10.   The % yields are reported by
Meehan and Straub (34) and Brookes and Osborne (38), respectively for binding to bases.   Trans and ' cis addition products were
not analyzed in detail.   Values from (38) were adjusted by 0.91
and 0.09 for the I(+) and I(-) isomers to correspond to the total
relative yields in (34).   The interpretation of experimental
yields for trans and cis adducts is consistent with the proposed
model. The external binding site (IIX) and quasi intercalated or
intercalative covalent site (IQ) are indicated.

resumes an overall B-DNA conformation except for the affected base.
The second results when the adduct is placed in the minor groove in
a receptor site kinked to yield a good steric fit.   These kinks occur when the B-DNA is bent to improve the steric fit of the adduct.
They differ from those used for intercalative covalent binding by
the possible minimum separation of the base pairs.   Of the two possible orientations of the affected base, adduct formation to N2(G)
is distinguished by placement of the pyrene moiety in the minor
(anti G) or major (syn G) grooves of the DNA.   They are illustrated
in Figure 11.   For both of these possibilities the orientation

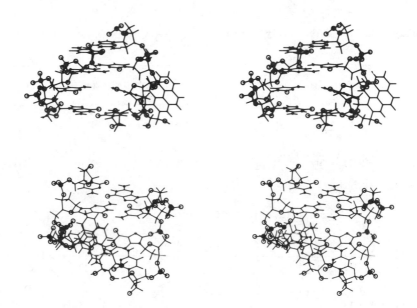

Figure 11. An externally bound BPDE I(+)-N2(G) adduct. (upper) The pyrene is placed in the major groove after an <u>anti</u> → <u>syn</u> rotation about the glycosidic bond of G by 200° in an otherwise B-DNA conformation. (lower) The pyrene moiety is placed in the minor groove in a DNA conformation with a -70° kink, α(BPDE) = 15° and γ(DNA = 35°.

angles, α(BPDE) and γ(DNA), are reported in Table XIII for several possible duplex structures. The most favorable fit of the externally bound BPDE-N2(G) adducts occurs for the <u>anti</u> orientation of G as the kink becomes more negative with the pyrene in the minor groove. At α_x = -70°, γ(DNA) = 35° and α(BPDE) = 15° and the unwinding angle is 36° -α_z ≈ 29°. These orientations are in good agreement with experimental values (42,45,50,51,58). However, caution must be stressed in this interpretation because the reactions were studied with the racemic I(±) adducts (42,45,58) and with the I(+) adduct (50,51) to DNA. The major product involving the <u>trans</u> I(+)-N2(G) adduct suggests that α(BPDE) and γ(DNA) are related to the theoretical structures. The measured unwinding angles (42,45) involve other adducts and intercalative covalently bound forms.

Both of these structures can arise after a denaturation and renaturation of the DNA. There is experimental evidence for the dynamic process of opening and closing of the DNA. In hydrogen exchange studies of the amino and, in Watson-Crick base pairing, buried imide groups, Teitelbaum and Englander (105,106) conclude that the G•C and A•T base pairs are open about 1% of the time, and that the opening rate constant is about 0.04 to 0.06 sec^{-1} in both cases. A study of intercalation by Gabbay et al. (107) of two molecules, a 1,8-naphthylimide with one bulky substituent and a 1,8,4,5-naphthylimide with bulky substituents on each end of the molecule to

Table XIII. Externally Bound I(+), I(−), II(+) and II(−) <u>trans</u>
             Adducts to N2(G) and N6(A) [a]

Adduct	base	sugars	α_x	α_z	γ(DNA)	α(BPDE)	$\delta\varepsilon_{BPDE}$
			<u>anti</u> → <u>syn</u> orientation of affected base				
I(+)−N2(G)[b]	<u>syn</u>	<u>s</u>	0	36.0	0	51	6.6
I(+)−N2(G)	<u>syn</u>	<u>s</u>	+10	36.1	5	49	9.6
II(−)−N6(A)[c]	<u>syn</u>	<u>s</u>	0	36.0	0	51	10^3
			<u>anti</u> orientation of all bases				
I(+)−N2(G)	<u>anti</u>	<u>s</u>	0	36.0	0	50	10^5
I(+)−N2(G)	<u>anti</u>	<u>s</u>	−20	22.5	10	33	141.7
I(+)−N2(G)	<u>anti</u>	<u>s</u>	−40	22.0	20	30	60.8
I(+)−N2(G)	<u>anti</u>	<u>a</u>	−30	26.0	15	29	63.7
I(+)−N2(G)	<u>anti</u>	<u>a</u>	−50	19.2	26	16	5.7
I(+)−N2(G)[b]	<u>anti</u>	<u>a</u>	−70	7.3	35	15	−10.1
I(+)−N2(G)[d]				1−14	29	15−43	exp
I(−)−N6(A)	<u>anti</u>	<u>s</u>	10	36.1	5	42	377.4
I(−)−N6(A)	<u>anti</u>	<u>s</u>	20	38.0	10	38	44.6
I(−)−N6(A)	<u>anti</u>	<u>s</u>	30	38.6	15	31	7.5
I(−)−N6(A)[e]						37−45	exp
II(+)−N2(G)	<u>anti</u>	<u>a</u>	−30	26.0	15	5	215.7
II(+)−N2(G)	<u>anti</u>	<u>a</u>	−50	19.2	25	7	−5.3
II(+)−N2(G)	<u>anti</u>	<u>a</u>	−70	7.3	35	15	−13.8
II(±)−N2(G)[f]						>65	exp
II(−)−N6(A)	<u>anti</u>	<u>s</u>	0	36.0	0	46	3×10^3
II(−)−N6(A)	<u>anti</u>	<u>s</u>	20	38.0	10	39	67.2
II(−)−N6(A)[c]	<u>anti</u>	<u>s</u>	30	38.6	15	31	12.7

[a] The sugars are C(2′)-<u>endo</u> to C(2′)-<u>endo</u> (<u>s</u>) and C(2′)-<u>endo</u> to
C(3′)-<u>endo</u> (<u>a</u>) in the 5′(p)3′ or 3′(sugar)5′ direction. Base
sequences: ↑G·C,C·G,BPT,G·C,C·G↓ and ↑G·C,T·A,BPT,A·T,C·G↓.
Energies for the BPDE−N2(G) and −N6(A) adducts are measured
relative to the 5′-orientation of the corresponding isomers in
Tables VIII and IX, respectively.
[b] See Figure 11.
[c] See Figure 13.
[d] Experimental results (<u>50,51,58</u>). α_z = 36° − unwinding angle
from (<u>42,45</u>).
[e] Experimental results (<u>51</u>). N6(A) adduct formation is assumed.
[f] Experimental results are for the racemic mixture (<u>58</u>).

prevent direct insertion into an intercalation site, also supports a dynamical structure of DNA. Their kinetic studies show that DNA complexes form in ca. <1 msec and >100 msec for these two compounds, respectively. The first compound can be inserted into DNA directly from one groove; however, for the second naphthylimide to be intercalated, the substituents must lie in opposite grooves. The slower time needed for complex formation suggests a denaturation and renaturation of the DNA to accommodate each substituent in each groove. Their results suggest that there are two modes available for intercalation: rapid (involving opening, RO) and slow (involving denaturation, SD) equilibrium processes, and conversely, that these two modes are available for the reverse process. For a covalently bound adduct which results during intercalation, the RO process will not permit the pyrene moiety to be dislodged from the intercalation site, whereas the SD process will. Therefore, the final state will depend on the direction of the equilibrium process.

The anti → syn reorientation of G, and A is illustrated in Figure 12. This transformation takes N2(G) from its position in the minor groove and places it in the major groove and quite far outside the helix. Watson-Crick pairing is lost. There are no poor steric contacts in this model, and energy can be recovered by hydrogen bonding with water. An anti → syn rotation of A displaces N6(A) only slightly farther into the major groove and similarly for O6(G). As already shown, the anti orientation of G yields the most favorable fit and the pyrene orientations are in agreement with experimental results for the I(+)-N2(G) adducts. Similar calculations were performed for the BPDE I(-)- and II(-)-N6(A) adducts.

The stereoselected Cda conformation of the BPDE I(-) and II(-) adducts to N6(A) were chosen for study in a reoriented complex with an externally bound pyrene moiety. In Figure 13, the adduct is shown in its optimum orientation in B-DNA with adenine after an anti → syn transformation for which the non-bonded contacts are poor, and with the normal anti base orientation with favorable contacts. The fit improves for the anti base as α_x → 30°. The orientation of the pyrene moiety is $\alpha(BPDE) = 31°$ and the local helical axis of the DNA is oriented at $\gamma(DNA) = 15°$. Calculations were not performed with externally bound BPDE-DNA adducts to O6(G) and N4(C). Calculations of externally bound BPDE I(-)-N6(A) adducts with kinked DNA with α_x → 30° yields an orientation $\alpha(BPDE) = 31°$ in good agreement with experimental results for the externally bound component (51).

The energies reported in Table XIII for the externally bound forms are measured relative to that for the intercalative covalently bound form. Thus, the trans BPDE I(+)-N2(G) adduct is 10.1 kcal/mole more stable and the trans BPDE II(-)-N6(A) adduct is 12.7 kcal/mole less stable in the externally bound form. Similarily, the trans BPDE II(+)-N2(G) adduct is -13.8 kcal/mole more stable and the trans BPDE I(-)-N6(A) adduct is 7.5 kcal/mole less stable. Therefore, site IQ (intercalative covalent) which is favored by the I(-) isomer (51) may be due to N6(A) and N4(C) adduct formation, specifically trans addition.

The distribution of BPDE I(-) adducts observed by Brookes et al. (38) as 59% N2(G), 21% O6(G), 18% N6(A) and 2% other must be addressed. Although the yield of N2(G) adduct is 59% compared to

that with other base atoms, it is actually 0.1 of the I(+)-N2(G) adduct.   The binding to N2(G) and the resulting form (externally or
internally bound) will predominant. In _trans_ addition the I(-) isomer is unfavored in the intercalative covalent step; however, _cis_
addition is possible as discussed in the previous section.   The _cis_
adduct should yield a stable complex analogous to the results for

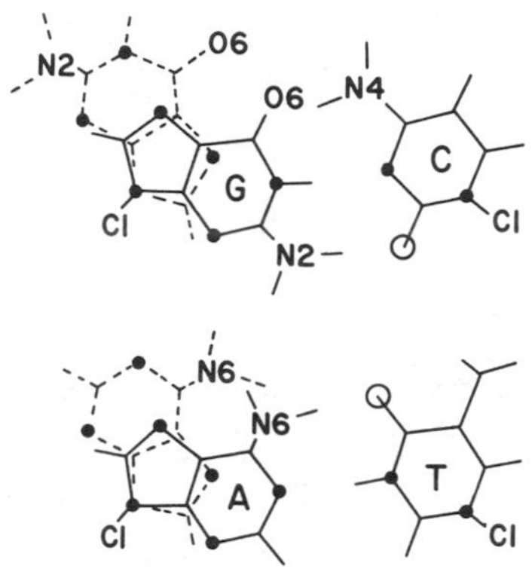

Figure 12.   _Anti_ (-) and _syn_ (---) reorientation of G and A about
their glycosidic bonds (C1′-O) in B-DNA.

the I(+) adduct in Table XIII.   The contribution to both IQ (internal) and IIX (external) binding sites by the I(-) isomer may be due
to the small amount of _cis_ adduct which binds to N2(G) (for IQ) and
low reactivity toward N6(A) (for site IQ).
        Two other models which orient the pyrene moiety externally have
been proposed. Aggarwal et al. (108) successfully fit both the I(+)
and I(-) into A-DNA which can arise from a local distortion of B-
DNA.   In the I(+) adduct, the chromophore is directed out of the
minor groove, whereas for the I(-) it fits snugly into the groove.
The angle subtended by the long axis of BP with respect to the helix
axis is 67° for I(+) and 63° for I(-).   Hingerty and Broyde (109)
have optimized the conformation of the dCpdG-BPDE I(+) adduct.   They
find a pyrene base stacked conformation with C and oriented by
α(BPDE) = 25°-30°.   If their complex can be incorporated into a
double helix, it may represent a local deformation of an externally
bound form with the pyrene oriented in the minor groove.   Thus, all
three models for the "final" externally bound BPDE I(+) to N2(G) in
the minor groove exhibit similar physical characteristics.   At the
time of writing this manuscript, stereoselectivity has not been
demonstrated with these latter two models (108,109).

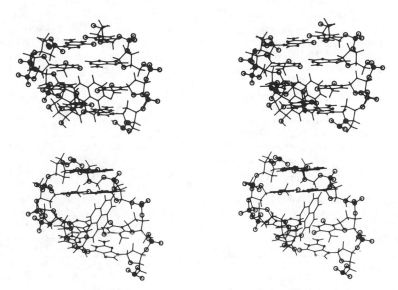

Figure 13.   An externally bound BPDE II(-)-N6(A) adduct with the
pyrene moiety   placed in the major groove.   The receptor sites are
(upper) B-DNA except for an <u>anti</u> → <u>syn</u> rotation by 180° about the
glycosidic bond of A, and (lower) DNA with an $\alpha_X = 30°$ kink.

Discussion and Conclusion

Theoretical results have been   presented with analytical data and
stereographic projections to support a mechanism given in Figure 14
in which stereoselectivity occurs   during intercalative covalent
binding.   We propose a molecular model and demonstrate that each
step plays the following roles: The intercalation step orients the
epoxide toward the major or minor groove with the reactive C10 atom
in the groove adjacent to the appropriate base atoms: N2(G), N6(A),
O6(G) and N4(C).   A proton catalyzed nucleophilic S_N2 reaction is
favored during intercalation because positive ions, especially H^+,
reside in the grooves and their presence assists in the activation
of the BPDEs.   However, during covalent bond formation, the DNA must
kink to orient the pyrene moiety within the kinked site for proper
bonding between C10 of the BPDEs and the reactive base atoms, i.e.,
to achieve a bond length of approximately 1.5 Å and tetrahedral hybridization on C10 of the BPDE and trigonal hybridization on the reactive base atoms.   In this   step the base pairs are deformed from
the parallel orientation of a classical intercalation site while the
BPDE remains inserted.   The DNA kinks or bends with a wedge opening
into the major or minor groove, respectively, while the adjacent
base pairs remain separated to accommodate the quasi intercalated
BPDE as it undergoes intercalative covalent binding to DNA.   It was

shown that stereoselectivity occurs during this step. Specifically,
BPDE I(+) and II(+) are the only isomers which fit in a kinked site
(+39°) when bound to N2 on guanine.   In contrast, BPDE I(-)   and
II(-) are the only isomers which fit in a kinked site (-30°) when
bound to atoms N6(A), O6(G) and it appears that binding to N4(C) may
not be stereoselective. The less favored proton addition via an $S_{N}1$
reaction forms the carbonium ion of BPDE.   This permits both <u>trans</u>
and <u>cis</u> addition, and minor products. For <u>cis</u> addition, the mirror
images are stereoselected, namely, I(-) and $\overline{II(}-)$ by N2(G), I(+) and
II(+) by N6(A) and O6(G) and II(-) and I(+) by N4(C). If both <u>trans</u>
and <u>cis</u> addition occurred to an equal extent, stereoselectivity
would not be observed.   Therefore, the favored <u>trans</u> addition as
well as the   steric fit of specific stereoisomers during intercalative covalent binding contribute to stereoselectivity. Possible re-
arrangements of the DNA to yield outside binding can occur in two
ways: First, an <u>anti</u> → <u>syn</u> rotation about the glycosidic bond of
the affected bases allows the remaining portion of the DNA to resume
its normal B-DNA conformation with an externally bound adduct that
fits well in the case of BPDE I(+) and II(+) bound to N2(G), but not
well for BPDE I(-) and II(-) bound to N6(A).   Second, a denaturation, rearrangement of the adduct and renaturation of the DNA allows
the adduct to lie in a groove in a slightly kinked DNA.

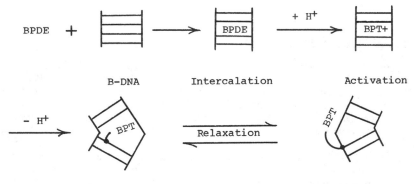

Intercalative covalent binding;   Outside binding;
stereoselectivity; final   final predominant
predominant structure for <u>trans</u>   structure for <u>trans</u>
addition to N6(A) and O6(A)   addition to N2(G) of
of BPDE I(-) and II(-).   BPDE I(+) and II(+).

Figure 14. Mechanism for BPDE-DNA adducts.

The main features of this proposed mechanism are (1) the stereoselectivity of the BPDEs by the DNA during intercalative covalent binding and (2) the final orientation of the bound pyrene which may be oriented internally (intercalative covalent) or externally (outside the helix). The stereoselectivity occurs during covalent bond formation and after intercalation. Relaxation of the DNA allows the adduct to adjust to its final orientation. If the experimental measurements are assumed to be made on the DNA-adducts after the final orientation has been achieved, then the following interpretations can be made.

The I(+) and II(+) isomers are stereoselected by N2(G), whereas the I(-) and II(-) isomers are stereoselected by the N6(A) and O6(G) during intercalative covalent steps with trans addition. The I(+)- and II(+)-N2(G) adducts are rearranged to an externally bound form with the pyrene in the minor groove, but the I(-)-N6(A) and II(-)-O6(G) adducts remain quasi intercalated. This is determined by the relative energy change between the two forms as we see from Table XIII. However, there is a superposition of the two types of sites, IQ and IIX (51,57,58), and BPDE I(-) DNA adducts exhibit both types of binding. By symmetry, the cis BPDE I(-)-N2(G) adduct is predicted to behave similarily to the trans I(+)-N2(G) adduct. It should be externally bound.

The N2(G) adducts are more stable than the N6(A) and O6(G) and N4(C) adducts. Because cis addition products are present, minor amounts of the other adducts are found. If only cis addition occurred, then the I(-) and II(-) isomers would be stereoselected by N2(G), and the I(+) and II(+) isomers would be stereoselected by N6(A) and O6(G). Although we did not perform calculations on the cis adducts, it can be seen from the stereographic projections that the change accompanied by a reflection of only the BPDE atoms through the plane of the pyrene changes the (+) into (-) isomers. Thus, the rules of stereoselectivity are reversed. However, the small amount of cis adduct yields these minor components; the I(-)-N2(G) adduct is most prevalent (38) for reactions of BPDE I(-) with DNA and we assume that this arises from the cis addition. If both trans and cis addition occurred equally, we predict that stereoselectivity would not be observed.

Based on the results in this paper, the following experimental data should be obtained for each of the diastereoisomers. (1) The relative yields of trans and cis addition products should be determined for adduct formation to each base atom. (2) Alternating and non-alternating homopolymers should be used to evaluate the base sequence specificity. (3) Binding to sites IQ and IIX should be correlated to trans and cis adducts and to stereoselectivity.

Acknowledgments

The authors acknowledge the grant of computer time from Rensselaer Polytechnic Institute and support by the National Institutes of Health under Grant CA-28924. Also, the authors wish to thank Chris Bonesteel for her technical assistance with the preparation of this manuscript.

Literature Cited

1.   Harvey, R. G. Acc. Chem. Res. 1981, 14, 218-226.
2.   Brookes, P.; Lawley, P. D. Nature 1964, 202, 781-784.
3.   Daudel, P.; Duquesne, M.; Vigny, P.; Grover, P. L.; Sims, P.
      FEBS Lett. 1975, 57, 250-253.
4.   Ivanovic, V.; Geacintov, N. E.; Weinstein, I. B.   Biochem.
      Biophys. Res. Commun. 1976, 70, 1172-1179.
5.   King, H. W. S.; Osborne, M. R.; Beland, F. A.; Harvey, R. G.;
      Brookes, P. Proc. Natl. Acad. Sci. (USA) 1976, 73, 2679-2681.
6.   Mager, R.; Huberman, E.; Yang, S. K.; Gelboin, H. V.; Sachs,
      L. Int. J. Cancer 1977, 19, 814-817.
7.   Moore, P. D.; Koreeda, M.; Wislocki, P. G.; Levin, W.; Conney,
      A. H.; Yagi, H.; Jerina, D. M. "Drug Metabolism Concepts"; D.
      M. Jerina, Ed.;   Am. Chem. Soc.:   Washington, D. C., 1977; p.
      127-154.
8.   Osborne, M. R.; Jacobs, S.; Harvey, R. G.; Brookes, P.
      Carcinogenesis 1981, 6, 553-558.
9.   Gamper, H. B.; Tung, A. S.-C.; Straub, K.; Bartholomew, J. C.;
      Calvin, M. Science 1977, 197, 671-674.
10.   Haseltine, W. A.; Lo, K. M.; D'Andrea, A. D.   Science 1980,
      209, 929-931.
11.   Gamper, H. B.; Bartholomew, J. C.; Calvin, M.   Biochemistry
      1980, 19, 3948-3956.
12.   Koreeda, M.; Moore, P. D.; Wilson, P. G.; Levin, W.; Conney, A.
      H.; Yagi, H.; Jeffrey, D. M. Science 1978, 199, 778-781.
13.   Weinstein, I. B.; Jeffrey, A. M.; Jennette, K. W.; Blobstein,
      S. H.; Harvey, R. G.; Harris, C.; Autrup, H.; Kasai, H.;
      Nakanishi, K. Science 1976, 193, 592-595.
14.   Jeffrey, A. M.; Weinstein, I. B.; Jennette, K. W.; Grzeskowiak,
      K.; Nakanishi, K.; Harvey, R. G.; Autrup, H.; Harris, C.
      Nature (London) 1977, 269, 348-350.
15.   Yagi, H.; Akagi, H.; Thakker, D. R.; Mah, H. D.; Koreeda, M.;
      Jerina, D. M. J. Am. Chem. Soc. 1977, 99, 2358-2359.
16.   Nakanishi, K.; Kasai, H.; Cho, H.; Harvey, R. G.; Jeffrey, A.
      M.; Jennette, K. W.; Weinstein, I. B. J. Am. Chem. Soc. 1977,
      99, 258-260.
17.   Ivanovic, V.; Geacintov, N. E.; Yamasaki, H.; Weinstein, I. B.
      Biochem. 1978, 17, 1597-1603.
18.   Jennette, K. W.; Jeffrey, A. M.; Blobstein, S. H.; Beland, F.
      A.; Harvey, R. G.; Weinstein, I. B. Biochem. 1977, 16, 932-
      938.
19.   Meehan, T.; Straub, K.; Calvin, M. Nature 1977, 269, 725-727.
20.   Jeffrey, A. M.; Jennette, K. W.; Blobstein, S. H.; Weinstein,
      I. B.; Beland, F. A.; Harvey, R. G.; Kasai, H.; Miura, I.;
      Nakanishi, K. J. Am. Chem. Soc. 1976, 98, 5714-5715.
21.   Jeffrey, A. M.; Grzeskowiak, K.; Weinstein, I. B.; Nakanishi,
      K.; Roller, P.; Harvey, R. G. Science 1979, 206, 1309-1311.
22.   Koreeda, M.; Moore, P. D.; Yagi, H.; Yeh, H. J. C.; Jerina, D.
      M. J. Am. Chem. Soc. 1976, 98, 6720-6722.
23.   Miller, E. C. Can. Res. 1978, 38, 1479-1496.
24.   Weisburger, J. H.   "Bioassays and Tests for Chemical Carcino-
      gens"; In "Chemical Carcinogens", C. E. S. Searle, Ed.; ACS

MONOGRAPH No. 173; American Chemical Society: Washington, D. C, 1976; p. 1-23.

25. Miller, J. A. Cancer Res. 1970, 30, 559-576.

26. Gelboin, H. V.; Kinoshita, N.; Wiebel, F. J. Fed. Proc. Fed. Am. Soc. Exp. Biol. 1972, 31, 1298-1309.

27. Cook, J. W.; Hewett, C. L.; Hieger, I. J. Chem. Soc. 1933, 395-405.

28. Sims, P.; Grover, P. L.; Swaisland, A.; Pal, K.; Hewer, A. Nature (London) 1974, 252, 326-327.

29. Huberman, E.; Sachs, L.; Yang, S. K.; Gelboin, H. V. Proc. Natl. Acad. Sci. (USA) 1976, 73, 607-611.

30. Meehan, T.; Straub, K.; Calvin, M. Proc. Natl. Acad. Sci. (USA) 1976, 73, 1437-1441.

31. Borgen, A.; Darvey, H.; Catagnoli, N.; Crocker, T. T.; Rasmussen, R. E.; Wang, I. Y. J. Med. Chem. 1973, 16, 502-506.

32. Osborne, M. R.; Harvey, R. G.; Brookes, P. Chem.-Biol. Interact. 1978, 20, 123-130.

33. King, H. W. S.; Osborne, M. R.; Brookes, P. Chem.-Biol. Interact. 1979, 24, 345-353.

34. Meehan, T.; Straub, K. Nature 1979, 277, 410-412.

35. Pulkrabek, P.; Leffler, S.; Grunberger, D.; Weinstein, I. B. Biochemistry 1979, 18, 5128-5134.

36. Wood, A. W.; Chang, R. L.; Levin, W.; Yagi, H.; Thakker, D. R.; Jerina; D. M.; Conney, A. H. Biochem. Biophys. Res. Commun. 1977, 77, 1389-1396.

37. Miller, K. J.; Taylor, E. R.; Dommen, J.; Burbaum, J. J. In "The Molecular Basis of Cancer", Part A; R. Rein, Ed.; Alan R. Liss, Inc.: New York, 1985; p. 187-197.

38. Brookes, P.; Osborne, M. R. Carcinogenesis 1982, 3, 1223-1226.

39. Lefkowitz, S. M.; Brenner, H. C.; Astorian, D. G.; Clarke, R. H. FEBS Letters 1979, 105, 77-80.

40. Prusik, T.; Geacintov, N. E.; Tobiasz, C.; Ivanovic, V.; Weinstein, I. B. Photochem. and Photobiol. 1979, 29, 223-232.

41. Prusik, T.; Geacintov, N. E. Biochem. Biophys. Res. Commun. 1979, 88, 782-790.

42. Drinkwater, N. R.; Miller, J. A.; Miller, E. C.; Yang, N. C. Cancer Res. 1978, 38, 3247-3255.

43. Miller, K. J.; Burbaum, J. J.; Dommen, J. In "Polynuclear Aromatic Hydrocarbons: Sixth International Symposium on Physical and Biological Chemistry"; M. Cooke, A. Dennis, Ed.; Battelle Press: Columbus, OH, 1981; p. 515-528.

44. Taylor, E. R.; Miller, K. J.; Bleyer, A. J. J. Biomol. Struct. and Dynam. 1983, 1, 883-904.

45. Gamper, H. B.; Straub, K.; Calvin, M.; Bartholomew, J. C. Proc. Natl. Acad. Sci. (USA) 1980, 77, 2000-2004.

46. Meehan, T.; Gamper, H.; Becker, J. F. J. Biol. Chem. 1982, 257, 10479-10485.

47. Geacintov, N. E.; Gagliano, A.; Ivanovic, V.; Weinstein, I. B. Biochemistry 17, 5256-5262 (1978).

48. Beland, F. A. Chem.-Biol. Interact. 22, 329-339 (1978).

49. Jeffrey, A. M., Kinoshita, T., Santella, R. M., Grunberger, D., Katz, L., Weinstein, I. B. In "Carcinogenesis: Fundamental Mechanisms and Environmental Effects"; Pullman, B., Tso, P. O. P., Gelboin, H., Eds.; 13th JERUSALEM SYMPOSIUM ON QUANTUM

CHEMISTRY AND BIOCHEMISTRY Vol. 13; Jerusalem, Israel; 1980; pp. 565-579.

50. Hogan, M. E.; Dattagupta, N.; Whitlock, Jr., J. P. J. Biol. Chem. 1981, 256, 4505-4513.
51. Geacintov, N. E.; Ibanez, V.; Gagliano, A. G.; Jacobs, S. A.; Harvey, R. G. J. Biol. Struct. and Dynam. 1984, 1, 1473-1484.
52. Geacintov, N. E.; Yoshida, H.; Ibanez, V.; Harvey, R. G. Biochem. Biophys. Res. Commun. 1981, 100, 1569-1577.
53. Geacintov, N. E.; Yoshida, H.; Ibanez, V.; Harvey, R. G. Biochem. 1982, 21, 1864-1869.
54. Geacintov, N. E.; Yoshida, H.; Ibanez, V.; Jacobs, S. A.; Harvey, R. G. Biochem. Biophys. Res. Commun. 1984, 122, 33-39.
55. MacLeod, M. C. & Selkirk, J. K. Carcinogenesis 1982, 3, 287-292.
56. Geacintov, N. E.; Ibanez, V.; Gagliano, A. G.; Yoshida, H.; Harvey, R. G. Biochem. Biophys. Res. Commun. 1980, 92, 1335-1342.
57. Geacintov, N. E.; Gagliano, A. G.; Ibanez, V.; Harvey, R. G. Carcinogenesis 1982, 3, 247-253.
58. Undeman, O.; Lycksell, P.-O.; Gräslund, A.; Astlind, T.; Ehrenberg, A.; Jerström, B; Tjerneld, F.; Norden, B. Cancer Research 1983, 43, 1851-1860.
59. Olson, W. Macromolecules 1975, 8, 272-275.
60. Smith, P. J. C.; Arnott, S. Acta Cryst. 1978, A34, 3-11.
61. Alden, C. J.; Arnott, S. Nucleic Acids Res. 1975, 2, 1701-1717.
62. Alden, C. J.; Arnott, S. Nucleic Acids Res. 1977, 4, 3855-3861.
63. Tumanyan, V. G.; Esipova, N. G. Biopolymers 1975, 14, 2231-2246.
64. Zhurkin, V. B.; Lysov, Yu. P.; Ivanov, V. I. Biopolymers 1978, 17, 377-412.
65. Miller, K. J. Biopolymers 1979, 18, 959-980.
66. Taylor, E. R.; Miller, K. J. Biopolymers 1984, 23, 2853-2878.
67. Miller, K. J. Proceedings of the Second SUNYA Conversation in the Discipline Biomolecular Stereodynamics; R. H. Sarma, ed.; New York, 1981; Vol. II, pp. 469-486.
68. Miller, K. J.; Pycior, J. F. Biopolymers 1979, 18, 2683-2719.
69. Miller, K. J.; Brodzinsky, R.; Hall, S. Biopolymers 1980, 19, 2091-2122.
70. Newlin, D. D.; Miller, K. J.; Pilch, D. F. Biopolymers 1984, 23, 139-158.
71. Taylor, E. R.; Olson, W. K. Biopolymers 1983, 22, 2667-2702.
72. Pack, G. R.; Loew, G. Biochim. Biophys. Acta 1978, 519, 163-172.
73. Nuss, M. E.; Marsh, F. J.; Kollman, P. A. J. Am. Chem. Soc. 1979, 101, 825-833.
74. Ornstein, R. L.; Rein, R. Biopolymers 1979, 18, 2821-2847.
75. Ornstein, R. L.; Rein, R. Biopolymers 1979, 18, 1277-1291.
76. Berman, H. M.; Neidle, S. In "Stereodynamics of Molecular Systems"; Sarma, R. H., Ed.; Pergamon: New York; 1979, pp. 367-382.

77. For example, Bingham, R. C.; Dewar, M. J. S.; Lo, D. H. J. Am. Chem. Soc. 1975, 97, 1285-1293 and subsequent papers.
78. Miller, K. J.; Pycior, J. F.; Moschner, K. F. QCPE BULLETIN 1981, 1, 67-70.
79. Neidle, S.; Subbiah, A.; Cooper, C. S.; Riberio, O. Carcinogenesis 1980, 1, 249-254.
80. Zacharias, D. E.; Glusker, J. P.; Fu, P. P.; Harvey, R. G. J. Am. Chem. Soc. 1979, 101, 4043-4051.
81. Silverman, B. D. Cancer Biochem. Biophys. 1983, 6, 131-136.
82. Lavery, R.; Pullman, B. Int. J. Quant. Chem. 1979, XV, 271-280.
83. Kikuchi, O.; Pearlstein, R.; Hopfinger, A. J.; Bickers, D. R. J. Pharm. Sci. 1983, 72, 800-808.
84. Politzer, P.; Daiker, K. C.; Estes, V. M. Int. J. Quant. Chem. Quant. Biol. Symp. 1979, 6, 47-53.
85. Klopman, G.; Grinberg, H.; Hopfinger, A. J. J. Theor. Biol. 1979, 79, 355-366.
86. Zielinski, T. J.; Breen, D. L.; Rein, R. J. Am. Chem. Soc. 1978, 100, 6266-6267.
87. Klopman, G.; Andreozzi, P.; Hopfinger, A. J.; Kikuchi, O. J. Am. Chem. Soc. 1978, 100, 6267-6268.
88. Neidle, S.; Subbiah, A.; Kuroda, R.; Cooper, C. S. Cancer Res. 1982, 42, 3766-3768.
89. Politzer, P.; Daiker, K. C.; Estes, V. M.; Baughman, M. Int. J. Quant. Chem.: Quant. Biol. Symp. 1978, 5, 291-299.
90. Loew, G, H.; Pudzianowski, A. T.; Czerwinski, A.; Ferrell, Jr., J. E. Intl. J. Quant. Chem.: Quant. Biol. Symp. 1980, 7, 223-244.
91. Malhotra, D.; Hopfinger, A. J. Nucleic Acids Res. 1980, 8, 5289-5304.
92. Subbiah, A.; Islam, S. A.; Neidle, S. Carcinogensis 1983, 4, 211-215.
93. Miller, K. J.; Macrea, J.; Pycior, J. F. Biopolymers 1980, 19, 2067-2089.
94. Pullman, B. J. Biomolec. Struct. and Dynam. 1983, 1, 773-794.
95. Clementi, E.; Corongiu, G. J. Biol. Phys. 1983, 11, 33-42.
96. Whalen, D. L.; Ross, A. M.; Montemarano, J. A.; Thakker, D. R.; Yagi, H.; Jerina, D. M. J. Am. Chem. Soc. 1979, 101, 5086-5088.
97. Geacintov, N. E.; Yoshida, H.; Ibanez, V.; Harvey, R. G. Biochem. Biophys. Res. Commun. 1981, 100, 1569-1577.
98. Geacintov, N. E.; Hibshoosh, H.; Ibanez, V.; Benjamin, M. J.; Harvey, R. G. Biophys. Chem. 1984, 20, 121-133.
99. Michaud, D. P.; Gupta, S. C.; Whalen, D. L.; Sayer, J. M.; Jerina, D. M. Chem.-Biol. Interactions 1983, 44, 41-52.
100. Kootstra, A.; Haas, B. L.; Slaga, T. J. Biochem. Biophys. Res. Commun. 1980, 94, 1432-1438.
101. Lavery, R.; Pullman, B. Intl. J. Quant. Chem. 1979, XVI, 175-188.
102. Lavery, R.; Pullman, A.; Pullman, B. Int. J. Quant. Chem.: Quant. Biol. Symp. 1978, 5, 21-34.
103. Lin, J.-h.; LeBreton, P. R.; Shipman, L. L. J. Phys. Chem. 1980, 84, 642-649.

104. Nakata, Y.; Malhotra, D.; Hopfinger, A. J.; Bickers, D. R.   J. Pharm. Sci. 1983, 72, 809-811.

105. Teitelbaum, H.; Englander, S. W.   J. Mol. Biol. 1975, 92, 55-78.

106. Teitelbaum, H.; Englander, S. W.   J. Mol. Biol. 1975, 92, 79-92.

107. Gabbay, E. J.; DeStefano, R.; Baxter, C. S.   Biochem. Biophys. Res. Commun. 1973, 51, 1083-1089.

108. Aggarwal, A. K.; Islam, S. A.; Neidle, S.   J. Biomol. Struct. and Dynam. 1983, 1, 873-881.

109. Hingerty, B.; Broyde, S.   J. Biomol. Struct. and Dynam. 1983, 1, 905-912.

RECEIVED April 18, 1985

One-Electron Oxidation in Aromatic Hydrocarbon Carcinogenesis

ERCOLE L. CAVALIERI and ELEANOR G. ROGAN

Eppley Institute for Research in Cancer, University of Nebraska Medical Center, Omaha, NE 68105

Two main pathways are involved in the carcinogenic activation of polycyclic aromatic hydrocarbons (PAH): one-electron oxidation and monooxygenation. One-electron oxidation produces PAH radical cations, which can react with cellular nucleophiles. Biochemical and biological data indicate that only PAH with relatively low ionization potentials (below ca. 7.35 eV) can be activated by one-electron oxidation. Furthermore, a carcinogenic PAH must have a relatively high charge localization in its radical cation to react effectively with target cellular macromolecules. Binding of benzo[a]pyrene (BP) to DNA in vitro and in vivo occurs predominantly at C-6, the position of highest charge density in the BP radical cation, and binding of 6-methylBP to mouse skin DNA yields a major adduct in which the 6-methyl is bound to the 2-amino of deoxyguanosine. PAH radical cations are also involved in the metabolic conversion of PAH to PAH diones. Carcinogenicity studies of PAH in rat mammary gland indicate that only PAH with ionization potential low enough for activation by one-electron oxidation induce tumors in this target organ. These results and others indicate that one-electron oxidation of PAH is involved in their tumor initiation process.

Covalent binding of chemical carcinogens to cellular macromolecules, DNA, RNA and protein, is well-accepted to be the first step in the tumor initiation process (1,2). Most carcinogens, including polycyclic aromatic hydrocarbons (PAH), require metabolic activation to produce the ultimate electrophilic species which react with cellular macromolecules. Understanding the mechanisms of activation and the enzymes which catalyze them is critical to elucidating the tumor initiation process.

Historically the process of activation has almost exclusively been studied by metabolizing compounds with liver preparations, leading most investigators in chemical carcinogenesis to think that

0097–6156/85/0283–0289$06.00/0
© 1985 American Chemical Society

oxygenation is the critical step to produce proximate and/or ultimate carcinogens. This emphasis has indeed been predominant for
PAH, in which formation of bay-region vicinal diol epoxides has been
described to be the most important, if not exclusive, pathway of
activation (3-5).

At present a variety of studies with PAH, as well as other
chemicals, suggest that metabolic activation in target tissues can
occur by one-electron oxidation (6,7). The electrophilic intermediate radical cations generated by this mechanism can react directly
with various cellular nucleophiles. In this paper, we will discuss
chemical, biochemical and biological evidence which indicates that
one-electron oxidation plays an important role in the metabolic
activation of PAH.

Chemical Properties of PAH Radical Cations

Nucleophilic Trapping of Radical Cations. To investigate some of
the properties of PAH radical cations these intermediates have been
generated in two one-electron oxidant systems. The first contains
iodine as oxidant and pyridine as nucleophile and solvent (8-10),
while the second contains $Mn(OAc)_3$ in acetic acid (10,11). Studies
with a number of PAH indicate that the formation of pyridinium-PAH
or acetoxy-PAH by one-electron oxidation with $Mn(OAc)_3$ or iodine,
respectively, is related to the ionization potential (IP) of the
PAH. For PAH with relatively high IP, such as phenanthrene,
chrysene, 5-methylchrysene and dibenz[a,h]anthracene, no reaction
occurs with these two oxidant systems. Another important factor
influencing the specific reactivity of PAH radical cations with
nucleophiles is localization of the positive charge at one or a few
carbon atoms in the radical cation.

For unsubstituted PAH, such as benzo[a]pyrene (BP), pyridinium
or acetoxy derivatives are formed by direct attack of pyridine or
acetate ion, respectively, on the radical cation at C-6, the position of maximum charge density (Scheme 1). This is followed by a
second one-electron oxidation of the resulting radical and loss of a
proton to yield the 6-substituted derivative. For methyl-substituted PAH in which the maximum charge density of the radical cation
adjacent to the methyl group is appreciable, as in 6-methylbenzo[a]-
pyrene (6-methylBP) (Scheme 2), loss of a methyl proton yields a
benzylic radical. This reactive species is rapidly oxidized by
iodine or Mn^{3+} to a benzylic carbonium ion with subsequent trapping
by pyridine or acetate ion, respectively.

For activation by one-electron oxidation, these properties of
PAH radical cations enable us to predict the position(s) at which
covalent binding of PAH to cellular targets may occur.

Synthesis of Radical Cation Perchlorates and Subsequent Coupling
with Nucleophiles. Syntheses of the radical cation perchlorates of
BP and 6-methylBP (12) were accomplished by the method reported
earlier for the preparation of the perylene radical cation (13,14).
More recently we have also synthesized the radical cation perchlorate of 6-fluoroBP (15). Oxidation of the PAH with iodine in benzene
in the presence of $\overline{AgClO_4}$ instantaneously produces a black precipitate containing the radical cation perchlorate adsorbed on AgI with

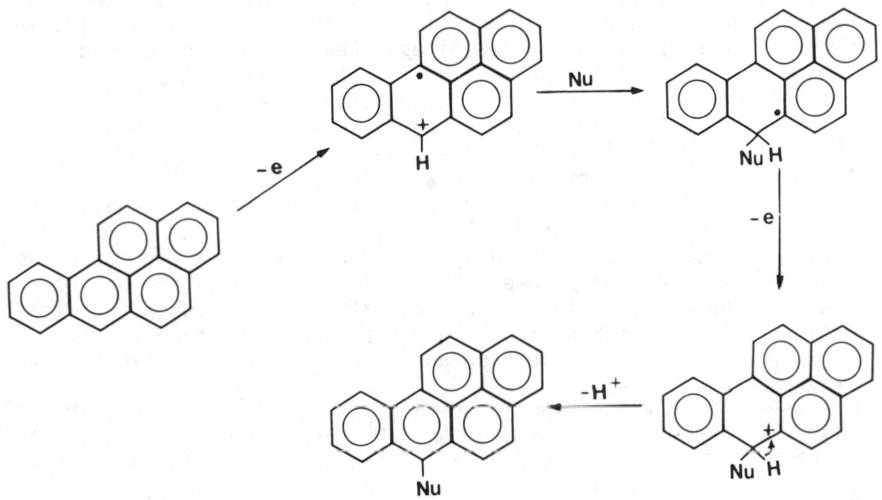

Scheme 1. Mechanism of trapping of BP radical cation by a nucleophile (Nu).

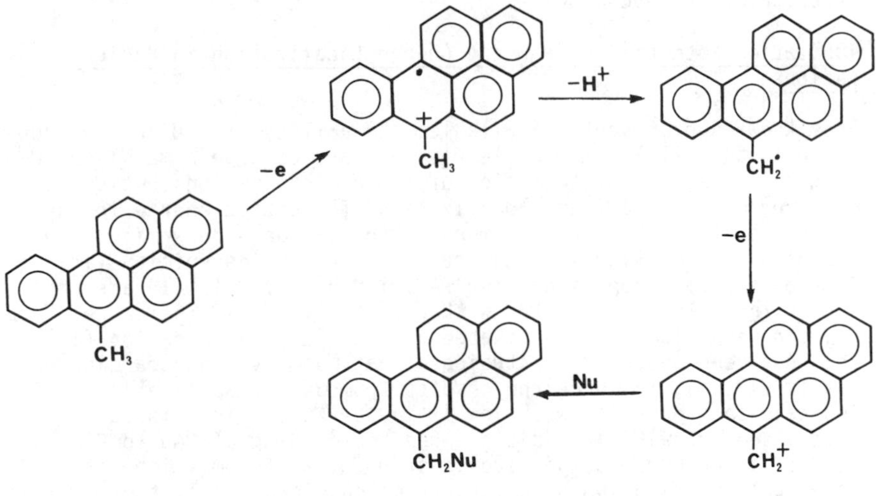

Scheme 2. Stepwise one-electron oxidation of 6-methylBP and subsequent trapping by a nucleophile (Nu).

yields of 28, 28 and 39% for $BP^{+}ClO_4^-$, 6-methylBP$^{+}ClO_4^-$ and 6-fluoro-BP$^{+}ClO_4^-$, respectively. The BP and 6-methylBP radical cations have been characterized by electron spin resonance spectroscopy ([12]) and by trapping with strong nucleophiles. Reaction of the BP radical cation with the two strong nucleophiles NaSCN and NaNO$_2$ yields 6-thiocyano- and 6-nitroBP, but also derivatives at C-1. Incidentally, in the BP radical cation, C-6 is the position of highest charge density, followed by C-1 and C-3. When the 6-methylBP and 6-fluoroBP radical cations react with NaNO$_2$ and NaSCN, only derivatives at the 1 and/or 3-position are obtained. Neither substitution at the 6-methyl group nor displacement of the fluorine atom is observed. These results generally indicate that strong nucleophiles display low selectivity toward the position in which the positive charge is better localized. Reaction of BP and 6-fluoroBP radical cations with the weak nucleophile H$_2$O affords a mixture of BP-1,6-, -3,6- and -6,12-dione. These products are the result of an initial attack of H$_2$O at C-6.

When the weak nucleophile acetate ion in water is used, $BP^{+}ClO_4^-$ yields specifically 6-acetoxyBP and the three diones, which are the result of reaction of the radical cation with H$_2$O. In the case of 6-fluoroBP$^{+}ClO_4^-$, BP diones are the predominant products, whereas only traces of 6-acetoxyBP are obtained. This indicates that the attack by the acetate ion is sterically hindered at the 6-position in the 6-fluoroBP$^{+}ClO_4^-$.

The overall conclusion from the reaction of BP and 6-substituted BP radical cations with nucleophiles of various strengths is that weak nucleophiles display higher selectivity toward the position of highest charge localization. Thus another important factor in the chemical reactivity of radical cations is represented by the strength of the nucleophile.

Ionization Potential of PAH and Charge Localization in Radical Cations

From knowledge presently available, the ability of PAH to bind covalently to cellular macromolecules appears to depend mainly on two factors: the ease of formation of PAH radical cations, which is measured by their IP, and localization of positive charge in the radical cation. The IP of numerous PAH have been determined and compared to a qualitative measure of their carcinogenicity ([16]). Some of the most representative PAH with high and low IP are presented in Table I. Only PAH with relatively low IP (below ca. 7.35 eV) can be biologically activated by one-electron oxidation ([16]). This has been observed in studies of rat mammary gland carcinogenesis ([10,17,18]), in which the results from direct application of PAH indicate that only PAH with low IP induce tumors in this target organ (see below). In addition when the binding of PAH to DNA is studied using horseradish peroxidase/H$_2$O$_2$, a system which catalyzes one-electron oxidation of a variety of chemicals, only those PAH with IP < ca 7.35 eV are significantly bound ([16]).

The carcinogenicity of PAH with relatively high IP, such as benzo[c]phenanthrene, benz[a]anthracene, chrysene, 5-methylchrysene and dibenz[a,h]anthracene (Table I), can be related to the formation of bay-region diol epoxides catalyzed by monooxygenase enzymes ([5]). However, the most potent carcinogenic PAH have IP < ca. 7.35 eV.

Table I. Structure, Ionization Potential, and Carcinogenicity of Selected PAH

Compound	Structure	Ionization[a] potential (eV)	Carcinogenicity[b]
Phenanthrene		8.19	−
Benzo[c]phenanthrene		7.93	+
Chrysene		CA. 7.8	±
5-Methylchrysene	CH₃	CA. 7.7	+++
Benzo[e]pyrene		7.62	−
Dibenz[a,h]anthracene		7.57	+++
Benz[a]anthracene		7.54	±
Pyrene		7.50	−
Anthracene		7.43	−
7-Methylbenz[a]anthracene	CH₃	7.37	+++

Continued on next page.

Table I. Continued.

Compound	Structure	Ionization[a] potential (eV)	Carcinogenicity[b]
Dibenzo[a,e]pyrene		7.35	+++
Dibenzo[a,l]pyrene		7.26	+++
Dibenzo[a,i]pyrene		7.25	++++
Benzo[a]pyrene		7.23	++++
6-Fluorobenzo[a]pyrene		7.23	++
7,12-Dimethylbenz[a]anthracene		7.22	+++++
3-Methylcholanthrene		7.12	++++
6-Methylbenzo[a]pyrene		7.08	+++
Perylene		7.06	—

Continued on next page.

Table I. Continued.

Compound	Structure	Ionization[a] Potential (eV)	Carcinogenicity
Dibenzo[a,h]pyrene		6.97	+ + + +
Anthanthrene		6.96	+

[a]Determined from absorption maximum of the charge-transfer complex of each compound with chloranil, with the exception of dibenz[a,h]-anthracene determined by polarographic oxidation (24).

[b]Extremely active, +++++; very active, ++++; active, +++; moderately active, ++; weakly active, +; very weakly active, ±; and inactive, —.

This list includes BP, 7,12-dimethylbenz[a]anthracene, 3-methylchol-
anthrene, dibenzo[a,i]pyrene and dibenzo[a,h]pyrene. These PAH can
be activated both by one-electron oxidation and/or monooxygenation.
There are a few PAH with low IP which are inactive (Table I), such
as perylene, or weakly active, such as anthanthrene. This indicates
that low IP is a necessary, but not sufficient factor for determining carcinogenic activity by one-electron oxidation. These inactive
or weakly active PAH have the highest density of positive charge
delocalized over several aromatic carbon atoms in their radical
cations, whereas the active PAH with low IP have charge mainly localized on one or a few carbon atoms in their radical cations.
These observations lead us to suggest that the second critical factor in binding of PAH radical cations is that the carcinogenic PAH
must have relatively high charge localization in their radical
cations to give them sufficient reactivity to bind with cellular
nucleophiles (6,7). Evidence on this point has been obtained by
one-electron oxidation of PAH with iodine (8-10) and Mn(OAc)$_3$
(10,11), although this concept of charge localization requires further study by more quantitative approaches.

Metabolic Formation of Quinones by an Initial One-Electron Oxidation of BP

Metabolism of BP mediated by the cytochrome P-450 monooxygenase
system forms three classes of products: phenols, dihydrodiols and
quinones. Formation of phenols and dihydrodiols is obtained by an
initial electrophilic attack of an enzyme-generated oxygen atom.
The same pathway of activation has been postulated in the formation
of quinones, although the putative 6-hydroxyBP precursor has never
been isolated (19,20). In this mechanism, formation of quinones
would proceed by autoxidation of 6-hydroxyBP (20). However, substantial evidence indicates that the first step in formation of
quinones does not involve the typical attack of the electrophilic
active oxygen to yield 6-hydroxyBP, but instead consists of the loss
of one electron from BP to produce the radical cation.
    The first line of evidence derives from the predominant forma-
tion of quinones when metabolism of BP is conducted under peroxidatic conditions, namely by prostaglandin H synthase (21) or by
cytochrome P-450 with cumene hydroperoxide as cofactor (22). Under
these metabolic conditions one-electron oxidation is the prepon-
derant mechanism of activation.
    Second, metabolism of 6-fluoroBP by rat liver microsomes yields
the same BP quinones obtained in the metabolism of BP (23). This
suggests that these products are formed by an initial attack of a
nucleophilic oxygen atom at C-6 in the 6-fluoroBP radical cation
with displacement of the fluoro atom. In fact, when 6-fluoroBP is
treated with the one-electron oxidant Mn(OAc)$_3$, the major products
obtained are 6-acetoxyBP and a mixture of 1,6- and 3,6-diacetoxyBP
(15), indicating that reaction occurs via an initial attack of acetate ion at C-6 of the 6-fluoroBP radical cation. On the other hand
electrophilic substitution of 6-fluoroBP with bromine or deuterium
ion shows no displacement of fluorine at C-6, although in both cases
substitution occurs at C-1 and/or C-3. These results indicate that

the only plausible chemistry in the metabolic formation of quinones from 6-fluoroBP is consistent with an initial one-electron oxidation of the compound to form 6-fluoroBP^{+}.

Finally, we have studied the metabolism of a series of PAH with decreasing IP. In these metabolic studies with Aroclor-induced rat liver microsomes, the formation of quinones was measured in the presence of NADPH or cumene hydroperoxide as cofactor.

As presented in Table II, no quinones are obtained with NADPH for dibenz[a,h]anthracene and benz[a]anthracene, whereas with cumene hydroperoxide a trace amount of benz[a]anthracene quinone is observed. For the PAH with low IP, quinones are formed in the presence of both cofactors. The relationship between IP and formation of quinones constitutes further evidence that these metabolites are obtained by an initial one-electron oxidation of the PAH with formation of its radical cation.

Table II. Metabolic Formation of Quinones for PAH of
Various Ionization Potentials

Compound	Ionization Potential (eV)[a]	Formation of Quinone by Aroclor-Induced Rat Liver Microsomes with	
		NADPH	Cumene Hydroperoxide
Dibenz[a,h]anthracene	7.57	-	-
Benz[a]anthracene	7.54	-	$\pm$[b]
Benzo[a]pyrene	7.23	+	+
Dibenzo[a,i]pyrene	7.20	+	+
Dibenzo[a,h]pyrene	6.97	+	+
Anthanthrene	6.96	$\pm$	+

[a]Determined from absorption maximum of the charge-transfer complex of each compound with chloranil, with the exception of dibenz[a,h]anthracene, which was determined by polarographic oxidation (24).

[b]$\pm$ indicates formation of a trace amount of quinone.

We propose that the first step in the formation of quinones, as shown in Scheme 3 for BP, involves an electron transfer from the hydrocarbon to the activated cytochrome P-450-iron-oxygen complex. The generated nucleophilic oxygen atom of this complex would react at C-6 of BP^{+} in which the positive charge is appreciably localized. The 6-oxy-BP radical formed would then dissociate to leave the iron of cytochrome P-450 in the normal ferric state. Autoxidation of the 6-oxy-BP radical in which the spin density is localized mainly on the oxygen, C-1, C-3 and C-12 (19,20) would produce the three BP diones.

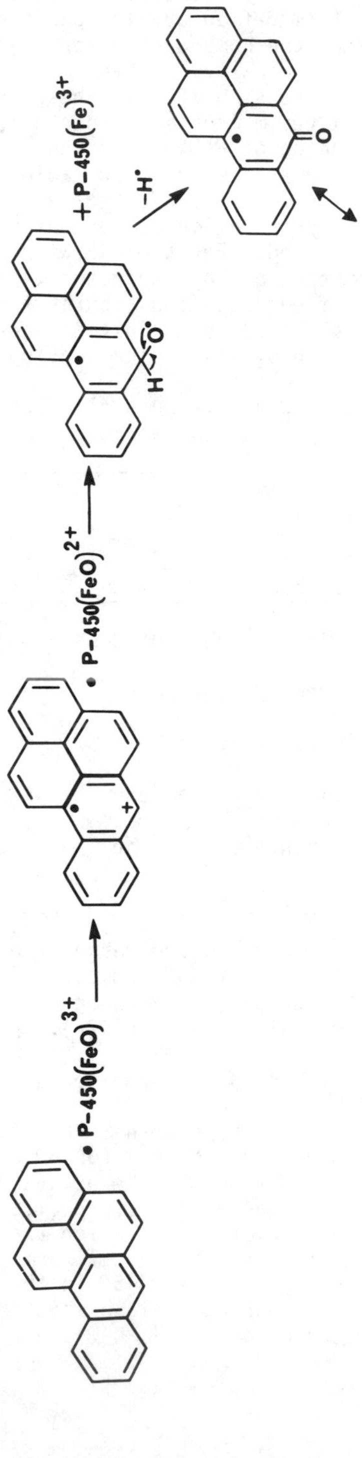

• P-450(FeO)3+
• P-450(FeO)2+
+ P-450(Fe)3+
-H•

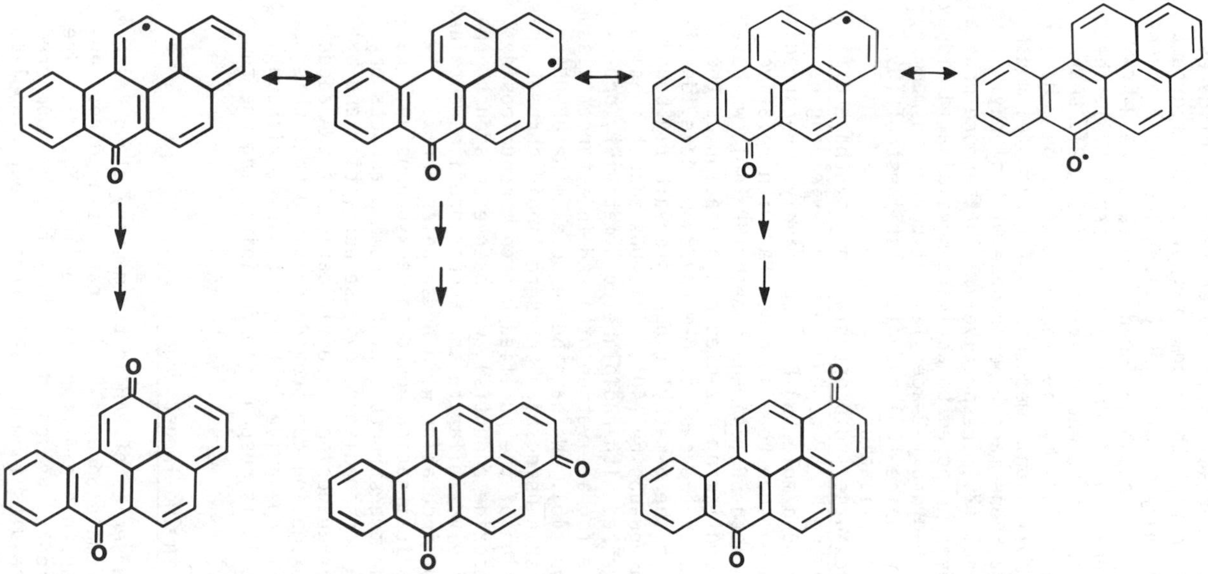

Scheme 3. Proposed mechanism of metabolic formation of BP diones by initial one-electron oxidation of BP.

Binding of PAH to DNA in vitro and in vivo

While most research on the enzymatic activation of chemical carcinogens has focused on monooxygenation by cytochrome P-450, it has
become increasingly clear that activation by cellular peroxidases,
including the prostaglandin H synthase complex, plays an important
role in the activation of many carcinogens (25). The model horseradish peroxidase/H_2O_2 system has been found to metabolize N-hy-
droxy-2-acetylaminofluorene (26,27), diethylstilbestrol (28), phenol
(29), aminopyrine (30), benzidine and derivatives (31, 32), tetramethylhydrazine (33) and BP (34) by one-electron oxidation. Mammalian peroxidases also follow this mechanism: for example, mouse
uterine peroxidase and rat bone marrow peroxidase with diethylstilbestrol (28) and phenol (29), respectively. Furthermore prostaglandin H synthase has been proposed to activate benzidine in kidney
carcinogenesis (35, 36), N-hydroxy-2-acetylaminofluorene in mammary
cells (37), tetramethylhydrazine (38) and diethylstilbestrol (39),
apparently by one-electron oxidation.
    Both horseradish peroxidase and prostaglandin H synthase effi-
ciently catalyze the binding of BP to DNA in vitro, yielding 89 + 5
and 310 + 64 μmole BP bound/mole DNA-P, respectively. Horseradish
peroxidase has already been seen to bind other PAH with relatively
low IP to DNA (16). For both BP (34) and 6-methylBP (40), we have
obtained clear evidence confirming one-electron oxidation as the
mechanism of activation. In the case of 6-methylBP we have identified a DNA adduct in which the 6-methyl group is covalently bound to
the 2-amino group of deoxyguanosine (40). This DNA adduct is also
present in mouse skin treated with radiolabeled 6-methylBP, providing the first evidence for activation of a PAH in a target tissue
by one-electron oxidation (40). We have begun to examine BP-DNA
adducts formed in mouse skin using high pressure liquid chromatography after enzymic digestion of the purified DNA to mononucleosides.
In addition to BP diol epoxide adduct(s), we observe an adduct profile which is qualitiatively similar to the adduct profiles obtained
from DNA with BP bound by incubation with horseradish peroxi-
dase/H_2O_2 and from BP radical cation bound to deoxyguanosine. We
are currently identifying the structure of the common adducts obtained on the skin, with horseradish peroxidase activation and by
reaction of BP^+ with deoxyguanosine. Identification of DNA adducts
formed by one-electron oxidation can provide evidence that this
mechanism of activation is operative in target tissues, although
this does not prove that it is responsible for initiating the tumor
process.

Carcinogenicity Studies in Two Target Organs

The carcinogenicity of a series of PAH in the mammary gland has been
examined in 50-day-old female Sprague-Dawley rats using direct application of the compound to the mammary tissue (10, 17, 18). The
results of these experiments, presented in Table III, are compared
to the carcinogenicity results in mouse skin from repeated application obtained in our laboratory and others. PAH were selected because they were or were not expected to be activated by one-electron
oxidation, based on the hypothesis that compounds with relatively
high IP cannot be activated by this mechanism. Furthermore, some

Table III. Comparative Carcinogenicity of PAH in Mouse Skin and Rat
Mammary Gland

Compound	Ionization[a] Potential (eV)	Carcinogenicity[b] in:	
		Mouse Skin	Rat Mammary Gland
Cyclopenta[cd]pyrene		+ +	-
Benzo[a]pyrene 7,8-dihydrodiol		+ + + +	-
5-Methylchrysene	ca. 7.7	+ + +	-
Dibenz[a,h]anthracene	7.57	+ + +	-
Benz[a]anthracene	7.54	$\pm$	-
7-Methylbenz[a]anthracene	7.37	+ + +	+
Benzo[a]pyrene	7.23	+ + + +	+ + +
7,12-Dimethylbenz[a]-anthracene	7.22	+ + + + +	+ + + +
10-Fluoro-3-methyl-cholanthrene	7.17	N.T.[c]	+ +
1,3-Dimethylcholanthrene	7.15	+ +	-
8-Fluoro-3-methylchol-anthrene	7.14	N.T.	+ +
2,3-Dimethylcholanthrene	7.13	N.T.	+ +
3-Methylcholanthrene	7.12	+ + + +	+ + + +
6-Methylbenzo[a]pyrene	7.08	+ + +	+

[a] Determined from absorption maximum of the charge-transfer complex of each compound with chloranil, with the exception of dibenz[a,h]-anthracene determined by polarographic oxidation (24).

[b] Extremely active, + + + + +; very active, + + + +; active, + + +, moderately active, + +; weakly active, +; very weakly active, $\pm$; inactive, -.

[c] N.T. = not tested.

PAH were chosen in which activation by monooxygenation or one-electron oxidation was blocked.

Compounds which have low IP and sufficient charge localization in the radical cation, namely 7-methylbenz[a]anthracene, BP, 7,12-dimethylbenz[a]anthracene, 10-fluoro-3-methylcholanthrene, 8-fluoro-3-methylcholanthrene, 2,3-dimethylcholanthrene, 3-methylcholanthrene, and 6-methylBP, are generally carcinogenic, both in mouse skin and rat mammary gland. However, 1,3-dimethylcholanthrene, which has a low IP, is inactive in rat mammary gland and active in mouse skin. This is presumably due to steric hindrance at C-1, the position of nucleophilic substitution in the 3-methylcholanthrene radical cation. Its carcinogenic activity in mouse skin can be attributed to activation by monooxygenation. In contrast 2,3-dimethylcholanthrene, in which the methyl substituent at C-2 does not prevent nucleophilic substitution at C-1 in the radical cation, is carcinogenic. PAH with relatively high IP, such as dibenz[a,h]anthracene and 5-methylchrysene, are not active when directly applied to the mammary gland. The carcinogenicity of 5-methylchrysene in mouse skin has been demonstrated to occur via a diol epoxide mechanism (41), and the potent activity of dibenz[a,h]-anthracene is presumably induced by the same mechanism (5). The inactivity of these two skin carcinogens suggests that diol epoxides are not formed in the mammary gland. No carcinogenic activity is observed in this target organ for the two mouse skin carcinogens BP 7,8-dihydrodiol(5) and cyclopenta[cd]pyrene (42), both of which require a simple epoxidation to become active.

From these experiments we can draw three main conclusions: 1) oxygenation of PAH by cytochrome P-450 monooxygenase enzymes does not seem to play a role in eliciting carcinogenicity in rat mammary gland; 2) the results in the mammary experiments support the hypothesis that one-electron oxidation might be the predominant mechanism of activation in this target organ; and 3) multiple mechanisms of activation appear to occur in mouse skin, although these experiments do not provide direct evidence on this point.

Conclusions

Based on present knowledge the carcinogenicity of PAH is best understood in terms of two major mechanisms of activation: one-electron oxidation and monooxygenation. The bay-region diol epoxides can be considered major ultimate carcinogenic intermediates when activation occurs by monooxygenation (3-5). One-electron oxidation of PAH with formation of radical cations can only play a role in biological systems when PAH have an IP below ca. 7.35 eV (Table I) (6,7). Thus carcinogenicity of compounds with relatively high IP (Table I), such as benzo[c]phenanthrene, chrysene, 5-methylchrysene and dibenz[a,h]-anthracene, can be attributed to monooxygenation with formation of bay-region diol epoxides. Most of the potent PAH, however, have IP below ca. 7.35 eV. This list includes BP, 7,12-dimethylbenz[a]anthracene, 3-methylcholanthrene, dibenzo[a,i]pyrene and dibenzo[a,h]-pyrene. These PAH can be activated by both one-electron oxidation and monooxygenation, depending on the enzymes present in the target organ. The ubiquity of peroxidases, in particular prostaglandin H synthase, in extrahepatic tissues which are responsive to PAH leads us to suggest that one-electron oxidation may be a major pathway of activation in most target tissues. Combined studies of enzymology, carcinogenicity and binding to cellular macromolecules should provide the information necessary to determine the role of the different mechanisms of PAH activation responsible for initiation of the cancer process in a certain target organ.

Acknowledgments

We appreciate the valuable collaboration of Drs. C. Warner, P. Cremonesi and A. Wong and of Mr. S. Tibbels. We are also grateful to Ms. M. Susman for excellent editorial assistance. Finally we thank the National Institutes of Health for supporting this research through grants R01 CA25176, R01 CA32376 and R01 ES02145.

Literature Cited

1. Miller, J.A. Cancer Res. 1970, 30, 559-76.
2. Miller, E.C.; Miller, J.A. Cancer, 1981, 47, 2327-45.
3. Nordqvist, M.; Thakker, D.R.; Yagi, H.; Lehr, R.E.; Wood, A.W.; Levin, W.; Conney, A.H.; Jerina, D.M. In "Molecular Basis of Environmental Toxicity"; Bhatnager, R.S., Ed.; Ann Arbor Science Publishers: Ann Arbor, Mich., 1979; pp. 329-357.
4. Sims, P.; Grover, P.L. In "Polycyclic Hydrocarbons and Cancer"; Gelboin, H.V.; Ts'o, P.O.P, Ed.; Academic: New York, 1978; Vol. 1, pp. 117-81.
5. Conney, A.H. Cancer Res. 1982, 42, 4875-917.
6. Cavalieri, E.L.; Rogan, E.G. In "Free Radicals in Biology"; Pryor, W.A., Ed.; Academic: New York, 1984; Vol. VI, pp. 323-369.
7. Cavalieri, E.; Rogan, E. In "Chemical Induction of Cancer"; by Woo, Y.-T; Lai, D.Y., Arcos, J.C.; Argus, M.F.; ; Academic: New York, 1984; in press.
8. Cavalieri, E.; Roth, R. J. Org. Chem., 1976, 41, 2679-84.
9. Cavalieri, E.; Roth, R.; Rogan, E.G. In "Polynuclear Aromatic Hydrocarbons: Chemistry, Metabolism and Carcinogenesis"; Freudenthal, R.I.; Jones, P.W., Eds.; Raven: New York, 1976; Vol. 1, pp. 181-190.
10. Cavalieri, E.; Rogan, E. In "Polynuclear Aromatic Hydrocarbons: Formation, Metabolism and Measurement"; Cooke, M.; Dennis, A.J., Eds.; Battelle Press, Columbus, Ohio, 1983; pp. 1-26.
11. Rogan, E.G.; Roth, R.; Cavalieri, E. In "Polynuclear Aromatic Hydrocarbons: Chemistry and Biological Effects"; Bjørseth, A.; Dennis, A.J., Eds.; Battelle Press: Columbus, Ohio, 1980; pp. 259-265.
12. Cavalieri, E.; Rogan, E.; Warner, C.; Bobst, A. In "Polynuclear Aromatic Hydrocarbons: Mechanisms, Methods and Metabolism"; Cooke, M.; Dennis, A.J., Eds.; Battelle Press, Columbus, Ohio, in press.
13. Sato, Y.; Kinoshita, M.; Sano, M.; Akamatu, H. Bull. Chem. Soc. Jap., 1969, 42, 3051-5.
14. Ristagno, C.V.; Shine, H.J. J. Org. Chem., 1971, 36, 4050-5.
15. Cavalieri, E.; Cremonesi, P.; Warner, C.; Tibbels, S.; Rogan, E. Proc. Am. Assoc. Cancer Res., 1984, 25, 124.
16. Cavalieri, E.L.; Rogan, E.G.; Roth, R.W.; Saugier, R.K.; Hakam, A. Chem.-Biol. Interact. 1983, 47, 87-109.

17. Cavalieri, E.; Sinha, D.; Rogan, E. In "Polynuclear Aromatic Hydrocarbons: Chemistry and Biological Effects"; Bjørseth, A.; Dennis, A.J., Eds.; Battelle Press: Columbus, Ohio, 1980; pp. 215-231.
18. Cavalieri, E.; Rogan, E. In "Polynuclear Aromatic Hydrocarbons: Physical and Biological Chemistry"; Cooke, M.; Dennis, A.J.; Fisher, G.L., Eds.; Battelle Press: Columbus, Ohio, 1982; pp. 145-155.
19. Nagata, C.; Kodama, M.; Ioki, Y.; Kimura, T. In "Free Radicals and Cancer"; Floyd, R.A., Ed.; Marcel Dekker: New York, 1982; pp. 1-62.
20. Lorentzen, R.J.: Caspary, W.J.; Lesko, S.A.; Ts'o, P.O.P. Biochemistry, 1975, 14, 3970-7.
21. Marnett, L.J.; Reed, G.A. Biochemistry, 1979, 18, 2923-9.
22. Renneberg, R.; Capdevila, J.; Chacos, H.; Estabrook, R.W.; Prough, R.A. Biochem. Pharmacol., 1981, 30, 843-8.
23. Buhler, D.R.; Ünlü, F.; Thakker, D.R.; Slaga, T.J.; Conney, A.H.; Wood, A.W.; Chang, R.L.; Levin, W.; Jerina, D.M. Cancer Res., 1983, 43, 1541-9.
24. Pish, E.S.; Yang, N.C. J. Am. Chem. Soc., 1963, 85, 2124-30.
25. Eling, T.; Boyd, J.; Reed, G.; Mason, R.; Sivarajah, K. Drug Metab. Rev., 1983, 14, 1023-53.
26. Bartsch, H.; Hecker, E. Biochim. Biophys. Acta, 1971, 237, 567-78.
27. Floyd, R.A.; Soong, L.M.; Culver, P.L. Cancer Res., 1976, 36, 1510-9.
28. Metzler, M.; McLachlan, J.A. Biochem. Biophys. Res. Comm., 1978, 85, 874-88.
29. Sawahata, T.; Neal, R.A. Biochem. Biophys. Res. Comm., 1982, 109, 988-94.
30. Griffith, B.W.; Ting, P.L. Biochemistry, 1978, 17, 2206-11.
31. Josephy, P.D.; Eling, T.; Mason, R.P. J. Biol. Chem., 1982, 257, 3669-75.
32. Josephy, P.D.; Mason, R.P.; Eling, T. Carcinogenesis, 1982, 3, 1227-30.
33. Kalyanaraman, B.; Mason, R.P. Biochem. Biophys. Res. Comm., 1982, 105, 217-24.
34. Rogan, E.G.; Katomski, P.A.; Roth, R.W.; Cavalieri, E.L. J. Biol. Chem., 1979, 254, 7055-9.
35. Mattammal, M.B.; Zenser, T.V.; Davis, B.B. Cancer Res., 1981, 41, 4961-6.
36. Josephy, P.D.; Eling, T.E.; Mason, R.P. J. Biol. Chem., 1983, 258, 5561-9.
37. Wong, P.K.; Hampton, M.J.; Floyd, R.A. In "Prostaglandins and Cancer: First International Conference", Powles, T.J.; Bockman, R.S.; Honn, K.V.; Ramwell, P., Eds.; Alan R. Liss: New York, 1982; pp. 167-179.
38. Kalyanaraman, B.; Sivarajah, K.; Eling, T.E.; Mason, R.P. Carcinogenesis, 1983, 4, 1341-3.
39. Degen, G.H.; Eling, T.E.; McLachlan, J.A. Cancer Res., 1982, 42, 919-23.
40. Rogan, E.G.; Hakam, A.; Cavalieri, E.L. Chem.-Biol. Interact., 1983, 47, 111-22.

41. Hecht, S.S.; Mazzarese, R.; Amin, S.; LaVoie, E.; Hoffmann, D. In "Polynuclear Aromatic Hydrocarbons. Third International Symposium on Chemistry and Biology--Carcinogenesis and Mutagenesis"; Ann Arbor Science Publishers: Ann Arbor, Mich., 1979; pp. 733-52.
42. Cavalieri, E.; Rogan, E.; Toth, B.; Munhall, A. Carcinogenesis, 1981, 2, 277-81.

RECEIVED March 5, 1985

Hydroperoxide-Dependent Oxygenation of Polycyclic Aromatic Hydrocarbons and Their Metabolites

LAWRENCE J. MARNETT

Department of Chemistry, Wayne State University, Detroit, MI 48202

Fatty acid hydroperoxides in the presence of heme complexes and heme proteins oxidize benzo(a)pyrene and
7,8-dihydroxy-7,8-dihydrobenzo(a)pyrene to quinones
and diol epoxides, respectively. The oxidizing agent
is a peroxyl radical derived from the fatty acid
hydroperoxide but not a higher oxidation state of a
mammalian peroxidase. The stereochemistry of (±)-BP-
dihydrodiol epoxidation is distinct from that catalyzed by mixed-function oxidases, which provides a
convenient method for discriminating the contributions
of the two systems to BP-7,8-dihydrodiol metabolism
in cell homogenates, cell or organ culture. Using
this method, epoxidation of BP-7,8-dihydrodiol has
been detected during prostaglandin biosynthesis, lipid
peroxidation, and xenobiotic oxygenation. Fatty acid
hydroperoxide-dependent oxidation constitutes a novel
pathway for metabolic activation of polycyclic hydrocarbons and other carcinogens which has widespread
potential *in vivo* significance.

Oxidation is intimately linked to the activation of polycyclic aromatic hydrocarbons (PAH) to carcinogens (1-3). Oxidation of PAH in
animals and man is enzyme-catalyzed and is a response to the introduction of foreign compounds into the cellular environment. The
most intensively studied enzyme of PAH oxidation is cytochrome P-450,
which is a mixed-function oxidase that receives its electrons from
NADPH via a one or two component electron transport chain (1). Some
forms of this enzyme play a major role in systemic metabolism of PAH
(4). However, there are numerous examples of carcinogens that
require metabolic activation, including PAH, that induce cancer in
tissues with low mixed-function oxidase activity (5). In order to
comprehensively evaluate the metabolic activation of PAH, one must
consider all cellular pathways for their oxidative activation.
    Peroxidases have been implicated in carcinogenesis by PAH,
aromatic amines, and estrogens *inter alia* (6-9). These enzymes
catalyze the reduction of hydrogen peroxide and organic

0097-6156/85/0283-0307$06.00/0
© 1985 American Chemical Society

hydroperoxides and use a wide variety of compounds as reducing
agents (Equation 1). Important observations on the oxidation of PAH

$$ROOH + DH_2 \rightarrow ROH + D + H_2O \qquad (1)$$

by peroxidases have been made by Cavalieri and Rogan and are
described in their chapter in this volume and elsewhere (6). Ten
years ago we reported that benzo(a)pyrene (BP) is oxidized during
the oxygenation of arachidonic acid by prostaglandin H (PGH) synthase
(10). PGH synthase is a widely distributed enzyme of polyunsatu-
rated fatty acid metabolism that possesses a peroxidase activity and
generates hydroperoxy endoperoxides as initial products of fatty
acid oxygenation (11-13). Its principal function is to biosynthesize
PGH_2, the endoperoxide intermediate of prostaglandin and thromboxane
biosynthesis (Figure 1) (11,14). The other enzyme of unsaturated
fatty acid oxygenation is lipoxygenase (15). It oxygenates unsaturated fatty acids to hydroperoxides that are reduced to alcohols or
converted to leukotrienes (Figure 1) (16). These two enzymes, PGH
synthase and lipoxygenase, represent the principal sources of
organic hydroperoxides in mammalian tissue (17). Our investigations
of the oxidation of PAH by the hydroperoxide products of PGH synthase
and lipoxygenase catalysis indicate that this pathway can generate
ultimate carcinogenic forms of PAH and that the mechanisms of oxidation are distinct from those of classic peroxidase-catalyzed oxidation. Fatty acid hydroperoxide-dependent oxidation, therefore,
represents a novel pathway for the metabolic activation of PAH.

Benzo(a)pyrene Oxidation

Incubation of BP with arachidonic acid and ram seminal vesicle microsomes, a rich source of PGH synthase, produces 1,6-, 3,6-, and 6,12-
quinones as the exclusive products of oxidation (Figure 2) (18).
These are the same quinones that are formed when 6-hydroxy-BP is
oxidized by air or microsomes (19). However, there is no definitive
evidence that 6-hydroxy-BP is an intermediate in their formation by
PGH synthase. Among all of the stable metabolites of BP, the
quinones are distinctive because, unlike phenols and dihydrodiols,
they are not derived from arene oxides. Thus, arene oxides do not
appear to be products of BP oxidation by PGH synthase (19,20).
Potent inhibition of PGH synthase-dependent BP oxidation by antioxidants suggests that the quinones are products of free radical
reactions (18).
      Addition of RNA or DNA prior to oxidation of BP by PGH synthase
results in substantial nucleic acid binding (17,21). Addition of
RNA five minutes after initiation of oxidation leads to no covalent
binding (17). This implies that the quinones do not bind to
nucleic acid but rather a short-lived intermediate in their formation does. Arachidonic acid oxygenation in ram seminal vesicle
microsomes is complete within two min, which suggests that the
reactive intermediate is generated concurrently with PGH_2. The
structures of the nucleic acid adducts have not been elucidated so
the identity of the reactive intermediate is unknown.
      Despite the high level of nucleic acid binding that is evident,
no mutagenic species can be detected when BP is incubated with ram

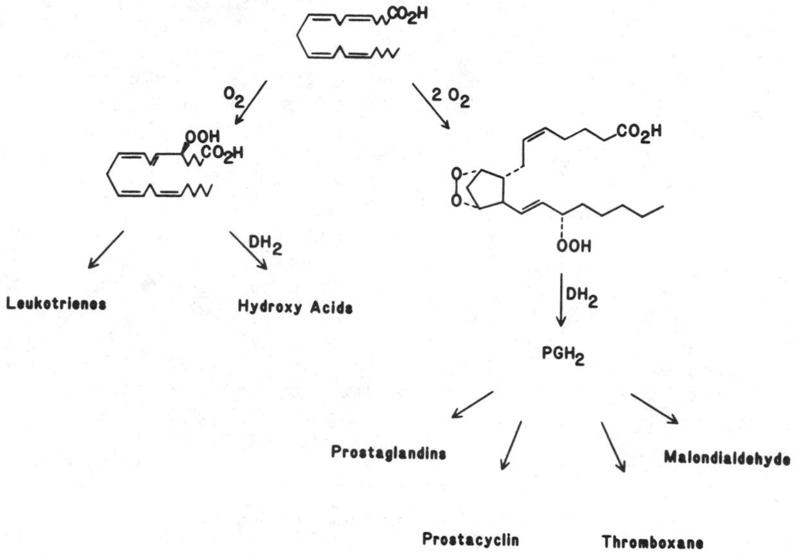

Figure 1. Pathways of oxygenation of unsaturated fatty acids in animal tissue.

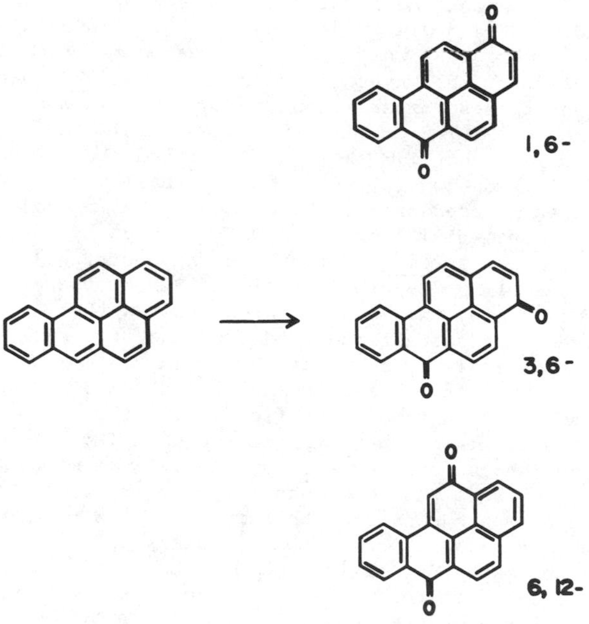

Figure 2. Products of BP oxidation by arachidonic acid and ram seminal vesicle microsomes.

seminal vesicle microsomes and arachidonic acid in the presence of
Salmonella typhimurium strains. The presence of the *Salmonella*
strains and nutrient broth in the incubations does not inhibit
quinone formation. Furthermore, one of the strains employed, TA98,
has been reported to detect 6-hydroxy-BP as a mutagen (23). It may
be that the intermediate responsible for nucleic acid binding is too
unstable to survive transit across the bacterial cell wall and membrane. Alternatively, the intermediate may bind to DNA but not
induce mutation. This is unlikely because the generation of bulky
adducts on a DNA molecule usually results in mutation. Although
some adducts of polycyclic hydrocarbons to DNA appear to be more
mutagenic than others, the differences are not greater than an order
of magnitude (24,25). Thus, it is unlikely that if adducts are
formed they are not mutagenic.

Two other polycyclic hydrocarbons, 3-methylcholanthrene and
7,12-dimethylbenzanthracene are oxidized during arachidonate
metabolism (21,26). Hydroxymethyl compounds that do not arise from
arene oxides appear to be the products formed from 7,12-dimethyl-
benzanthracene.

7,8-Dihydroxy-7,8-Dihydrobenzo(a)pyrene Oxidation

In contrast to the results with BP, incubation of BP-7,8-dihydrodiol
with ram seminal vesicle microsomes and arachidonate generates a
species that is strongly mutagenic to *Salmonella* strains TA98 and
TA100 (Figure 3) (22). Formation of the mutagen is inhibited by
indomethacin indicating the involvement of PGH synthase. Similar
experiments with BP-4,5-dihydrodiol and BP-9,10-dihydrodiol do not
generate potent mutagens, which suggests that activation is specific
for the precursor of the bay-region diol epoxide (22). The obvious
interpretation of these experiments is that PGH synthase catalyzes
the epoxidation of BP-7,8-dihydrodiol to the ultimate carcinogen
BP-diol epoxide. To confirm this we identified the products of BP-
7,8-dihydrodiol oxidation (27,28). Theoretically, the epoxide
oxygen can be introduced from either side of the molecule giving
rise to syn- or anti-diol epoxides. Each epoxide hydrolyzes
rapidly to a mixture of cis and trans tetrahydrotetraols (Figure 4).
When incubations of BP-7,8-dihydrodiol and PGH synthase are allowed
to proceed for 15 min, two products are obtained that we identified
as the cis and trans tetraols derived from the anti-diol epoxide
(27,28,29). Hydrolysis products of the syn-diol epoxide were not
detected. When incubations were terminated after 3 min, a new
product was detected that we identified as a methyl ether that is
formed by methanolysis of the anti-diol epoxide (Equation 2) (29).

This reaction can only have occurred after termination of the
reaction because there was no methanol in the incubation mixture.
Additional experiments confirmed that methanolysis occurs during
chromatography (reverse phase, methanol-water gradients). The
detection of the methyl ether is important because it confirms that
a diol epoxide is generated, survives solvent extraction, and then
undergoes solvolysis on the HPLC column. This provides direct
evidence for the formation of the anti-diol epoxide as a product of
PGH synthase-dependent cooxidation of BP-7,8-dihydrodiol. The correlation of the rate of BP-7,8-dihydrodiol oxidation, anti-diol
epoxide formation, and mutagen generation are shown in Figure 5 (30).

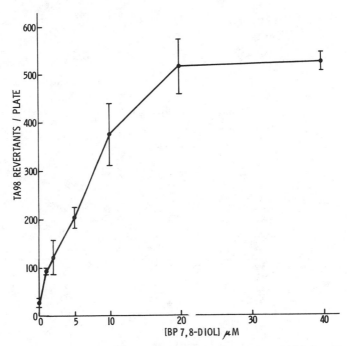

Figure 3.   Induction of mutation in <u>S. typhimurium</u> TA98 by BP-7,8-
            dihydrodiol, arachidonic acid, and ram seminal vesicle
            microsomes. Concentration dependence on BP-7,8-dihydro-
            diol. (Reproduced with permission from Ref. 22. Copyright
            1978 Academic.)

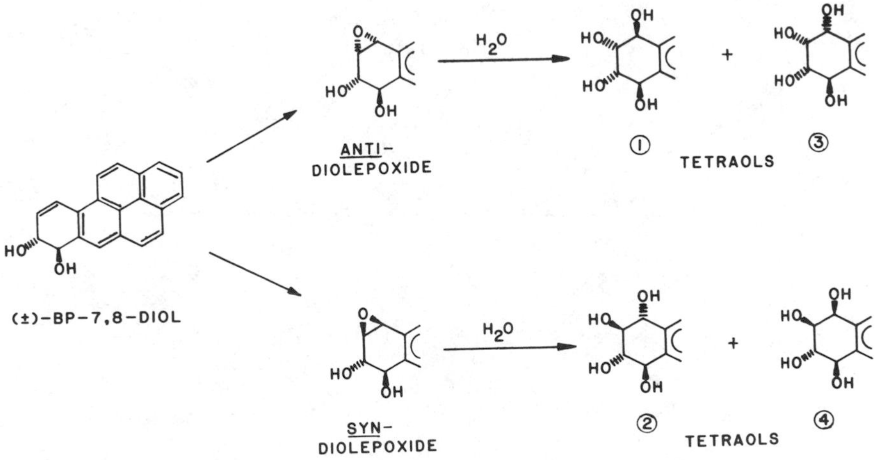

Figure 4.   Diol epoxide products of BP-7,8-dihydrodiol oxidation and
            their hydrolysis products.

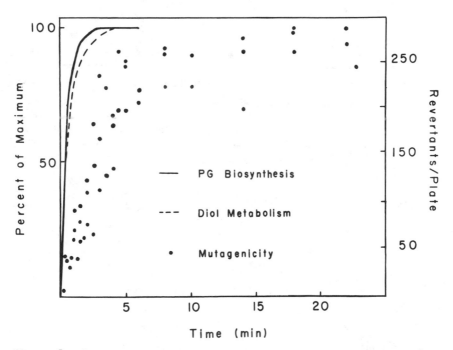

$$(2)$$

Figure 5.   Comparison of the time course of PGH$_2$ biosynthesis, BP–
7,8–dihydrodiol metabolism, and generation of a mutagen
from BP–7,8–dihydrodiol by RSVM. (Reproduced with
permission from Ref. 30. Copyright 1982 Marcel Dekker.)

Further support for epoxidation of BP-7,8-dihydrodiol to the anti-diol epoxide is the identification of RNA and DNA adducts formed as a result of incubation of BP-7,8-dihydrodiol, PGH synthase, and polyguanylic acid or DNA (30,31). Following digestion of the nucleic acid, the major guanosine and deoxyguanosine adducts were identified as arising by addition of the exocyclic amino group of guanosine to the benzylic carbon of the anti-diol epoxide (Table I) (31). These experiments also defined the stereochemistry of epoxidation. Both enantiomers of BP-7,8-dihydrodiol are epoxidized at equal rates to enantiomers of the anti-diol epoxide. The direction of oxygen introduction is from the same side of the molecule as the hydroxyl group at carbon-8 of BP-7,8-dihydrodiol.

Table I. Relative Yields of Diastereomeric Adducts From Anti-diol Epoxide Plus Polyguanylic Acid Compared to Adducts Generated During Metabolism of BP-7,8-dihydrodiol by Ram Seminal Vesicles in the Presence of Arachidonic Acid

Incubation	% Radioactivity as (-)-cis and (-)-trans diastereomers	% Radioactivity as (+)-cis diastereomer	% Radioactivity as (+)-trans diastereomer
[^{14}C]-anti-diol epoxide	50	12	38
[^{3}H]-BP-7,8-dihydrodiol	51	7	42

When the 7,8-hydroxyl groups are missing, epoxide introduction occurs from both sides of the pyrene ring. Thus 7,8-dihydrobenzo(a)-pyrene is cooxidized by PGH synthase to a potent mutagen that is identified by product and nucleic acid binding studies as 9,10-epoxy-7,8,9,10-tetrahydrobenzo(a)pyrene (Equation 3) (32). The structures of the guanosine adducts formed in incubations containing poly-guanylic acid indicate that equal amounts of epoxide are formed by introduction of oxygen from above and below the plane of the pyrene ring (Equation 3).

These findings indicate that PGH synthase in the presence of arachidonate can catalyze the terminal activation step in BP carcinogenesis and that the reaction may be general for dihydrodiol metabolites of polycyclic hydrocarbons. Guthrie et. al. have shown that PGH synthase catalyzes the activation of chrysene and benzanthracene dihydrodiols to potent mutagens (33). As in the case with BP, only the dihydrodiol that is a precursor to bay region diol epoxides is activated. We have recently shown that 3,4-dihydroxy-3,4-dihydro-benzo(a)anthracene is oxidized by PGH synthase to tetrahydrotetraols derived from the anti-diol epoxide (Equation 4) (34).

Nature of Oxidants Generated From Fatty Acid Hydroperoxides

PGH synthase contains two heme-requiring activities (13). The cyclo-oxygenase component oxygenates arachidonic acid to the hydroperoxy endoperoxide, PGG$_2$, and the peroxidase component reduces PGG$_2$ to the hydroxy endoperoxide, PGH$_2$. The cyclooxygenase is inhibited by non-steroidal antiinflammatory agents such as aspirin and indomethacin,

but the peroxidase is not (35,36). Both components are contained
on the same 70,000 Dalton protein (13). The presence of a peroxidase as an integral component of PGH synthase implies that hydroperoxide-dependent oxidations are catalyzed by this component (37).
As a first approximation one might expect that the mechanisms of
these oxidations would be analogous to those of other heme peroxidases.

Extensive studies have established that the catalytic cycle for
the reduction of hydroperoxides by horseradish peroxidase is the one
depicted in Figure 6 (38). The resting enzyme interacts with the
peroxide to form an enzyme-substrate complex that decomposes to
alcohol and an iron-oxo complex that is two oxidizing equivalents
above the resting state of the enzyme. For catalytic turnover to
occur the iron-oxo complex must be reduced. The two electrons are
furnished by reducing substrates either by electron transfer from
substrate to enzyme or by oxygen transfer from enzyme to substrate.
Substrate oxidation by the iron-oxo complex supports continuous
hydroperoxide reduction. When either reducing substrate or hydroperoxide is exhausted, the catalytic cycle stops.

We have developed an assay to identify peroxidase reducing substrates based on their ability to stimulate reduction of 1-hydro-
peroxy-5-phenyl-4-pentene (Equation 5) (39). The hydroperoxide is
incubated with limiting concentrations of peroxidase in the presence
or absence of a potential reducing substrate. In the absence of
reductant catalytic reduction cannot occur and negligible quantities
of alcohol are produced (the hydroperoxide and alcohol are quantitated after separation by HPLC). In the presence of a good reducing
substrate catalytic turnover occurs and quantities of alcohol are
produced that are stoichiometric with reducing substrate oxidized.
The assay appears to be general for all plant and animal, heme and
non-heme, peroxidases. One can rank the relative efficacy of
reducing substrates using this assay. Aromatic amines, phenols,
catechols, β-dicarbonyls, nitrogen heterocycles, and aromatic sul-
fides are good to excellent reducing substrates (39). In contrast,
polycyclic hydrocarbons and dihydrodiol metabolites of PAH are very
poor to non-reducing compounds. Because BP and BP-7,8-dihydrodiol
do not stimulate hydroperoxide reduction they cannot be oxidized by
higher oxidation states of the peroxidase (iron-oxo complexes) The
concentrations of hydroperoxide. PGH synthase, and BP or BP-7,8-
dihydrodiol are analogous to those in which BP or BP-7,8-dihydrodiol
oxidation can be detected in ram seminal vesicle microsomes.
Therefore, we conclude that the oxidizing agent that converts BP to
quinones or BP-7,8-dihydrodiol to diol epoxides is not an iron-oxo
intermediate of peroxidase turnover.

Support for this conclusion is provided by the hydroperoxide
specificity of BP oxidation. The scheme presented in Figure 6
requires that the same oxidizing agent is generated by reaction of
H_2O_2, peroxy acids, or alkyl hydroperoxides with the peroxidase.
Oxidation of any compound by the iron-oxo intermediates should be
supported by any hydroperoxide that is reduced by the peroxidase.
This is clearly not the case for oxidation of BP by ram seminal
vesicle microsomes as the data in Figure 7 illustrate. Quinone
formation is supported by fatty acid hydroperoxides but very poorly
or not at all by simple alkyl hydroperoxides or H_2O_2. The fact that

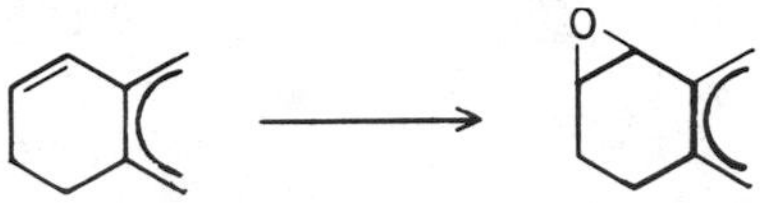

(3)

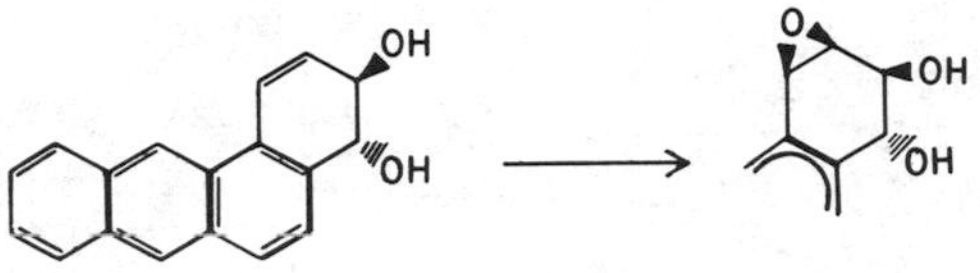

(4)

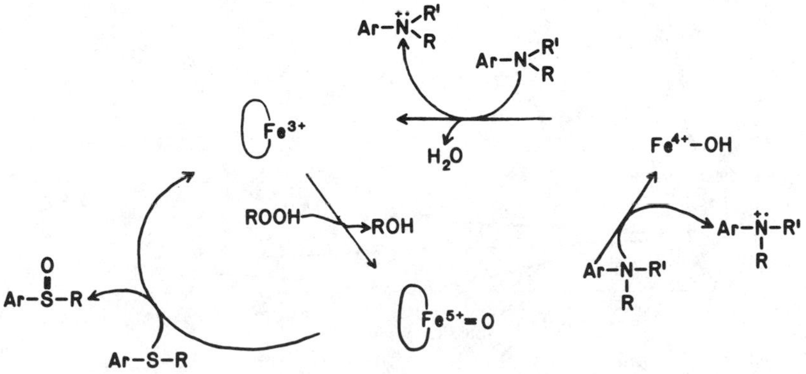

Figure 6. Catalytic cycle of horseradish peroxidase.

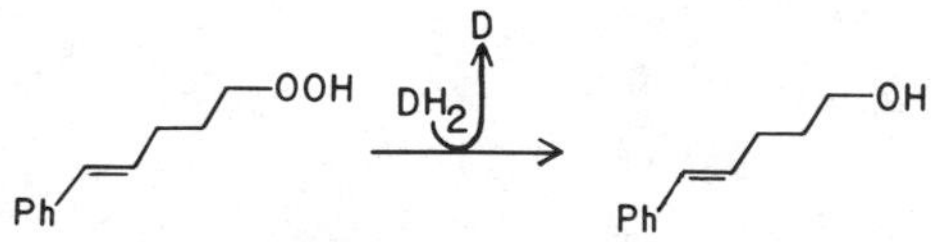

(5)

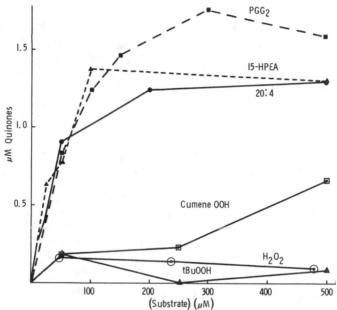

Figure 7.   Dependence of BP oxidation by ram seminal vesicle micro-
            somes on the concentration of different hydroperoxides.
            Abbreviations used are 20:4, arachidonic acid; 15-HPEA,
            15-hydroperoxy-eicosatetraenoic acid; t-BuOOH, t-butyl
            hydroperoxide.   The structure is PGG$_2$ is given in Figure 1.

H_2O_2 does not support oxidation is especially significant because the same concentrations of H_2O_2 support vigorous oxidation of reducing substrates such as aromatic amines and phenylbutazone. Therefore, we conclude that BP and BP-7,8-dihydrodiol are oxidized by a species that is not a functional intermediate of peroxidase catalysis.

The oxidizing agent that is responsible for the oxygenation of BP and BP-7,8-dihydrodiol appears to be a free radical. Reaction of fatty acid hydroperoxides with metal complexes generates alkoxyl and peroxyl radicals that can oxidize organic molecules (40-43). Incubation of fatty acid hydroperoxides with certain hemeproteins or their prosthetic group, hematin, causes oxidation of BP to quinones and BP-7,8-dihydrodiol to diol epoxides (17,40). In the case of BP-7,8-dihydrodiol epoxidation, the source of the epoxide oxygen is molecular oxygen; epoxidation is potently inhibited by antioxidants, and epoxidation is supported by unsaturated but not saturated fatty acid hydroperoxides (Table II)(40,44). These observations are analogous to the results of microsomal incubations and are consistent with a free radical mechanism of hydroperoxide-dependent epoxidation. BP oxidation to quinones occurs during autoxidation of lipids initiated by enzymes or γ-irradiation (45,46).

In the case of the hematin-catalyzed reaction we have proposed that peroxyl radicals are the epoxidizing agents (40). The mechanism is illustrated in Figure 8. Hematin reduces the hydroperoxide to an alkoxyl radical that cyclizes to the adjacent double bond. The incipient carbon-centered radical couples with O_2 to form a peroxyl radical that we propose epoxidizes BP-7,8-dihydrodiol. Peroxyl radicals are well-known in chemical systems to epoxidize isolated double bonds such as the 9,10-double bond of BP-7,8-dihydrodiol (Equation 6) (47). However, they have been largely ignored as potential oxidizing agents in biochemical systems although their half-lives (0.1-10 sec) suggest they can serve as diffusible, selective oxidants (48). The mechanism outlined in Figure 8 is consistent with all of the experimental observations and explains the requirement for a double bond in the vicinity of the hydroperoxide (Table II). The ability of peroxyl radicals to epoxidize double bonds appears to depend upon the ability of the peroxyl radical-olefin adduct to stabilize the carbon-centered radical. Thus, 3,4-dihydroxy-3,4-dihydrobenzo(a)anthracene is oxidized to 1/6 the extent of BP-7,8-dihydrodiol and aflatoxin B_1 is epoxidized to only a slight extent (34,49).

Peroxyl radicals are the species that propagate autoxidation of the unsaturated fatty acid residues of phospholipids (50). In addition, peroxyl radicals are intermediates in the metabolism of certain drugs such as phenylbutazone (51). Epoxidation of BP-7,8-dihydrodiol has been detected during lipid peroxidation induced in rat liver microsomes by ascorbate or NADPH and during the peroxidatic oxidation of phenylbutazone (52,53). These findings suggest that peroxyl radical-mediated epoxidation of BP-7,8-dihydrodiol is general and may serve as the prototype for similar epoxidations of other olefins in a variety of biochemical systems. In addition, peroxyl radical-dependent epoxidation of BP-7,8-dihydrodiol exhibits the same stereochemistry as the arachidonic acid-stimulated epoxidation by ram seminal vesicle microsomes. This not only provides additional

Table II.   Epoxidation of Unsaturated and Saturated
Fatty Acid Hydroperoxides

HYDROPEROXIDE	O_2 UPTAKE (uM)	7,8-DIOL OXIDATION VI (uM/min.)
n-$C_{16}H_{33}$OOH	0	0.16±0.02
	61±3	6.5±0.6
	160±16	12.3±0.9
	160±13	12±1.3
	160±5	12±1.6
	0	0
	65±1	7.0±1.1
	240±15	16±1.8

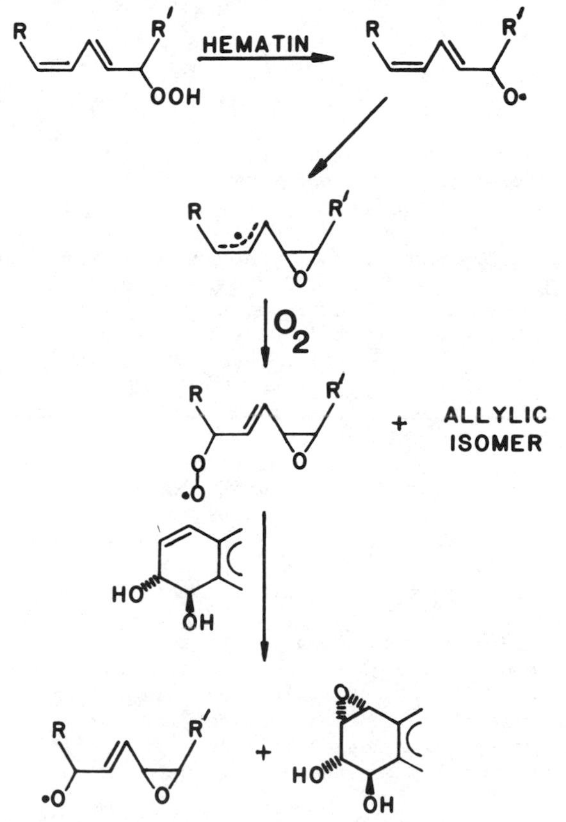

Figure 8.   Proposed mechanism of the generation of peroxyl radicals
            by reaction of hematin with unsaturated fatty acid hydro-
            peroxides.

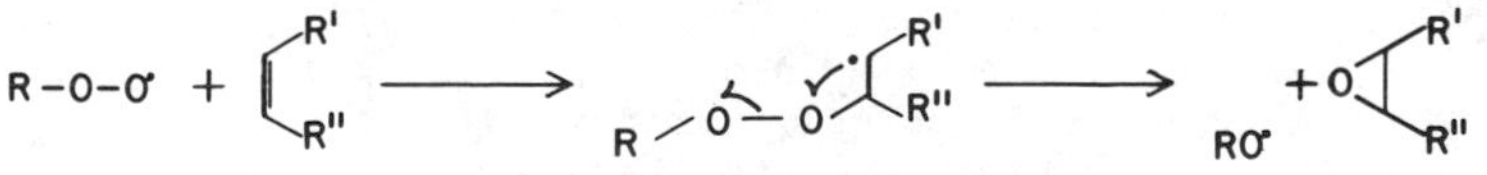

(6)

evidence that the oxidizing agent in the enzymatic reaction is a
peroxyl radical but also suggests that the stereochemistry of BP-
7,8-dihydrodiol oxidation is an important and general diagnostic
probe to differentiate epoxidation by mixed-function oxidases and by
peroxyl radicals.

Significance of Fatty Acid Hydroperoxide-Dependent PAH Oxidation

What is the significance of arachidonic acid-dependent xenobiotic
metabolism? Experiments described above firmly establish that it
can cause metabolic activation *in vitro*. Dihydrodiol metabolites of
polycyclic hydrocarbons are oxidized to diol epoxides that represent
the ultimate carcinogenic forms of the parent hydrocarbons. Interestingly, only dihydrodiols that form bay region diol epoxides are
activated by PGH synthase; no activation of other PAH dihydrodiols
occurs. Arachidonate-dependent cooxidation is essentially an activation pathway specific for generation of bay-region diol epoxides.
Work described elsewhere indicates that aromatic amines can also be
oxidized to mutagenic derivatives although the identity of the mutagenic derivative is, at present, uncertain (54).
     Is it possible to quantitate the relative contribution of
hydroperoxide-dependent and mixed-function oxidase-dependent oxidation of PAH *in vitro*, in cells and organs, and *in vivo*? Adding
arachidonic acid or NADPH to support oxidation *in vitro* gives a good
estimate of oxidative potential but its relation to cellular oxidation is not straightforward. Likewise, "specific" inhibitors can be
helpful in *in vitro* experiments but their use can be compromised in
cellular, organismal, or *in vivo* experiments by overlapping specificities or altered potencies. For example, many compounds that inhibit lipoxygenase activity at low concentration in microsomal or
cytoplasmic fractions are ineffective when they are employed in
cellular experiments. The reason for the differential effect is un-
clear but the implication for the use of such compounds as *in vivo*
inhibitors is obvious.
     A potentially powerful probe for sorting out the contribution
of hydroperoxide-dependent and mixed-function oxidase-dependent
polycyclic hydrocarbon oxidation is stereochemistry. Figure 9 summarizes the stereochemical differences in epoxidation of (±)-BP-7,8-
dihydrodiol by hydroperoxide-dependent and mixed-function oxidase-
dependent pathways (31,55,56). The (-)-enantiomer of BP-7,8-
dihydrodiol is converted primarily to the (+)-anti-diol epoxide by
both pathways whereas the (+)-enantiomer of BP-7,8-dihydrodiol is
converted primarily to the (-)-anti-diol epoxide by hydroperoxide-
dependent oxidation and to the (+)-syn-diol epoxide by mixed-function
oxidases. The stereochemical course of oxidation by cytochrome
P-450 isoenzymes was first elucidated for the methycholanthrene-
inducible form but we have detected the same stereochemical profile
using rat liver microsomes from control, phenobarbital-, or methylcholanthrene-induced animals (32). The only difference between the
microsomal preparations is the rate of oxidation.
     The findings summarized in Figure 9 provide a practical diag-
nostic tool for distinguishing the two routes of oxidation.
Reactions can be performed with cellular or subcellular preparations
and (±)- or (+)-BP-7,8-dihydrodiol and the tetraol hydrolysis

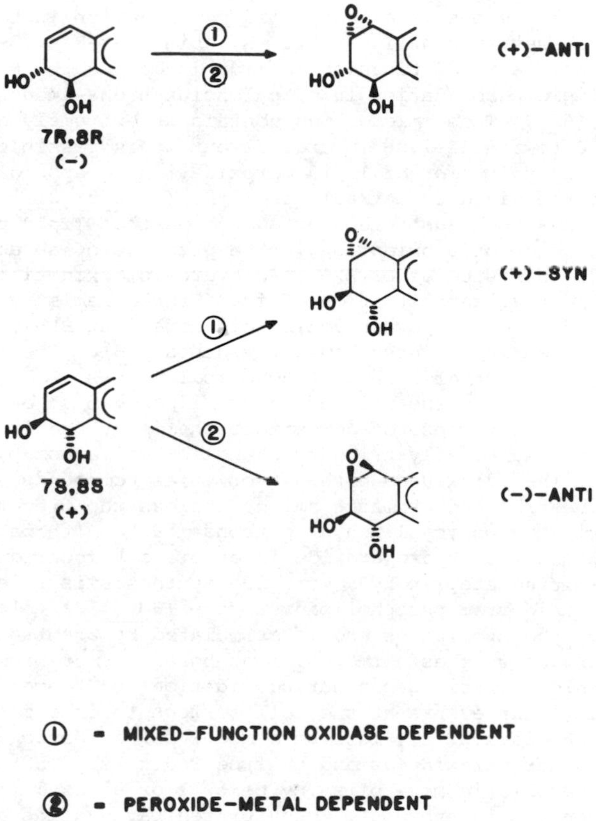

① = MIXED-FUNCTION OXIDASE DEPENDENT

② = PEROXIDE-METAL DEPENDENT

Figure 9. Stereochemical differences between fatty acid hydro-
peroxide- and mixed-function oxidase-dependent oxidation
of (±)-BP-7,8-dihydrodiol.

products of the diol epoxides separated by HPLC and quantitated
(57). When the substrate is (±)-BP-7,8-dihydrodiol an anti/syn
ratio in excess of 2.5 is seen for peroxide-dependent oxidation and
an anti/syn ratio of 1 for mixed-function oxidase-dependent oxidation. When the substrate is (+)-BP-7,8-dihydrodiol the anti/syn
ratio for the mixed-function oxidase-dependent reaction decreases
to ∿0.3. The tenfold difference in the anti/syn ratio between
peroxide- and cytochrome P-450-dependent epoxidation makes it an
extremely sensitive indicator of the pathway of oxidation. We have
exploited it to demonstrate that lipid peroxidation in rat liver
microsomes causes epoxidation (52). By using (+)-BP-7,8-dihydrodiol,
we have been able to distinguish epoxidation caused by NADPH-
dependent lipid peroxidation in methylcholanthrene-induced rat liver
microsomes (52). These microsomes contain an extremely active cytochrome P-450 toward BP-7,8-dihydrodiol but it is possible to differentiate the contribution of lipid peroxidation to epoxidation by
determining the yield of tetraols from the (-)-anti-diol epoxide.
Although it has been suspected for some time that lipid peroxidation
could cause xenobiotic oxidation in the presence of an active cytochrome P-450, our studies of BP-7,8-dihydrodiol oxidation provided
the first clearcut demonstration of it. Stereochemistry has also
been employed to detect arachidonic acid-dependent BP-7,8-dihydro-
diol epoxidation in cultured hamster trachea (58). These examples
illustrate the power of such stereochemical probes.

PGH synthase and the related enzyme lipoxygenase occupy a
position at the interface of peroxidase chemistry and free radical
chemistry and can clearly trigger metabolic activation by both
mechanisms. The peroxidase pathway activates compounds such as
diethylstilbestrol and aromatic amines whereas the free radical
pathway activates polycyclic hydrocarbons (59). Both pathways
require synthesis of hydroperoxide in order to trigger oxidation.
The rate-limiting step in hydroperoxide synthesis is release of
arachidonic acid from phospholipid storage (60,61). Release is
catalyzed by phospholipases and is stimulated by agents that act at
the cell surface such as hormones, ionophores, tumor promoters, etc
(62). Arachidonic acid-dependent cooxidation is, therefore, a
pathway that links events at the cell surface to intracellular oxidation of xenobiotics. It is also a model for oxidation of xenobiotics by other peroxidases and by free radicals. There are few
reports of xenobiotic metabolism by peroxyl or alkoxyl free radicals
but the potential is enormous. Unsaturated fatty acids are present
in all cells to some extent and, in fact, are quite abundant in most
cells. For example, ventricular myocardial muscle contains 14.1
μmol linoleic and arachidonic acids per gram wet weight tissue (63).
Both fatty acids are quite susceptible to lipid peroxidation which
generates peroxyl radicals capable of oxidizing certain xenobiotics,
e.g., BP-7,8-dihydrodiol. Inhibition of lipid peroxidation is
obviously a task that must be constantly performed by cells to prevent tissue destruction and xenobiotic metabolism. The turnover of
only 0.1% of the unsaturated fatty acid residues of cells could
generate a very significant amount of peroxyl radicals inside membrane regions of cells where many xenobiotics are dissolved.

Metabolism of aromatic amines and BP-7,8-dihydrodiol has been
detected during arachidonate oxygenation in intact cells and in
cultured trachea (64,58). Exogenous arachidonate was added to

stimulate hydroperoxide synthesis and cooxygenation in most of the studies. Recently, though, Amstad and Cerutti reported that the levels of aflatoxin B_1-DNA adducts formed in C3H 10T½ fibroblasts were decreased by treatment of the cells with indomethacin or eicosatetraynoic acid, inhibitors of arachidonate oxygenation (65). They concluded that a significant fraction of total aflatoxin epoxidation by 10T½ cells occurs as a result of arachidonate-dependent cooxygenation. This implies that cooxygenation takes place in cells and that it is triggered by release of arachidonate from endogenous stores.

To what extent does cooxygenation occur *in vivo* and is it important in chemical carcinogenesis? This is a very difficult question to answer at the present time. Recent results demonstrate that aromatic amines and diamines can be cooxidized *in vivo* (66,67). In the case of β-napthylamine it is estimated that 30% of the adducts that form to DNA in the dog bladder, a target organ for napthylamine carcinogenesis, arise as a result of arachidonate-dependent cooxidation (66). This conclusion is based on the detection of unique peroxidase adducts to DNA that are structurally distinct from mixed-function oxidase-generated adducts. In contrast, pretreatment of A/HeJ mice with aspirin or indomethacin does not lower the levels of DNA adducts formed from BP in lung nor does it reduce the incidence of lung neoplasms induced by BP (68). Control experiments indicate that aspirin treatment abolishes PGH synthase activity *in vivo* (68). This suggests that PGH synthase-dependent cooxidation does not play a role in lung tumorigenesis by benzo(a)pyrene in the adenoma model. This may be related to the high levels of the endogenous antioxidant, vitamin E, in rodent lung (69). However, administration of aspirin to guinea pigs does not lower the levels of protein or DNA adducts formed from BP in several different tissues, so the levels of vitamin E may not be a determinant of BP cooxidation (70).

The tissue distribution of PGH synthase suggests that it does not play a major role in systemic drug metabolism because most of the tissues where it is present in high concentration do not receive a significant proportion of cardiac output (12). However, several of these tissues, e.g., kidney and uterus, are target organs for carcinogens that require metabolic activation. In order to detect arachidonate-dependent metabolic activation in these tissues, it will be necessary to develop unique and specific probes. If systemic metabolism of a given compound proceeds with a unique pattern of stereochemistry (e.g., BP-7,8-dihydrodiol) or produces unique DNA adducts (β-napthylamine) then it should be possible to quantitate the extent to which unique stereoisomers or DNA adducts are formed. *In vitro* studies define these distinctive features of cooxidative metabolism and guide the intelligent design of critical experiments. Hopefully, by using such diagnostic probes it will be possible to provide quantitative answers to questions about the extent to which cooxidation of polycyclic hydrocarbons and other carcinogens occurs *in vivo*.

Acknowledgments

This research has been generously supported by grants from the American Cancer Society (BC244) and the National Institutes of Health (GM23642). LJM is a recipient of an American Cancer Society Faculty Research Award (FRA243).

Literature Cited

1. Conney, A. H. Cancer Res. 1982, 42, 4875-4917.
2. Harvey, R. G. Acc. Chem. Res. 1981, 14, 218-26.
3. Sims, P.; Grover, P.L. In "Polycyclic Hydrocarbons and Cancer";
    Gelboin, H.V.; Ts'o, P. O. P., Eds.; Academic: New York, NY,
    1978; Vol. 1, pp. 117-81.
4. Levin, W.; Lu, A. Y. H.; Ryan, D.; Wood, A. W.; Kapitulnik, J.;
    West, S.; Huang, M.-T.; Conney, A. H.; Thakker, D. R.; Holder,
    G.; Yagi, H.; Jerina, D. M. In "Origins of Human Cancer";
    Hiatt, H. H.; Watson, J. D.; Winsten, J. A., Eds.; Cold Spring
    Harbor: Cold Spring Harbor, 1977; pp. 659-82.
5. Rydstrom, J.; Montelius, J.; Bengtsson, M. "Extrahepatic Drug
    Metabolism and Chemical Carcinogenesis"; Elsevier: New York,
    1983.
6. Cavalieri, E. L.; Rogan, E. G. In "Free Radicals in Biology";
    Pryor, W. A., Ed.; Academic: New York, 1984; Vol. 6, pp. 323-69.
7. Bartsch, H.; Hecker, E. Biochim. Biophys. Acta 1971, 237, 567-
    78.
8. Floyd, R. A.; Soong, L. M.; Culver, P. L. Cancer Res. 1976, 36,
    1510-19.
9. Metzler, M.; McLachlan, J. A. Biochem. Biophys. Res. Comm.
    1978, 85, 874-84.
10. Marnett, L. J.; Wlodawer, P.; Samuelsson, B. J. Biol. Chem.
    1975, 250, 8510-17.
11. Hamberg, M.; Svensson, J.; Wakabayashi, T.; Samuelsson, B.
    Proc. Natl. Acad. Sci. USA 1974, 71, 345-49.
12. Christ, E. J.; Van Dorp, D. A. Biochim. Biophys. Acta 1972,
    270, 537-45.
13. Ohiki, S.; Ogino, N.; Yamamoto, S.; Hayaishi, O. J. Biol. Chem.
    1979, 254, 839-46.
14. Nugteren, D. H.; Hazelhof, E. Biochim. Biophys. Acta 1973,
    326, 448-61.
15. Hamberg, M.; Samuelsson, B. Proc. Natl. Acad. Sci. USA 1974,
    71, 3400-04.
16. Samuelsson, B. Science (Wasnington, D.C.) 1983, 220, 5689-
    5775.
17. Marnett, L. J.; Reed, G. A. Biochemistry 1979, 18, 2923-29.
18. Marnett, L. J.; Reed, G. A.; Johnson, J. T. Biochem. Biophys.
    Res. Comm. 1977, 79, 569-76.
19. Lorentzen, R. J.; Caspary, W. J.; Lesko, S. A.; Ts'o, P. O. P.
    Biochemistry 1975, 14, 3970-77.
20. Lesko, S.; Caspary, W.; Lorentzen, R.; Ts'o, P. O. P.
    Biochemistry 1975, 14, 3978-84.
21. Sivarajah, K.; Anderson, M. W.; Eling, T. Life Sci. 1978, 23,
    2571-78.
22. Marnett, L. J.; Reed, G. A.; Dennison, D. J. Biochem. Biophys.
    Res. Comm. 1978, 82, 210-16.
23. Wislocki, P. G.; Wood, A. W.; CHang, R. L.; Levin, W.; Yagi, H.;
    Hernandez, O.; Dansette, P. M.; Jerina, D. M.; Conney, A. H.
    Cancer Res. 1976, 36, 3350-57.
24. Fahl, W. E.; Scarpelli, D.; Gill, K. Cancer Res. 1981, 41,
    3400-06.
25. Brooks, P.; Osborne, M. R.; Carcinogenesis 1982, 3, 1223-26.
26. Reed, G. A., unpublished data.

27. Marnett, L. J.; Johnson, J. T.; Bienkowski, M. J.   FEBS Letts. 1979, 106, 13-16.
28. Sivarajah, K.; Mukhtar, H.; Eling, T.   FEBS Letts. 1979, 106, 17-20.
29. Marnett, L. J.; Bienkowski, M. J.   Biochem. Biophys. Res. Comm. 1980, 96, 639-47.
30. Marnett, L. J.; Panthananickal, A.; Reed, G. A.   Drug Metab. Rev. 1982, 13, 235-47.
31. Panthananickal, A.; Marnett, L. J.   Chem. Biol. Interact. 1981, 33, 239-52.
32. Panthananickal, A.; Weller, P.; Marnett, L. J.   J. Biol. Chem. 1983, 258, 4411-18.
33. Guthrie, J.; Robertson, I. G. C.; Zeiger, E.; Boyd, J. A.; Eling, T. E.   Cancer Res. 1982, 42, 1620-23.
34. Dix, T. A.; Buck, J.; Marnett, L. J., manuscript in preparation.
35. Egan, R. W.; Gale, P. H.; Baptista, E. M.; Kennicott, K. L.; VandenHeuvel, W. J. A.; Walker, R. W.; Fagerness, P. E.; Kuehl, F. A., Jr.   J. Biol. Chem. 1981, 256, 7352-61.
36. Marnett, L. J.; Siedlik, P. H.; Fung, L. W.-M.   J. Biol. Chem. 1982, 257, 6957-64.
37. Pagels, W. R.; Sachs, R. J.; Marnett, L. J.; Dewitt, D. L.; Day, J. S.; Smith, W. L.   J. Biol. Chem. 1983, 258, 6517-23.
38. Dunford, H. B.   Coord. Chem. Rev. 1976, 19, 187-251.
39. Weller, P.; Markey, C.; Marnett, L., manuscript in preparation.
40. Dix, T. A.; Marnett, L. J.   J. Amer. Chem. Soc. 1981, 103, 6744-46.
41. Gardner, H. W.; Eskins, K.; Grams, G. W.; Inglett, G. E.   Lipids 1972, 7, 324-34.
42. Gardner, H. W.; Weisleder, D.; Kleiman, R.   Lipids 1978, 13, 246-52.
43. Hamberg, M.   Lipids 1975, 10, 87-92.
44. Dix, T. A.; Fontana, R.; Panthananickal, A.; Marnett, L. J., submitted for publication.
45. Morgenstern, R.; DePierre, J. W.; Lind, C.; Guthenberg, C.; Mannervik, B.; Ernster, L.   Biochem. Biophys. Res. Comm. 1981, 99, 682-90.
46. Gower, J. D.; Wills, E. D.   Carcinogenesis 1984, 5, 1183-89.
47. Mayo, R.   Acc. Chem. Res. 1968, 1, 193-201.
48. Pryor, W. A.   In "Free Radicals in Biology and Aging"; Armstrong, D., Ed.; Raven: New York, in press.
49. Battista, J. R.; Marnett, L. J.   Carcinogenesis, submitted.
50. Porter, N. A.   Met. Enzymol. 1984, 105, 273-82.
51. Marnett, L. J.   In "Free Radicals in Biology"; Pryor, W. A., Ed.; Academic: New York, 1984; Vol. 6, pp. 63-94.
52. Dix, T. A.; Marnett, L. J.   Science 1983, 221, 77-79.
53. Reed, G. A.; Brooks, E. A.; Eling, T. E.   J. Biol. Chem. 1984, 259, 5591-95.
54. Robertson, I. G. C.; Sivarajah, K.; Eling, T. E.; Zeiger, E.   Cancer Res. 1983, 43, 476-80.
55. Thakker, D. R.; Yagi, H.; Akagi, H.; Koreeda, M.; Lu, A. Y. H.; Levin, W.; Wood, A. W.; Conney, A. H.; Jerina, D. M.   Chem. Biol. Interact. 1977, 16, 281-300.

56. Deutsch, J.; Vatsis, K. P.; Coon, M.; Leutz, J. C.; Gelboin, H. V. Mol. Pharmacol. 1979, 14, 1011-18.

57. Dix, T. A.; Marnett, L. J. Met. Enzymol. 1984, 105, 347-52.

58. Reed, G. A.; Grafstrom, R. C.; Krauss, R. S.; Autrup, H.; Eling, T. E. Carcinogenesis 1984, 5, 955-60.

59. Marnett, L. J.; Eling, T. E. In "Reviews in Biochemical Toxicology"; Hodgson, E.; Bend, J. R.; Philpot, R. M., Eds.; Elsevier/North Holland: New York, 1983; Vol. 5, pp. 135-72.

60. Lands, W. E. M.; Samuelsson, B. Biochim. Biophys. Acta 1968, 164, 426-29.

61. Vonkeman, H.; Van Dorp, D. A. Biochim. Biophys. Acta 1968, 164, 430-32.

62. Galli, C.; Galli, G.; Porcellati, G. "Advances in Prostaglandin and Thromboxane Research"; Raven: New York, 1978; Vol. 3.

63. Fletcher, R. Lipids 1972, 7, 728-732.

64. Wong, P. K.; Hampton, M. J.; Floyd, R. A. In "Prostaglandins and Cancer: First International Conference"; Powles, T. J.; Bockman, R. S.; Honn, K. V.; Ramwell, P., Eds.; Liss: New York, 1982; pp. 167-79.

65. Amstad, P.; Cerutti, P. Biochem. Biophys. Res. Comm. 1983, 112, 1034-40.

66. Yamazoe, Y.; Miller, D. W.; Gupta, R. C.; Zenser, T. V.; Weis, C. C.; Kadlubar, F. F. Proc. Amer. Assoc. Cancer Res. 1984, 25, 91.

67. Zenser, T. V.; Mattammal, M. B.; Brown, W. W.; Davis, B. B. Kidney International 1979, 16, 688-94.

68. Adriaenssens, P. I.; Sivarajah, K.; Boorman, G. A.; Eling, T. E.; Anderson, M. W. Cancer Res. 1983, 43, 4762-67.

69. Kornbrust, D. J.; Mavis, R. D. Lipids 1980, 15, 315-22.

70. Garattini, E.; Coccia, P.; Romano, M.; Jiritano, L.; Noseda, A.; Salmona, M. Cancer Res. 1984, 44, 5150-5155.

RECEIVED May 13, 1985

The Mutational Consequences of DNA Damage Induced by Benzo[a]pyrene

ERIC EISENSTADT

Department of Cancer Biology and Laboratory of Toxicology, Harvard School of Public Health, Boston, MA 02115

Induced mutagenesis in Escherichia coli is an active process involving proteins with DNA replication, repair, and recombination functions. The available evidence suggests that mutations are generated at sites where DNA has been damaged and that they arise via an error-prone repair activity. In an attempt to understand what specific contributions to mutagenesis are made by DNA lesions, we have studied the mutational specificity of some carcinogens, such as benzo[a]pyrene and aflatoxin B$_1$, whose chemical reactions with DNA are well-studied. Our results, obtained by monitoring the distribution of lacI nonsense mutations in E. coli, suggest that the major mutational events induced by benzo[a]pyrene and aflatoxin B$_1$ are base substitutions.

The base substitutions are primarily transversions at G:C base pairs and the available evidence suggests that these mutations are induced by apurinic sites which are generated as secondary consequences of the initial alkylation event. The significance of these results in the context of carcinogenesis is briefly considered.

The high fidelity with which genomes are replicated in vivo and passed on to daughter cells is achieved by a repertoire of activities which function during replication, repair, and recombination (1,2). These activities, which collectively maintain the structural and informational integrity of the DNA molecule, are severely tested when the DNA template is damaged and becomes non-replicable. Under these circumstances, which obtain, for example, when cells are exposed to such human carcinogens as UV-light (3) or polycyclic aromatic hydrocarbons (4), it is commonly observed that the frequency of mutation is enhanced by many orders of magnitude. The correlation between the mutagenic and carcinogenic activity of many physical and chemical agents has been well-documented (5). Recent observations even suggest the possibility that one step in tumorigenesis might literally involve the mutational alteration of specific chromosomal genes (6–8).

0097–6156/85/0283–0327$06.00/0
© 1985 American Chemical Society

In this chapter I will review some aspects of mutagenesis mechanisms and the mutational consequences of DNA damage generated by benzo[a]pyrene. The focus will be on knowledge derived from investigations involving the bacterium Escherichia coli.

Mutagenesis is an active process

E. coli and eukaryotic cells can respond to DNA damage by inducing
the synthesis of specific gene products (9-12). The phenomenon of
gene induction by DNA damage has been most thoroughly described for
E. coli and has recently been reviewed by Walker (9). Among the inducible responses to DNA damage is the mutagenic repair process,
whose existence was first suggested over 30 years ago by the experiments of Weigle (13).
    Weigle showed that UV-light was mutagenic to bacteriophage lamb-
da only if the UV-irradiated lambda were grown on bacteria which had
also been irradiated with UV-light. In other words, the UV treatment
was not mutagenic per se. Furthermore, he demonstrated that irradiated lambda phage could be reactivated by growing the phage on pre-
irradiated bacteria. His results suggested the possibility that bacteria had an inducible system for DNA repair and mutagenesis which
acted on UV-irradiated lambda phage. The genetics of what is now
called Weigle or W-reactivation and W-mutagenesis is now very well
understood.
    Some twenty genes in E. coli -- known collectively as din genes
(damage inducible; 9,14) are coordinately regulated by the products
of the genes recA and lexA. The LexA protein represses din gene expression by binding to the operator region of each gene and prevent-
ing its transcription into RNA by RNA polymerase. Treatments which
damage the cell's DNA or otherwise interfere with DNA synthesis, activate the RecA protein; activated RecA protein then promotes the proteolytic inactivation of LexA repressor (15). Genes whose transcription had been repressed by LexA protein can now be transcribed and
new proteins can be synthesized. The overall response of E. coli to
DNA damage, which is genetically regulated by the recA and lexA loci,
is known as the SOS-response (16,17).
    Mutations in either recA or lexA can abolish the SOS-response
and eliminate both W-reactivation and W-mutagenesis. These mutations
also eliminate the mutability of the bacteria by UV-irradiation (16).
The observation that UV mutagenesis depended on the SOS-response established that mutations were not inevitable outcomes of DNA damage
and that DNA damage required processing by cellular mechanisms in order for mutations to be recovered. What specific processes regulated
by the SOS-response are responsible for mutagenesis?
    A major contribution towards answering this question was made by
the isolation of mutations which specifically eliminated the mutability of E. coli without affecting any of the other components of the
SOS-response. Mutations at the umuDC locus were independently discovered by Kato and Shinoura (18) and Steinborn (19) to abolish the
mutational affects of DNA damage by UV-irradiation. These mutants
were also shown to be defective in W-mutagenesis (18,20) and W-reactivation (18,20). The biochemical nature of the activity performed by
the umuDC gene products is not known. However, several observations
suggest that the function of the umuDC proteins is to enhance some
mode of DNA repair, either directly or indirectly: 1) E. coli carry-

ing the umuC36 allele is more sensitive to the lethal effects of UV-
irradiation (18) and angelicin plus near-UV (21); 2) Plasmid borne
analogs of the umuDC locus (mucAB; 12) enhance W-reactivation and the
resistance of bacteria to the lethal effects of UV-irradiation (23);

3) As previously noted, W-reactivation in Uvr⁻ bacteria is eliminated
by mutations at the umuDC locus (18). Of the approximately twenty
genes induced by DNA damage, the umuDC genes and their products are
the best candidates for direct participants in the biochemical processing of DNA lesions to mutations. The processing of DNA damage in
E. coli via the umuDC gene products and the associated proteins regulated by the SOS-response is called SOS-processing (9) or, sometimes,
error-prone repair (16,17). Mutagenesis in E. coli, therefore, appears to be a genetically and biochemically active process requiring
the participation of inducible proteins.

Not all mutagenesis in E. coli is dependent on SOS-processing.
Mutations may arise quite simply during DNA replication if a base is
substituted by or converted to another, incorrect, base. Consider
the consequence of oxidative deamination of the base 5-methylcytosine
to thymine. Replication followed by daughter strand segregation will
result in a G:C base pair having been mutated to an A:T base pair.
Sites containing 5-methylcytosine are hotspots for G:C to A:T transi-
tions in E. coli (24).

Alkylation of some bases at the exocyclic oxygen atoms can lead
to chemically stable alterations in the base pairing properties of a
base and, thereby, directly induce base mis-pairing by DNA polymerase. A well-studied example of this is the consequence of alkylating
guanine at the O-6 position (25-27). This has the effect of freezing
guanine in its (rare) enol tautomer permitting the G:T mismatch to
form in place of the usual G:C base pair. A subsequent round of DNA
replication leads to the generation of a G:C to A:T transition mutation. These exceptions notwithstanding, most DNA damaging agents induce mutations in E. coli via SOS-processing.

How universal is the SOS-processing system of E. coli?

The dependence of mutation on functions involving DNA repair seems to
be widespread among organisms. Many prokaryotic species are inherently non-mutable by UV-light but become mutable when plasmids encod-
ing for functions analogous to the umuDC functions are introduced
(e.g. 28). Non-mutable mutants of the yeast Saccharomyces cerevisiae
have been isolated and shown to possess defects which implicate DNA
repair and recombination processes (see 29 for a recent review).
Furthermore, there are many examples of DNA repair strategies which
are common to prokaryotic and eukaryotic organisms (nucleotide exci-
sion repair, DNA glycosylases, apurinic/apyrimidinic endonucleases,
O^6-methylguanine-DNA-methyl transferase; 2,30). Purified DNA polymerases from mammalian cells and viruses behave similarly in vitro to
E. coli DNA polymerase when DNA damage is encountered --i.e. replication ceases at the site of the lesion (31). Of course, even if analogues of SOS-processing are identified in eukaryotes, the regulation
of these activities might differ in detail from the scheme which ob-
tains in E. coli (e.g. analogous functions may be constitutively expressed). Nonetheless, Ruby and Szostak (10) have demonstrated the
existence of DNA damage inducible loci in S. cerevisiae and estimate

that there may exist as many as 80 such genes (11). Shorpp et al. (12) have recently reported that UV light enhances the synthesis of at least eight proteins in human fibroblast cells.

Are mutations distributed at sites of DNA damage?

The dependence of mutagenesis on SOS processing raised questions about the role(s) played by DNA lesions in mutagenesis. Do DNA lesions simply trigger the SOS response by interfering with DNA replication thereby generating mutations indirectly via an error-prone form of DNA replication? Or do mutations arise directly at the sites in DNA where damage has been generated?

The observations that mutation frequencies are elevated several-fold above normal levels in mutants which constitutively express their SOS-functions (32) and that the mutation frequency of unirradiated phage is elevated by growing them on irradiated (i.e. SOS-induced) bacteria (33), have been invoked to argue for the notion that mutagenesis via SOS-processing may be indirect.

On the other hand, the observation that 95% of the UV induced base substitution mutations arose at the very sites (pyrimidine–pyrimidine sequences) where the major fraction of UV damage is deposited suggested that at least the UV induced mutations were targeted (24).

Drake and Baltz (34) and Witkin and Wermundsen (35) presented arguments in favor of the notion that, for the most part, SOS mutagenesis was occurring at sites of DNA damage. More recent evidence, based on analyzing the distribution of mutations within the lacI gene of E. coli, strongly suggests that mutations arising via SOS-processing are occurring at the sites of DNA damage (36,37). Briefly, when one examines the spectrum of mutations induced by a variety of mutagens whose activity is dependent on SOS-processing, one finds that both where the mutations are induced and which mutations are induced depends on the mutagen. The observed differences among mutagens apply both to the mutational events that are distributed non-randomly at only a few sites (hotspots) and to events that are distributed randomly at many different sites within the gene (low frequency occurrences or LFO events) (36). Since each mutagenic treatment leaves behind its own characteristic distribution of mutations within the gene (37), mutations generated by SOS-processing of damaged DNA must be occurring at the sites of damage. Furthermore, the recent study by Miller and Low (38) on the distribution of mutations generated by turning on the SOS-response without DNA-damaging treatments shows that even these mutations are generated at specific sites in a gene as if they arose at sites where spontaneously generated lesions occur with a high frequency.

The well characterized reactions of carcinogens such as benzo-[a]pyrene and aflatoxin B_1 with DNA (39–50) suggested to us that an analysis of the kinds of mutations these agents induced could shed light on the contribution of specific DNA lesions to mutagenesis. Such an analysis could, in turn, provide clues as to which specific DNA lesions generated by these agents were mutagenic.

I would like to describe our investigations, performed in collaboration with Jeffrey Miller. To begin, I will briefly outline the genetic system developed by Miller which permits a rapid, rigorous determination of the position and kinds of mutants induced in a particular gene by DNA damaging agents.

The lacI system for analyzing nonsense mutations in E. coli

The lacI system has been described in detail by Miller (51). The
lacI gene product is the repressor of the lac operon. Cells which
have normal repressor activity are repressed for the synthesis of the
lacZ gene product, β-galactosidase, and the other products of the lac
operon. Cells which carry mutations in lacI which lead to synthesis
of a defective repressor protein will constitutively synthesize β-
galactosidase. Such mutants can be selected by demanding growth of
bacteria on a medium containing a galactoside analog such as phenyl-
β-D-galactoside (P-gal). P-gal is not itself an inducer of the lac
operon. Thus, it is a simple matter to treat a population of bacterial cells with a DNA damaging agent, grow them out non-selectively to
permit processing of DNA damage and phenotypic expression, and then
plate them on P-gal to select for cells carrying mutations in lacI.
    A large class of base substitution mutants can be analyzed di-
rectly by screening for suppressible mutations among the collection
of lacI mutants. The suppressible mutations are due to wild-type
codons having been mutated to TAA, TAG, or TGA. These nonsense codons are normally signals for the termination of protein synthesis by
ribosomes and can arise via all single base pair substitution mutations with the exception of the A:T to G:C transition. Thus, all
base pair substitutions, except for the one transition, can be monitored by collecting nonsense mutations in lacI. There are over 60
sites in lacI at which a single base pair substitution will generate
a nonsense codon.
    LacI nonsense mutants can be identified using classical bacter-
ial genetic methods. The entire gene has been sequenced (52). The
site, and therefore, the base pair which has been mutated can be i-
dentified simply by mapping the position of the nonsense mutation.
This can be accomplished by using an extensive set of lacI deletion
mutants (53). Mapping the mutation allows one to determine which
base pair substitution has been generated by a particular treatment.
Thus, by identifying many nonsense mutations induced by a mutagen, a
picture emerges of both the mutagens site specificity (where, within
the gene the mutations arise) and its mutagenic specificity (which
particular base substitutions are generated). To determine classes
of mutation other than base pair substitutions, it is possible to
genetically cross a given lacI allele onto small plasmid or phage
molecules and determine the sequence of the mutant allele (54,55).
    We have applied the genetic system for analyzing lacI nonsense
mutants to the investigation of the mutagenic specificity of benzo-
[a]pyrene (56) and aflatoxin B$_1$ (57). The results of our studies
have provided some important clues as to the chemical nature of the
mutagenic lesions induced by benzo[a]pyrene. Before I discuss these
results, I will briefly summarize previous investigations on the mutagenicity of BPDE.

The mutagenicity of benzo[a]pyrene diolepoxide -- previous
investigations

Carcinogens first began to be evaluated directly for mutagenic activity against microorganisms 35 years ago by Barrett and Tatum (58). However, the systematic use of sensitive microbial mutation assays to monitor the biological activity of carcinogens was not achieved until the realization that metabolic activation of carcinogens was essential (reviewed in 59). By the use of sub-cellular fractions derived from liver homogenates, it became possible to detect the mutagenic activity of benzo[a]pyrene and many other polycyclic aromatic hydrocarbons.

The mutagenicity of benzo[a]pyrene for bacteria was demonstrated by Ames et al. (60). They found that in the presence of rat liver homogenates benzo[a]pyrene induced both frameshift and base-pair substitution mutations. When the chemistry of benzo[a]pyrene activation had been worked out and the ultimate carcinogenic form identified as a diolepoxide, BPDE (reviewed in 61-62), several investigators (63-66) showed that BPDE was an extremely potent mutagen, also capable of inducing both frameshift and base-pair substitution mutations. McCann et al. (67) had shown that benzo[a]pyrene was mutagenic for S. typhimurium only if the bacteria carried the mutation enhancing plasmid pKM101 whose activity was later shown by Walker (23) to be entirely dependent on bacterial recA and lexA controlled functions. This provided early evidence that the mutagenicity of carcinogens such as benzo[a]pyrene was dependent on SOS-repair. Later, Ivanovic and Weinstein (68) directly showed that benzo[a]pyrene was mutagenic for E. coli only if the bacteria were both $recA^+$ and $lexA^+$.

Two questions that are raised by these observations are:

1) what is the mutagenic specificity of BPDE, i.e. what kinds of mutations are induced by treating cells with BPDE?

2) what is(are) the pre-mutational lesion(s) generated by BPDE which is(are) responsible for mutations?

The mutagenic specificity of BPDE

We have obtained important clues to these questions by determining the spectrum of 185 nonsense mutations induced in the lacI gene of E. coli by BPDE. The results of this investigation (56) are summarized in Tables I to III and in Figure 1. The results were striking. They showed that transversion mutations at G:C base pairs were the dominant induced event, although, other substitutions, in particular A:T to T:A transversion were clearly induced, but at lower frequencies. The specificity of induction of G:C to T:A was most clearly seen by examining the mutations occurring at the TAC codons for tyrosine (Table III). At these sites, both G:C to T:A mutations (yielding TAA, ochre nonsense mutants) and G:C to C:G (yielding TAG, amber nonsense mutants) transversions are monitorable. Table III clearly shows that at the two tyrosine codons where mutations were well-induced there is a striking preference for one mutational event over the other.

We have not directly determined the relative frequencies of frameshift mutations and other mutational events in comparison to the base substitution mutations. However, based on the high frequency of nonsense mutations (11%) among all lacI mutants induced by BPDE and because nonsense mutations are monitorable at less than one-fifth of the lacI codons and, even then, only via certain base pair substitutions, we believe that base substitutions account for a major fraction of mutations induced by BPDE.

Table I. Summary of Base Substitution Events Generated by BPDE

Substitution	No. of Available Sites	No. of Sites Found	Total No. of Occurrences	% of Analyzed Mutations
G:C to A:T	26	12	22	12
G:C to T:A	23	21	123	66
A:T to T:A	15	9	33	18
A:T to C:G	5	3	4	2
G:C to C:G	3	2	3	2
Total	72	47	185	
Amber	36	26	96	
Ochre	36	21	89	

What is (are) the pre-mutational lesion(s) induced by BPDE?

BPDE reacts at several different sites on DNA to generate several kinds of lesions at the N2 (43-45) and N7 (46,47) positions of guanine, apurinic sites (48,49), and strand breaks (50). Which of these lesions are responsible for the transversion mutations at G:C sites? Evidence derived from a number of experiments suggests the hypothesis that apurinic sites generated by BPDE reactions with DNA are responsible for the transversion mutations:

1. When we examined the mutagenic specificity aflatoxin B_1, a carcinogen which specifically reacts with the N7 atom of guanine (39-42), we found virtually only G:C to T:A transversions were induced (57); N7 purine adducts can induce depurination by destabilizing the N-glycosylic bond (69).

2. The work of Loeb and Kunkel and their colleagues (70-72) has clearly established that apurinic sites in DNA are mutagenic; they specifically cause transversion mutations, due to a strong preference for the incorporation of adenine residues during bypass of apurinic sites in template DNA. Thus, A:T to T:A and G:C to T:A transversions are the major mutagenic outcome generated by depurination of DNA.

3. Recently, Sage and Haseltine (49) have quantitatively determined the spectrum of DNA lesions induced by reactions of BPDE with DNA. They found that alkali-labile lesions account for about 40% of the DNA adducts. There was a striking correlation between the mutation frequencies induced by BPDE in lacI and the frequencies of alkali sensitive lesions at G, A, and C residues. Apurinic/apyrimidinic sites are common alkali-sensitive lesions. Earlier work by Drinkwater et al. (48) had also shown that treatment of DNA with BPDE generated apurinic/apyrimidinic sites.

Table II. Distribution of lacI Nonsense Mutations Induced by BPDE

Base Substitutions	Site	No. Independent Occurrences	Site	No. Independent Occurrences
G:C – A:T	A5	0	09	0
	*A6	5	010	0
	A9	0	011	0
	*A15	0	013	0
	A16	0	017	0
	A19	1	021	3
	A21	2	024	3
	A23	0	027	2
	A24	1	028	1
	A26	0	029	1
	A31	1	034	0
	A33	1	035	0
	*A34	1		
	A35	0		
G:C – T:A[#]	A2	7	03,A1[#]	12,1 (13)
G:C – C:G[#]	A7	5	06	0
	A10	5	07	0
			08,A8[#]	4,0 (4)
	A12	18	014	2
	A13	5	015	7
	A17	5	019	4
	A20	5	020	12
	A25	4	025	0
	A27	5	026	4
	A28	3	030,A29[#]	1,2 (3)
			031	2
			032	1
			036	12
A:T – T:A[@]	A11	2	01	2
A:T – C:G[@]	A18	4	02	5
	A32	1	04,A3[@]	0 (0)
	A36	8	05,A4[@]	2 (2)
			012	0
			016	6
			018,A14[@]	1 (1)
			022	4
			023,A22[@]	0 (0)
			033,A30[@]	1 (1)

Sites at which nonsense mutations are detected are identified by their amber (A) or ochre (0) alleles (Coulondre and Miller, 1977). The 8 tyrosine codons in lacI each have two nonsense alleles, one amber and one ochre. The amber alleles at these sites are marked by the symbols # and @.
*Sites containing 5-methylcytosines (CCAGG). These are spontaneous lacI hotspots.

Table III.   BPDE-Induced Mutations at the Three TAC Tyrosine Codons
in the lacI Gene

Site	coding position	# of independent occurrences of:	
		TAC - TAA (GC - TA)	TAC - TAA (GC - CG)
A1,03	tyr 7	12	1
A8,08	tyr 47	4	0
A29,030	tyr 273	1	2
Total		17	3

Both amber and ochre mutations can be generated at these sites,
allowing both G:C to T:A and G:C to C:G transversions to be
monitored.

Thus, while BPDE and aflatoxin B_1 might generate G:C to T:A
transversions via different pathways, it is reasonable to consider
the hypothesis that there is a common mechanism by which they induce
this mutation and that, therefore the transversion mutations induced
by BPDE result, not from the major adduct to the N2 atom of guanine
but from the generation of apurinic sites in DNA. These secondary
lesions might be generated spontaneously or via the activity of DNA
glycosylases (2, 30). An alternative hypothesis is that bulky lesions in general, the N2 adduct among them, may be noninformational
sites opposite which adenines are preferentially inserted during replication after DNA damage.

Other molecular genetic studies on the mutagenicity of BPDE

The genetic system we used to study the mutagenic specificity of BPDE
limits one to analyzing base substitution mutations. What is known
about the ability of BPDE to induce other categories of mutation? As
mentioned above, results from the Ames test revealed that benzo[a]py-
rene and its diol epoxide were capable of inducing frameshift mutations (60,63,64). More recently, Mizusawa and co-workers (73-76)
have investigated the mutational consequences of modifying plasmid
DNA in vitro with BPDE. In a series of studies they have shown that:
    1. plasmid molecules are inactivated (become non-replicable) in
Uvr⁻ bacteria by 1 covalent adduct (73,76) per molecule; this result
is in perfect agreement with an earlier report by Hsu et al. (77)
which had demonstrated that one molecule of bound BPDE was sufficient

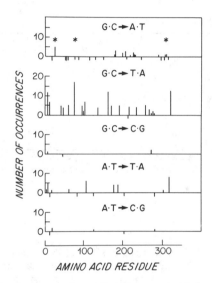

Figure 1. The frequencies of amber mutations in the lacI
gene induced by BPDE. Solid bars, individual sites at
which we detected mutations; open bars, sites at which we
did not detect mutations; asterisks (in a), sites at which
the target codon contains 5-methylcytosine.

to inhibit the replication of single molecule of the single stranded
bacteriophage ∅X174.

    2. BPDE-induced mutations in plasmid borne genes can be depen-
dent on umuC (75);

    3. mutations induced by BPDE include transversion, transition,
and frameshift mutations (74,76).

    Wei et al. (78) characterized mutations resulting from alkyla-
tion of a 10-base pair oligonucleotide with BPDE. Deletion mutations
were the major mutational event detected.

    The number of mutations analyzed in each of these investigations
was too small (only 7 to 8) to permit drawing firm conclusions about
mutational and site specificities. However, the results suggest
that, under some circumstances, BPDE can induce many different kinds
of mutations.

Future studies on the genetic effects of BPDE

To rigourously establish the genetic consequences resulting from BP
adduct to the N2 position of guanine, the approach taken by Essigman
and his colleagues (79,80) will be required. They have been developing techniques for placing defined chemical lesions into plasmid DNA
at pre-determined sites at which mutations can be monitored. If a
BP adduct can be "built" into DNA at the N2 of guanine, its biological and genetic effects can be determined.

    It would be interesting to know if the mutational consequences
of DNA lesions in mammalian cells were the same as those which obtain
in bacteria. Methods for retrieving and sequencing mutations in mammalian cells and their viruses are now being developed (81-83). If
yeast, a eukaryotic microorganism, can be considered representative
of higher eukaryotes, then judging from the observations that the
mutational spectra for UV-irradiation and 4-nitroquinoline-1-oxide
treatment are identical for yeast (84) and bacteria (85), the spectrum of mutations induced by BPDE in mammalian cells could well resemble those induced in E. coli.

Is the mutagenicity of BPDE directly responsible for its carcinogenicity?

Though one is far from being able to make a definitive statement,
there are some indications that the answer to this question might be
no. Recent experiments with mammalian cell culture systems, which
allow one to study the progression of cells from a state where their
growth is normally regulated to a tumorigenic state have revealed
that the transformation process requires at least two steps (86-88).
The first step, which is induced following exposure to a carcinogen
(X-rays: 86,88; 3-methylcholanthrene: 87), occurs with a very high
frequency and seemingly affects every exposed cell in the treated
population. The frequency of the initial event implies that it is
not a mutational event but rather an epigenetic one related, perhaps,
to the responses induced by DNA damage in E. coli and S. cerevisiae.
The second event is a very rare event consistent with the possibility
that it might be mutational in nature. However, since the second e-
vent occurs many cell generations after the exposure to DNA damaging
agents, it seems highly improbable that it occurs as a direct consequence of repairing the initial DNA damage. Thus, the link between

carcinogenesis and mutagenesis may simply be that the two processes originate from the same starting point, namely DNA damage.

Acknowledgments

Work in my laboratory has been supported by grants from the NIH. I am indebted to many of my present and former colleagues, in particular to Drs. A.J. Warren and P.L. Foster for their work on the mutagenic specificity of chemical carcinogens and to Dr. J.H. Miller for his collaborative effort in studying the mutagenic specificity of benzo[a]pyrene and aflatoxin B_1.

Literature Cited

1. Loeb, L.A.; Kunkel, T.A. Annu. Rev. Biochem. 1982, 52, 429–57.
2. Lindahl, T. Annu. Rev. Biochem. 1982, 51, 61–87.
3. Robbins, J.H.; Kraemer, K.H.; Lutzner, M.A.; Festoff, B.W.; Coon, H.G. Ann. Int'l. Med. 1974, 80, 221–48.
4. IARC Monographs on the evaluation of the carcinogenic risk of chemicals to humans, Vol 32, International Agency for Research on Cancer, France, 1983.
5. McCann, J.; Choi, E.; Ames, B.N. Proc. Natl. Acad. Sci. USA 1975, 72, 5135–9.
6. Tabin, C.J.; Bradley, S.M.; Bargman, C.I.; Weinberg, R.A.; Papageorge, A.G.; Scolnick, E.M.; Dhar, R.; Lowy, D.R.; Chang, E.H. Nature 1982, 300, 143–9.
7. Reddy, E.P.; Reynolds, R.K.; Santos, E.; Barbacid, M. Nature 1982, 300, 149–52.
8. Taparowsky, E.; Suard, Y.; Fasano, J.; Shimizu, K.; Goldfarb, M.; Wigler, M. Nature 1982, 300, 762–5.
9. Walker, G.C. Microbiol. Rev. 1984, 48, 60–93.
10. Ruby, S.W.; Szostak, J.W.; Murray, A.W. Methods in Enzymology 1983, 101, 253–68.
11. Ruby, S.W.; Szostak, J.W. Mol. Cell. Biol. 1985, 5, 75–84.
12. Schorpp, M.; Mallick, U.; Rahmsdorf, H.J.; Herrbick, P. Cell 1984, 37, 861–8.
13. Weigle, J.J. Proc. Natl. Acad. Sci. USA 1953, 39, 628–36.
14. Kenyon, C.J.; Walker, G.C. Proc. Natl. Acad. Sci. USA 1980, 77, 2819–23.
15. Little, J.W. Proc. Natl. Acad. Sci. USA 1984, 81, 1375–9.
16. Defais, M.; Fauquet, P.; Radman, M.; Errera, M. Virology 1971, 43, 495–503.
17. Witkin, E. Bacteriol. Rev. 1976, 40, 869–907.
18. Kato, T.; Shinoura, Y. Molec. Gen. Genet. 1977, 156, 121–31.
19. Steinborn, G. Molec. Gen. Genet. 1978, 165, 87–93.
20. Walker, G.C.; Dobson, P.P. Mol. Gen. Genet. 1979, 172, 17–24.
21. Miller, S.; Eisenstadt, E., unpublished observation
22. Perry, K.L.; Walker, G.C. Nature 1982, 300, 278–81.
23. Walker, G.C. Mol. Gen. Genet. 1977, 152, 93–103.
24. Coulondre, C.; Miller, J.H.; Farabaugh, P.J.; Gilbert, W. Nature 1978, 274, 775–80.
25. Gerchman, L.L.; Ludlum, D.B.; Biochim. Biophys. Acta. 1973, 308, 310–16.
26. Singer, B.; Fraenkel-Conrat, H.; Kusmierek, J.T. Proc. Natl. Acad. Sci. USA 1978, 75, 1722–6.

27. Snow, E.T.; Foote, R.S.; Mitra, S. J. Biol. Chem. 1984, 259, 8095-100.
28. Hofemeister, J.; Kohler, H.; Filippov, V.D. Mol. Gen. Genet. 1979, 176, 265-73.
29. Haynes, R.H.; Kunz, B.A. In "The Molecular Biology of the Yeast Saccharomyces"; Strathern, J.N.; Jones, E.W.; Broach, J.R., Ed.; Cold Spring Harbor Laboratory: Cold Spring Harbor, 1981; pp. 371-414.
30. Linn, S.M.; Roberts, R.J., Ed.; "Nucleases", Cold Spring Harbor Laboratory: Cold Spring Harbor, 1982.
31. Strauss, B.; Rabkin, S.; Sagher, D.; Moore, P. Biochimie 1982, 64, 829-38.
32. Witkin, E.M. Proc. Natl. Acad. Sci. USA 1974, 71, 1930-4.
33. Quillardet, P.; Devoret, R. Biochimie 1982, 64, 789-96.
34. Drake, J.W.; Baltz, R.H. Annu. Rev. Biochem. 1976, 45, 11-37.
35. Witkin, E.M.; Wermundsen, I.E. Cold Spring Harbor Symp. Quant. Biol. 43, 881-6.
36. Foster, P.L.; Eisenstadt, E.; Cairns, J. Nature 1982, 299, 365-7.
37. Miller, J.H. Cell 1982, 31, 5-7.
38. Miller, J.H.; Low, K.B. Cell 1984, 37, 675-82.
39. Essigman, J.M.; Croy, R.G.; Nadzan, A.M.; Busby, W.F., Jr.; Reinhold, V.N.; Buchi, G.; Wogan, G.N. Proc. Natl. Acad. Sci. 1977, 74, 1870-4.
40. Lin, J.K.; Miller, J.H.; Miller, E.C. Cancer Res. 1977, 37, 4430-8.
41. Martin, C.N.; Garner, R.C. Nature 1977, 267, 863-5.
42. Croy, R.G.; Essigman, J.M.; Reinhold, V.N.; Wogan, G.N. Proc. Natl. Acad. Sci USA 1978, 75, 1745-9.
43. Jeffrey, A.M.; Weinstein, I.B.; Jennette, K.W.; Grzeskowiak, K.; Nakanishi, K.; Harvey, R.G.; Autrup, H.; Harris, C. Nature 1977, 269, 348-50.
44. Jeffrey, A.M.; Grzeskowiak, K.; Weinstein, I.B.; Nakanishi, K.; Roller, P.; Harvey, R.G. Science 1979, 206, 1309-11.
45. Meehan, T.; Straub, K.; Calvin, M. Nature 1977, 269, 725-7.
46. King, H.W.S.; Osborne, M.R.; Brookes, P. Chem. Biol. Interact. 1979, 24, 345-53.
47. Osborne, M.R.; Jacobs, S.; Harvey, R.G.; Brookes, P. Carcinogenesis 1981, 2, 553-8.
48. Drinkwater, N.R.; Miller, E.C.; Miller J.A. Biochemistry 1980, 19, 5087-92.
49. Sage, E.; Haseltine, W.A. J. Biol. Chem. 1984, 259, 11098-102.
50. Haseltine, W.A.; Lo, K.M.; D'Andrea, A.D. Science 1980, 209, 929-31.
51. Miller, J.H. In "The Operon"; Miller, J.H.; Reznikoff, W.S., Ed.; Cold Spring Harbor Laboratory: Cold Spring Harbor, 1980; pp. 31-88.
52. Farabaugh, P.J. Nature 1978, 274, 765-9.
53. Schmeissner, U.; Ganem, D.; Miller, J.H. J. Mol. Biol. 1977, 109, 303-26.
54. Calos, M.P.; Johnsrud, L.; Miller, J.H. Cell 1978, 13, 411-8.
55. Schaaper, R.; Glickman, B., personal communication.
56. Eisenstadt, E.; Warren, A.J.; Porter, J.; Atkins, D.; Miller, J.H. Proc. Natl. Acad. Sci. USA 1982, 79, 1945-9.

57. Foster, P.L.; Eisenstadt, E.; Miller, J.H. Proc. Natl. Acad. Sci. USA 1983, 80, 2695-8.
58. Barrett, R.W.; Tatum, E.L. Cancer Research 1951, 11, 234.
59. Miller, E. Cancer Res. 1978, 38, 1479-96.
60. Ames, B.N.; Durston, W.E.; Yamasaki, E.; Lee, F.D. Proc. Natl. Acad. Sci. USA 1973, 70, 2281-5.
61. Conney, A.H. Cancer Res. 1982, 42, 4875-4917.
62. Harvey, R.G. Acc. Chem. Res. 1981, 14, 218-26.
63. Wood, A.W.; Wislocki, P.G.; Chang, R.L.; Levin, W.; Lu, A.Y.H.; Yagi, H.; Hernandez, O.; Jerina, D.M.; Conney, A.H. Cancer Res. 1976, 36, 3358-66.
64. Malaveille, C.; Kuroki, T.; Sims, P.; Grover, P.L.; Bartsch, H. Mutat. Res. 1977, 313-26.
65. Huberman, A.; Sachs, L.; Yang, S.K.; Gelboin, H.V. Proc. Natl. Acad. Sci. USA 1976, 73, 607-11.
66. Newbold, R.F.; Brookes, P. Nature (London) 1976, 261, 52-54.
67. McCann, J.; Spingarn, N.E.; Kobori, J.; Ames, B.N. Proc. Natl. Acad. Sci. USA 1975, 72, 979-83.
68. Ivanovic, V.; Weinstein, I.B. Cancer Res. 1980, 40, 3508-11.
69. Shapiro, R. Prog. Nucleic Acid Res. Mol. Biol. 1968, 8, 73-112.
70. Schaaper, R.M.; Loeb, L.A. Proc. Natl. Acad. Sci. USA 1981, 78, 1773-7.
71. Schaaper, R.M.; Glickman, B.W.; Loeb, L.A. Cancer Res. 1982, 42, 3480-2.
72. Kunkel, T.A. Proc. Natl. Acad. Sci. USA 1984, 81, 1494-1498.
73. Mizusawa, H.; Lee, C-H.; Kakefuda, T. Mutat. Res. 1981, 82, 47-57.
74. Mizusawa, H.; Lee, C-H.; Kakefuda, T.; McKenney, K.; Shimatake, H.; Rosenberg, M. Proc. Natl. Acad. Sci. USA 1981, 78, 6817-20.
75. Mizusawa, H.; Chakrabarti, S.; Seidman, M. J. Bacteriol. 1983, 156, 926-30.
76. Chakrabarti, S.; Mizusawa, H.; Seidman, M. Mutat. Res. 1984, 126, 127-37.
77. Hsu, W-T.; Lin, E.J.S.; Harvey, R.G.; Weiss, S.B. Proc. Natl. Acad. Sci. USA 1977, 74, 3335-9.
78. Wei, S-J.C.; Desai, S.M.; Harvey, R.G.; Weiss, S.B. Proc. Natl. Acad. Sci. USA 1984, 81, 5936-3940.
79. Fowler, K.W.; Büchi, G.; Essigman, J.M. J. Am. Chem. Soc. 1982, 1051-4.
80. Green, C.L.; Loechler, E.L.; Fowler, K.W.; Essigman, J.M. Proc. Natl. Acad. Sci. USA 1984, 81, 13-7.
81. Razzaque, A.; Mizusawa, H.; Seidman, M.M. Proc. Natl. Acad. Sci. USA 1983, 80, 3010-4.
82. Calos, M.; Lebkowski, J.S.; Botehan, M.R. Proc. Natl. Acad. Sci. USA 1983, 80, 3015-9.
83. Bourre, F.; Sarasin, A. Nature 1983, 305, 68-70.
84. Prakash, L.; Sherman, F. J. Mol. Biol. 1973, 79, 65-82.
85. Coulondre, C.; Miller, J. J. Mol. Biol. 1977, 117, 577-606.
86. Kennedy, A.R.; Fox, M.; Murphy, G.; Little, J.B. Proc. Natl. Acad. Sci. USA 1980, 77, 7262-6.
87. Fernandez, A.; Mondal, S.; Heidelberger, C. Proc. Natl. Acad. Sci. USA 1980, 77, 7272-6.
88. Kennedy, A.R.; Cairns, J.; Little, J.B. Nature 1984, 307, 85-5.

RECEIVED May 2, 1985

Chemical Properties of Ultimate Carcinogenic Metabolites of Arylamines and Arylamides

FRED F. KADLUBAR and FREDERICK A. BELAND

National Center for Toxicological Research, Jefferson, AR 72079

A number of arylamines and arylamides are carcinogenic in a variety of tissues of several species including the urinary bladder of man. These compounds undergo metabolic activation to ultimate carcinogens through a number of enzymatic and nonenzymatic pathways. In this review, these activation mechanisms are considered in detail and their relative contribution to the observed carcinogenicity of these compounds is discussed.

The metabolism of carcinogenic arylamines and arylamides results in a broad spectrum of reactive, electrophilic metabolites that form covalent adducts with cellular constituents. These activation pathways are summarized in Figure 1. Arylamides and primary arylamines are readily interconverted by N-acetyltransferases and N-deacetylases (reviewed in 1) and they are initially activated by cytochrome P-450- and flavin-containing monooxygenases to form N-hydroxy arylamides and N-hydroxy arylamines, respectively (reviewed in 2). These N-hydroxy metabolites, which can also be interconverted by enzymatic N-deacetylation/N-acetylation (1), are proximate carcinogens since they are generally more carcinogenic and mutagenic than their parent compounds. Further enzymatic or non-enzymatic processes lead to ultimate carcinogens, which are usually defined by their electrophilic reactivity with nucleic acids or proteins (3).

N-Hydroxy arylamides are converted to ultimate carcinogens through conjugation with sulfuric, acetic or glucuronic acids (reviewed in 1,4). Sulfuric acid conjugation is catalyzed by 3'-phosphoadenosine-5'-phosphosulfate (PAPS)-dependent sulfotransferases and yields N-sulfonyloxy arylamides (I); while N-acetoxy arylamides (II) are formed through nonenzymatic esterification with acetyl coenzyme A or by a peroxidase-mediated, one-electron oxidation and subsequent dismutation of a nitroxyl radical (5,6). N-Glucuronyloxy arylamides (III) are also formed by enzymatic conjugation and they can undergo subsequent N-deacetylation to N-glucuronyloxy arylamines (IV). An additional pathway by which N-hydroxy arylamides are activated is through an intramolecular rearrangement to N-acetoxy arylamines (V) which is catalyzed by cytosolic N,O-acyltransferases.

This chapter not subject to U.S. copyright.
Published 1985, American Chemical Society

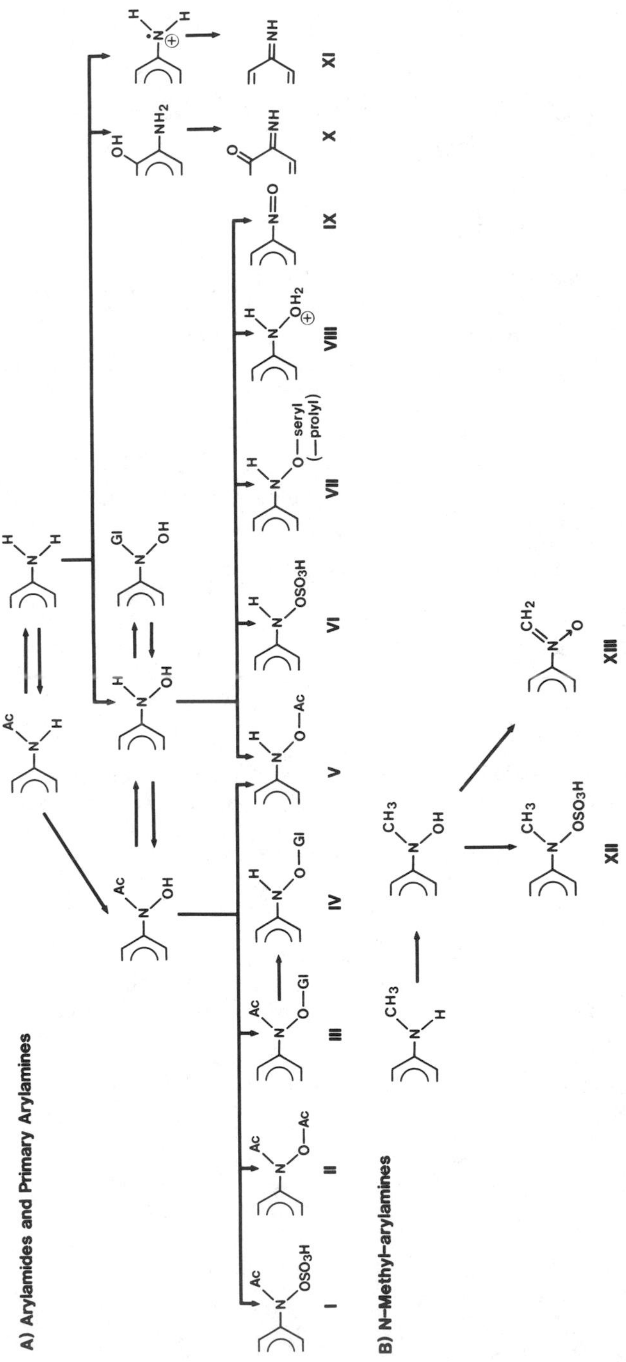

Figure 1. Metabolic Activation Pathways for Carcinogenic Arylamides, Primary Arylamines, and N-Methyl Arylamines. Putative ultimate carcinogenic metabolites are designated I–XIII. Ac, acetyl; Gl, glucuronyl.

N-Hydroxy arylamines are also converted to N-acetoxy arylamines (V), but apparently by an acetyl coenzyme A-dependent enzymatic O-esterification (7,8). Similarly, N-sulfonyloxy arylamines (VI) are thought to arise by a PAPS-dependent enzymatic O-sulfonylation of N-hydroxy arylamines (9,10); while O-seryl or O-prolyl esters (VII) are formed by their corresponding aminoacyl tRNA synthetases in a ATP-dependent reaction (11,12).

N-Hydroxy arylamines readily form glucuronide conjugates, but in contrast to the N-hydroxy arylamides, these are N-glucuronides which are unreactive and stable at neutral pH. The N-glucuronides are readily transported to the lumens of the urinary bladder and intestine where they can be hydrolyzed to the free N-hydroxy arylamines by mildly acidic urine or by intestinal bacterial β-glucuronidases (13,14). Non-enzymatic activation of N-hydroxy arylamines can occur in an acidic environment by protonation (15,16) of the N-hydroxy group (VIII) as well as by air oxidation (reviewed in 17) to a nitrosoarene (IX).

Alternative metabolic pathways involve ring-oxidation and peroxidation of arylamines. Although ring-oxidation is generally considered a detoxification reaction, an electrophilic iminoquinone (X) can be formed by a secondary oxidation of the aminophenol metabolite (18,19). Lastly, reactive imines (XI) can be formed from the primary arylamines by peroxidase-catalyzed reactions that involve free radical intermediates (reviewed in 20).

Only a limited number of activation pathways appear to be available to N-methyl arylamines. Following enzymatic N-hydroxylation to secondary N-hydroxy arylamines (21,22), these compounds are converted into reactive electrophiles through enzymatic esterification (9) to N-sulfonyloxy-N-methyl arylamines (XII) or by further oxidation to N-arylnitrones (XIII).

In this review, the chemical properties of these electrophilic metabolites (I-XIII) are discussed in terms of their metabolic formation and reactivity with nucleophiles, solvolysis and redox characteristics, reaction mechanisms, and their role as ultimate carcinogenic metabolites.

N-Sulfonyloxy Arylamides (I)

The metabolic formation of N-sulfonyloxy-N-acetyl-2-aminofluorene (N-sulfonyloxy-AAF) and its observed electrophilic reactivity, provided the first evidence for the importance of enzymatic conjugation reactions in chemical carcinogenesis (23,24). This reaction was shown to be catalyzed by PAPS-dependent sulfotransferases that are located predominantly in liver cytosol and has been subsequently demonstrated for N-hydroxy arylamide metabolites of several other carcinogens, including N-acetyl-4-aminobiphenyl (AABP), benzidine, N-acetyl-2-aminophenanthrene and phenacetin.

Accordingly, the contribution of this metabolic activation pathway to the formation of covalently-bound adducts of arylamides with cellular proteins and nucleic acids has been the subject of numerous investigations, and has been reviewed extensively by Mulder (25). From these and more recent data (4,26,27) it is apparent, particularly in the case of N-hydroxy-AAF (N-OH-AAF), that in vivo formation of reactive N-sulfonyloxy derivatives is primarily responsible for the carcinogen adducts with hepatic protein, RNA, DNA and glutathione (GSH) that retain the N-acetyl group. With N-OH-AAF, for example, these N-acetylated adducts account for 70-80%, 60-80%, and 15-30% of the total binding to rat liver protein, RNA, and DNA, respectively (28,30); and GSH-AAF adducts excreted in the bile account for about 10% of the dose given (26). In comparable studies with N-hydroxy-AABP (N-OH-AABP), N-acetylated adducts represent about 10% and 20% of the RNA and DNA binding, respectively (30); and N-acetylated adducts derived from 4'-fluoro-N-OH-AABP and N,N'-diacetylbenzidine amount to 10-20% of the total DNA-bound products (31,32). In contrast, only deacetylated adducts are detectable in rat hepatic DNA after administration of N-acetyl-4-aminostilbene (33) or N-acetyl-7-fluoro-2-aminofluorene (34), both of which induce tumors in the liver and other tissues. Similarly, only deacetylated DNA adducts are found in rat liver after treatment with the extrahepatic arylamide carcinogen, N-acetyl-2-aminophenanthrene (35), or with the N-hydroxy-N-acetyl derivative of the colon carcinogen, 3,2'-dimethyl-4-aminobiphenyl (36).

Structural identification of the N-acetylated adducts found in vivo has shown that binding to protein or GSH involves predominantly ortho-ring substitution of the arylamide with the sulfur atom in methionine or cysteine, respectively. In contrast, arylamide binding to nucleic acids in vivo involves both N-substitution at the C-8 position of guanine and ortho-ring substitution with the exocyclic N^2 atom of guanine (26,29-31,37,38).

Several synthetic N-sulfonyloxy arylamides have been prepared in order to compare their reactivity with nucleophiles to that observed in vivo and in in vitro metabolic systems. Synthetic N-sulfonyloxy-AAF reacts appreciably with both protein or methionine to give high yields of ortho-methylmercapto derivatives that are identical to those formed in vivo. Similarly, methionine has been shown to trap 65-85% of N-sulfonyloxy-AAF generated in incubations containing PAPS, N-OH-AAF, and hepatic cytosolic sulfotransferase (9). N-Sulfonyloxy-AAF also reacts with GSH in vitro to give 1-, 3-, 4-, and 7-AAF ring-substituted glutathion-S-yl adducts (39), of which two (1-, 3-) are major biliary metabolites (26). N-(Guanosin-8-yl)-AAF, a major in vivo adduct with hepatic RNA, can be prepared by reaction of guanosine with N-sulfonyloxy-AAF or by in vitro sulfotransferase activation of N-OH-AAF in the presence of RNA or guanosine (40). Reaction of N-sulfonyloxy-AAF with DNA yields both N-(deoxyguanosin-8-yl)-AAF and 3-(deoxyguanosin-N^2-yl)-AAF, which are identical to the N-acetylated adducts found in vivo (30,41). However, a similar reaction with deoxyguanosine in an aqueous medium gives only the C8-substituted product; while both C8- and N^2-substituted adducts can be prepared by reaction of N-sulfonyloxy-AAF with deoxyguanosine in anhydrous dimethylsulfoxide/triethylamine (41).

Though much less reactive than N-sulfonyloxy-AAF, N-sulfonyloxy esters of N-OH-AABP and its 4'-fluoro derivative have been prepared and shown to react with methionine to give ortho-substituted methylmercapto arylamides and with DNA to give C8- and N^2-substituted deoxyguanosine-arylamide adducts (reviewed in 42). Again, only C8-substituted guanine derivatives are obtained on reaction of N-sulfonyloxy-AABP with deoxyguanosine, guanosine, or RNA. N-Sulfonyloxy-N-acetyl-2-aminophenanthrene has been prepared and shown to react to a limited extent with methionine, deoxyguanosine and deoxyadenosine to give 1-methylmercapto, N-(deoxyguanosin-8-yl), and 1-(deoxyadenosin-N⁶-yl) derivatives, respectively. Metabolic formation of N-sulfonyloxy phenacetin has also been proposed since hepatic sulfotransferase-catalyzed activation of N-hydroxy phenacetin leads to the formation of adducts with protein, nucleic acids and GSH (25,43).

From these studies and those involving N-acetoxy arylamides (vide infra), it is clear that any proposed reaction mechanism must account for the ability of different nucleophiles to direct substitution to the N-, ortho- and meta-ring positions of the arylamide and should be consistent with reaction kinetics and with solvolysis or rearrangement products found in the reaction medium. In this regard, studies with model compounds such as methanesulfonate esters of N-hydroxy acetanilides (44,45) and N-sulfonyloxy acetanilides (46) have been particularly useful. These data indicate that reactive N-sulfonyloxy derivatives undergo heterolytic cleavage of the N-O bond to form an intimate ion pair consisting of a partially delocalized singlet nitrenium/carbenium cation and the sulfate anion (Figure 2), as originally proposed by Scribner et al. (47) and more recently supported by molecular orbital calculations (48). Collapse of the ion pair by internal return results in an o-sulfonyloxy acetanilide while reducing agents convert it to the parent acetanilide. Evidence has also been presented that hydrolysis of the ion pair may proceed through an imine intermediate which would account for meta- and possibly N-substituted products (45,46). In addition, earlier studies with the metabolically generated N-sulfonyloxy ester of phenacetin (p-ethoxyacetanilide) indicate that N-acetyl benzoquinone imine is formed as a reactive intermediate (49).

Recently, the decomposition of N-sulfonyloxy-AAF under aqueous conditions has been further examined and appears to be consistent with this overall mechanism (50). That is, the major products appear to be 1- and 3-sulfonyloxy-AAF with small amounts of AAF, 4-hydroxy-AAF, and a dimer formed by addition of the electrophile onto the aromatic ring of another AAF molecule (51). Furthermore, the relative yields of AAF could be increased by addition of the reducing agent, ascorbic acid (52).

The involvement of the nitrenium/carbenium cation-sulfate anion pair as the major electrophilic reactant from arylamide carcinogens is also consistent with the nature of the products formed with cellular nucleophiles (vide supra) and is in accord with the Pearson hard/soft acid-base concept of electrophilic substitution (53). Thus, ortho-substitution of the arylamine is favored by soft nucleophiles (RSCH₃, RSH, RNH₂) which tend to advance ion pair separation resulting in greater charge delocalization in the aromatic ring (48); while N-substitution is favored with hard nucleophiles that attack a tight ion pair with a positive nitrogen center (54). Both types of substitution represent an S_N1 reaction mechanism which is determined by the strength of the sulfate leaving group, along with formation of the ion pair whose overall reactivity or selectivity (N vs. ortho substitution) can be influenced by changes in reaction medium and by the nature of the nucleophile (41,42,55).

Although metabolically-formed N-sulfonyloxy arylamides are strong electrophiles, bind to cellular macromolecules, and have long been considered ultimate carcinogens, their precise role in aryl-

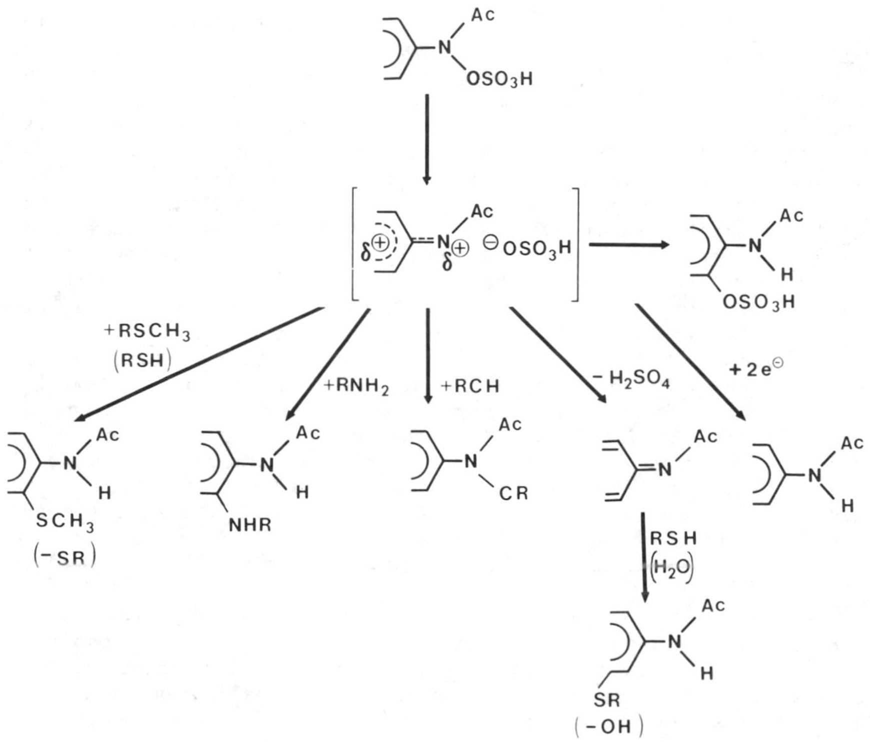

Figure 2.   Reaction Mechanism for N-Sulfonyloxy Arylamides (I).   Ac,
            acetyl; RSCH$_3$, methionine; RSH, glutathione or cysteine;
            RNH$_2$, N^2-guanine- and/or N^6-adenine-nucleosides, -nucleo-
            tides, or -nucleic acids; RCH, C8-guanine-nucleosides,
            -nucleotides, or -nucleic acids, or C7-AAF.

amide tumorigenesis is not certain. For example, N-sulfonyloxy-AAF is not a direct-acting, local carcinogen (56), even though it is highly toxic (57). It is mutagenic when reacted with purified B. subtilis transforming DNA (58), but does not serve as a direct-acting mutagen in the S. typhimurium test system (51,52,59,60). During chronic administration of a carcinogenic dose of AAF, hepatic sulfotransferase activity is greatly diminished (61) and deacetylated arylamine-DNA adducts eventually account for 97-100% of the total adducts (62). In addition, extrahepatic tissues which have little or no sulfotransferase activity and contain only deacetylated adducts, are also susceptible to AAF or N-OH-AAF carcinogenesis (63,64). However, sensitivity to hepatic tumor induction by AAF correlates well with hepatic sulfate availability and with sex, strain, and species differences in hepatic sulfotransferase levels (reviewed in 4,25). Thus, it has been proposed that N-sulfonyloxy arylamides may not be responsible for initiating hepatic tumorigenesis, but may rather serve to promote fixation of an initiating lesion through a cytotoxic response that induces cell replication (25,60).

N-Acetoxy Arylamides (II)

N-Acetoxy arylamides have been widely used as synthetic models to study electrophilic reactivity with cellular constituents and they yield reaction products similar to those observed with the N-sulfonyloxy esters. Furthermore, since they are highly carcinogenic at local sites of application (56,65,66) they have also been regarded as ultimate carcinogens (47). However, the N-acetoxy esters are generally less reactive than the corresponding sulfonyloxy derivatives, they exhibit much longer half-lives in aqueous solution (41,55,57,67-71) and their reaction mechanism is decidedly more complex. They react, at least in part by an S_N1 mechanism involving ion pair formation similar to that shown in Figure 2. This is supported by: a) their lower electrophilic reactivity and selectivity in comparison to N-sulfonyloxy esters which is due to the decreased strength and hardness of the acetate leaving group (41,55,67,72); b) their thermal rearrangement to ortho-acetoxy arylamides (69); c) their facile reduction to the parent arylamide (68,73); d) their conversion to reactive imines (45,74); and e) their reactivity with nucleophiles to give N-, ortho- and meta-substituted products (42,75). Yet heterolytic cleavage at the N-O bond must occur to only a minor extent because unlike N-sulfonyloxy arylamides (46,50), N-acetoxy arylamides have been shown to undergo preferentially cleavage of ester linkage to form a hydroxamate anion and presumably an acetyl cation which would account for the observed acetylation of lysine in proteins and ribose in nucleic acids (66,72,76,77). Evidence has also been presented that N-acyloxy arylamides may decompose homolytically to yield free radicals that could arylamidate DNA bases and also result in DNA-protein crosslinks (78-81). However, in view of the relative stability of N-acetoxy arylamides in aqueous media, their rapid reaction with added nucleophiles, and the failure to detect racemization to N-[(G-^{18}O)-acetoxy]-arylamides on prolonged incubation of N-[(carbonyl-^{18}O)-acetoxy]-arylamides in the absence of nucleophiles (82), it appears that an S_N2 reaction involving bimolecular displacement of acetate or hydroxamate by the attacking nucleophile represents a more probable mechanism to account for the major arylamidated or acetylated products obtained (27,75,76,82).

The role of N-acetoxy arylamides as metabolically formed ultimate carcinogens in vivo also appears to be limited. Their enzymatic formation via peroxidation of N-hydroxy arylamides can be excluded since tissues containing high levels of peroxidases such as the rat mammary gland (83) and the dog urinary bladder (84) do not form acetylated carcinogen-DNA adducts in vivo (63). Their non-enzymatic formation by reaction of acetyl coenzyme A with N-hydroxy arylamides (6) cannot be excluded; however, even if formed, their direct reaction with cellular DNA appears unlikely as treatment of cultured cells with synthetic N-acetoxy AAF (85,86) results primarily in deacetylated arylamine-DNA adducts, apparently due to rapid N-deacetylation to form the reactive N-acetoxy arylamine (V).

N-Glucuronyloxy Arylamides (III) and Arylamines (IV)

Metabolic conjugation of N-hydroxy arylamides to form N-glucuronyl-oxy ethers (III) represents a major pathway for biliary and urinary excretion of aromatic amine carcinogens (87,88). While these conjugates are generally considered to be stable detoxification products, the N-glucuronyloxy derivatives of AAF, N-acetyl-4-aminostilbene, phenacetin, but not of AABP or N-acetyl-2-aminophenanthrene, have been shown to react slowly either with protein, nucleic acids, or their constituents (89-91). Since reaction of N-glucuronyloxy-AAF with methionine and guanosine yields ortho-methylmercapto and N-(guanosin-8-yl) derivatives (89), respectively, a reaction mechanism involving formation of a nitrenium/carbenium cation - glucuronyl lactonate anion pair can be envisaged (Figure 3, path a). Studies on the mechanism of decomposition of N-glucuronyloxy phenacetin (92) are consistent with this hypothesis as ortho-glucuronyloxy phenacetin was the major rearrangement product, and evidence for an imine intermediate (45,92) leading to a meta-substituted derivative and to N-acetyl benzoquinone imine and its reaction products was obtained. The reduction product, phenacetin, was also obtained although its formation was not increased by ascorbate. However, an internal redox process yielding phenacetin and saccharic acid is plausible.

Conversion of N-glucuronyloxy-AAF to an N-glucuronyloxy aryl-amine (IV) has also been demonstrated (Figure 3, paths b and c). This can occur spontaneously at alkaline pH by migration of the N-acetyl group to the 2'-hydroxyl of the glucuronyl moiety (93) or in tissues by enzymatic N-deacetylation (94). N-Glucuronyloxy-2-aminofluorene (AF) is highly electrophilic, directly mutagenic, and reacts with nucleic acids and with methionine and guanosine (and 5'-guanylic acid) to give the corresponding ortho-methylmercapto and N-(guan-8-yl) derivatives (89,95,96), presumably via an S_N1 mechanism. Interestingly, enzymatic formation of N-glucuronyloxy aryla-mines by direct O-glucuronidation of N-hydroxy arylamines does not appear to occur, as only stable N-hydroxy arylamine N-glucuronides are obtained in in vitro hepatic microsomal incubations (16).

N-Glucuronyloxy arylamides do not appear to be important in hepatocarcinogenesis as their increased metabolic formation does not result in increased hepatic macromolecular binding (4,25).

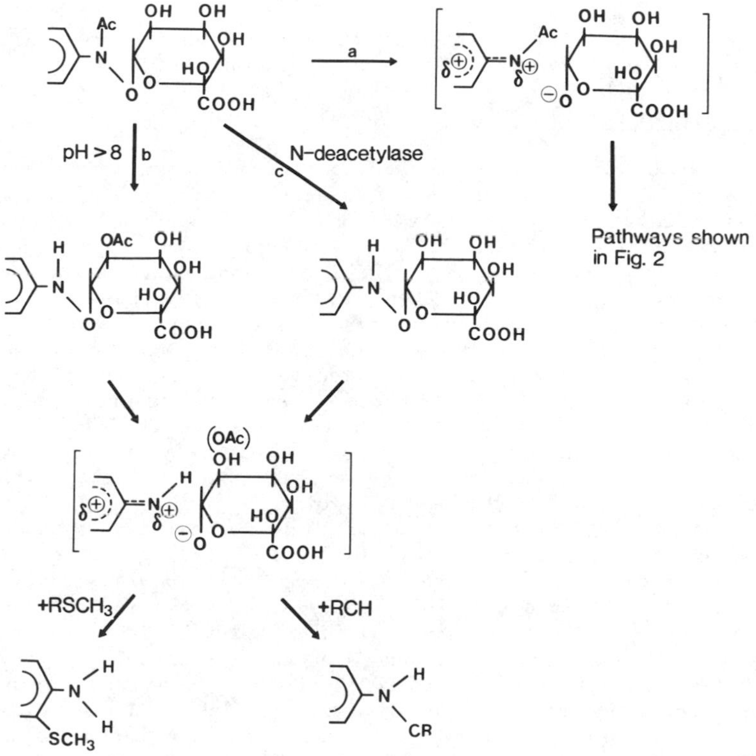

Figure 3.   Reaction Mechanism for N-Glucuronyloxy Arylamides (III) and
Arylamines (IV).   Ac; acetyl; RSCH₃, methionine; RCH,
C8-guanine-nucleosides, -nucleotides, or -nucleic acids.
Pathways a, b, and c are discussed in the text.

Prolonged residence in the intestine or urinary bladder lumen could
allow time for significant reaction with tissue components; however,
N-glucuronyloxy-AAF was only weakly carcinogenic at local subcu-
taneous sites of application (89). Enzymatic deacetylation to
N-glucuronyloxy-AF has been detected in hepatic tissue but this
activity in different species does not correlate with their relative
susceptibility to AAF hepatocarcinogenesis (94). On the other hand,
the alkaline pH-induced conversion to a reactive derivative may play
an important role in urinary bladder carcinogenesis (87) by AAF and
other arylamides in those species or individuals where normal urine
pH is alkaline (e.g. normal rabbit urine pH is 8.5-9.0).

N-Acetoxy Arylamines (V)

Early studies on the in vitro metabolic activation of carcinogenic
N-hydroxy arylamines indicated that N-acetoxy arylamines (V) are
formed as highly reactive intermediates that yield adducts with
proteins and nucleic acids (40,97). With N-hydroxy arylamides as
substrates, an enzyme mechanism involving intramolecular N,O-acetyl-
transfer was proposed (98); while an intermolecular process could be
demonstrated using N-hydroxy arylamines as substrates and N-hydroxy
arylamides as acetyl donors (99). Since that time, this acyltrans-
ferase has been extensively characterized (1) and purified to homo-
geneity from hepatic and extrahepatic tissues of several species
(reviewed in 100). More recently, Flammang et al. (7,101) have
shown that acetyl coenzyme A can serve effectively as an acetyl
donor for this enzyme, catalyzing the apparent direct O-acetylation
of several carcinogenic N-hydroxy arylamines.

Because of their instability and high reactivity, synthetic
N-acetoxy arylamines have never been isolated (97,99). However, NMR
spectral evidence for the existence of N-acetoxy-4-aminoquinoline-
1-oxide has been obtained (102,103); and N-acetoxy-4-aminoazobenzene
(104), N-acetoxy-2-amino-6-methyldipyrido [1,2-a:3',2'-d]imidazole
(N-acetoxy-Glu-P-1); ref. 105), and N-acetoxy-3-amino-1-methyl-5H-
pyrido[4,3-b]indole (N-acetoxy-Trp-P-2; ref. 106) have been prepared
as intermediates and then reacted with nucleosides or nucleic acids
to afford N-(guan-8-yl) products. In each of these cases, a more
stable imino tautomer can exist (Figure 4). Similar attempts at
preparation of N-acetoxy-4-aminobiphenyl have not been successful
(107); however, N-acetoxy-N-trifluoroacetyl-4-aminobiphenyl has been
prepared and shown to react rapidly in aqueous buffer with guanosine
(or 5'-guanylic acid) to give N-(guan-8-yl)-4-aminobiphenyl
derivatives, apparently by sequential detrifluoroacetylation and
generation of an electrophilic N-acetoxy arylamine (108). Evidence
for the formation of other N-acetoxy arylamines in situ has been
obtained by treatment of N-hydroxy arylamines with acetic anhydride
in buffered aqueous solutions containing N-acetylmethionine which
yielded the corresponding ortho-methylmercapto arylamines (97).
With in vitro metabolic activation systems, enzymatically generated
N-acetoxy arylamines have also been shown to react with
N-acetylmethionine or 2-mercaptoethanol to yield ortho-alkylmercapto
arylamines (68,97,99) and with nucleosides or nucleic acids to give
N-(guan-8-yl)- and ortho-(guan-N^2-yl)-arylamines (97,101,103,104).

From their high reactivity and nucleophilic selectivity, it
seems likely that N-acetoxy arylamines readily undergo heterolytic

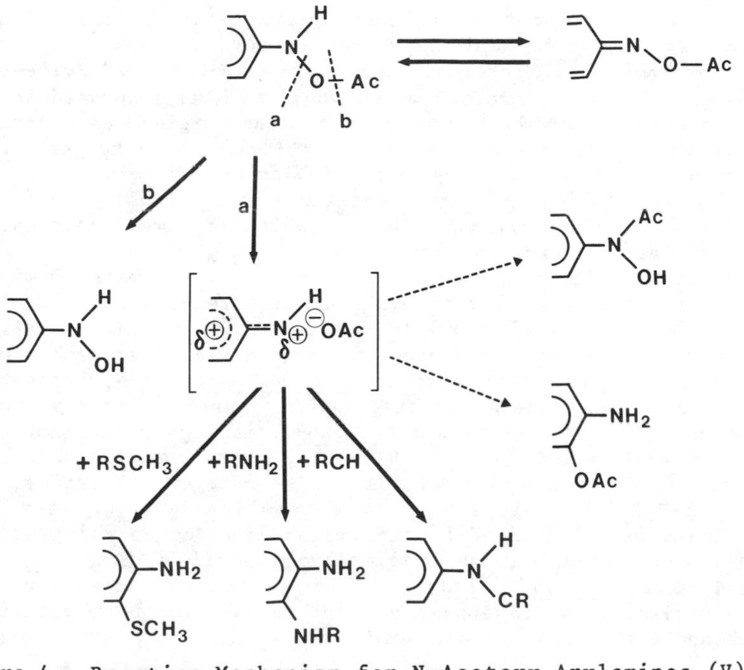

Figure 4. Reaction Mechanism for N-Acetoxy Arylamines (V). Ac, acetyl; RSCH₃ methionine; RNH₂, N²-guanine-nucleosides, -nucleotides, or -nucleic acids; RCH, C8-guanine-nucleosides, -nucleotides, or -nucleic acids. Pathways and heterolytic cleavages a and b are discussed in the text. Dashed arrows indicate proposed pathways.

cleavage to form singlet nitrenium/carbenium cation-acetate anion pairs (Figure 4, path a). Nucleophilic attack by RSCH$_3$, RNH$_2$, or RCH would then give the observed ortho- and N-substituted products. Under acidic conditions, hydrolysis of N-acetoxy-4-aminoquinoline-1-oxide to 4-hydroxyaminoquinoline-1-oxide (Figure 4, path b) has also been observed (102). Although the identification of decomposition products of chemically or enzymatically-generated N-acetoxy arylamines in neutral aqueous solution has not been reported, model studies with N-benzoyloxy-4-aminophenanthrene (109) suggest that internal rearrangement to an ortho-acetoxy arylamine and an N-hydroxy arylacetamide should occur (Figure 4, dashed arrows). The latter conversion has important implications for enzyme mechanisms. Thus, for N-hydroxy arylamide N,O-acetyltransferase, conversion to an N-acetoxy arylamine and internal return to an N-hydroxy arylacetamide represents a cyclic process which would terminate upon addition of a nucleophile and may be responsible for the suicide inactivation of the enzyme (99). For N-hydroxy arylamine O-acetylase, the rearrangement of the initial N-acetoxy arylamine intermediate to an N-hydroxy arylacetamide product represents an overall enzymatic N-acetylation of an N-hydroxy arylamine, which is a well-documented metabolic pathway for aromatic amines (1).

An important role for N-acetoxy arylamines as ultimate chemical carcinogens seems likely in view of their high reactivity, the wide tissue and species distribution (7,98) of enzyme(s) that catalyze their formation, and the prevalence of non-acetylated arylamine-DNA adducts in carcinogen-target tissues (110). In addition, synthetic N-acetoxy-4-aminoquinoline-1-acetate, which generates the acetoxy arylamine on reaction with thiols (102,103), is highly carcinogenic at sites of application (111). Recently, Saito et al. (8) have shown that the N-hydroxy metabolites of the mutagenic heterocyclic amines, Trp-P-2 and Glu-P-1, are metabolically activated to ultimate mutagens by an acetyl coenzyme A-dependent enzyme present within the test bacterium, as originally proposed by Sakai et al. (112) and McCoy et al. (113). Thus, metabolic formation of N-acetoxy arylamines would appear a major pathway for both mutation induction and initiation of carcinogenesis.

N-Sulfonyloxy Arylamines (VI)

For certain carcinogenic primary N-hydroxy arylamines, metabolic O-sulfonylation to a reactive ester has been demonstrated. With rat hepatic sulfotransferase preparations, the PAPS-dependent activation of N-hydroxy derivatives of 4-aminobiphenyl, 4-aminoazobenzene, 1-naphthylamine, and 2-naphthylamine yielded electrophilic intermediates that formed adducts with methionine or nucleic acids; while N-hydroxy-4-aminostilbene, N-hydroxy-3,2'-dimethyl-4-aminobiphenyl, N-hydroxy-N'-acetybenzidine and N-hydroxy-AF were not activated in this in vitro system (9,37,101,114). By comparison, mouse hepatic sulfotransferase has recently been shown to catalyze the activation of both N-hydroxy-AF (10) and N-hydroxy-4-aminoazobenzene (115) to intermediates that react with guanosine to yield N-(guan-8-yl) products.

Like the N-acetoxy arylamines, a reaction mechanism for N-sulfonyloxy esters would be expected to involve formation of a nitrenium/carbenium cation-sulfate anion pair which then reacts with methionine in proteins and guanine (or adenine) in nucleic acids to give <u>ortho-</u> and N-substituted products (Figure 5). Interestingly, the simple electrophilic ester, hydroxylamine-O-sulfonic acid, also reacts with guanosine under acidic conditions to give the C8-substituted amino-guanosine apparently by a similar mechanism (<u>116</u>). Furthermore, this mechanism is consistent with decomposition products identified from sulfotransferase incubations of N-hydroxy-2-naphthylamine (<u>9</u>) and with rearrangement products observed with synthetically-prepared N-sulfonyloxy aniline and N-sulfonyloxy-2-naphthylamine (<u>117</u>). That is, the <u>ortho-</u>sulfonyloxy arylamine was a major product and this could arise by internal return upon collapse of the ion pair (<u>cf</u>. N-sulfonyloxy arylamides). N-Hydroxy-N-sulfonyl-2-naphthylamine may also be formed as an intermediate rearrangement product as it has been reported (<u>117,118</u>) to decompose to <u>ortho-</u>sulfonyloxy-2-naphthylamine and 2-amino-1-naphthol, the latter of which was also detected in the sulfotransferase incubation with N-hydroxy-2-naphthylamine (<u>9</u>). The electrophilic ion pair also appeared to undergo a facile reduction to 2-naphthylamine. In this metabolic activation system, this proceeded at the expense of N-hydroxy-2-naphthylamine, which was oxidized to 2,2'-azoxynaphthalene; however, other reducing agents may serve this purpose <u>in vivo</u> and effectively detoxify the reactive ester. Similar redox processes could occur with N-acetoxy arylamines and other primary arylamine O-esters but this has not yet been investigated.

The role of N-sulfonyloxy arylamines as ultimate carcinogens appears to be limited. For N-hydroxy-2-naphthylamine, conversion by rat hepatic sulfotransferase to a N-sulfonyloxy metabolite results primarily in decomposition to 2-amino-1-naphthol and 1-sulfonyloxy-2-naphthylamine which are also major urinary metabolites; and reaction with added nucleophiles is very low, which suggests an overall detoxification process (<u>9,17</u>). However, for 4-aminoazobenzene and N-hydroxy-AAF, which are potent hepatocarcinogens in the newborn mouse, evidence has been presented that strongly implicates their N-sulfonyloxy arylamine esters as ultimate hepatocarcinogens in this species (<u>10,104</u>). This includes the inhibition of arylamine–DNA adduct formation and tumorigenesis by the sulfotransferase inhibitor pentachlorophenol, the reduced tumor incidence in brachymorphic mice that are deficient in PAPS biosynthesis (<u>10,115</u>), and the relatively low O-acetyltransferase activity of mouse liver for N-hydroxy-4-aminoazobenzene and N-OH-AF (<u>7,114,115</u>).

0-Seryl (0-Prolyl) Esters (VII) of N-Hydroxy Arylamines

The formation of O-seryl or O-prolyl esters (Figure 1) of certain N-hydroxy arylamines has been inferred from the observations that highly reactive intermediates can be generated <u>in vitro</u> by incubation with ATP, serine or proline, and the corresponding aminoacyl tRNA synthetases (<u>11,12,119</u>). For example, activation of N-hydroxy-4-aminoquinoline-1-oxide (<u>119,120</u>), N-hydroxy-4-aminoazobenzene (<u>11</u>) and N-hydroxy-Trp-P-2 (<u>121</u>) to nucleic acid-bound products was demonstrated using seryl–tRNA synthetase from yeast or rat ascites hepatoma cells. More recently, hepatic cytosolic prolyl-, but not seryl-, tRNA synthetase was shown to activate N-hydroxy-Trp-P-2 (<u>12</u>); however, no activation was detectable for the N-hydroxy metabolites of AF, 3,2'-dimethyl-4-aminobiphenyl, or N'-acetylbenzidine (<u>122</u>).

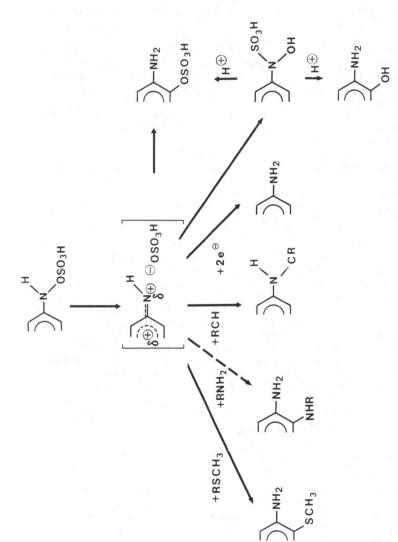

Figure 5.   Reaction Mechanism for N-Sulfonyloxy Arylamines (VI). RSCH₃, methionine; RNH₂, N²-guanine- and N⁶-adenine-nucleosides or -nucleic acids; RCH, C8-guanine-nucleosides or -nucleic acids. The dashed arrow indicates a proposed pathway.

The identification of C8-guanyl and N^6-adenyl adducts of 4-aminoquinoline-1-oxide (102,103) in DNA modified by the metaboli-cally-generated O-seryl ester and the similarity of the adduct pro-file with that obtained on reaction of DNA with N-acetoxy-4-amino-quinoline-1-oxide suggest an electrophilic reaction mechanism similar to that for the N-acetoxy or N-sulfonyloxy arylamines (Figures 4 and 5). However, N-seryloxy or N-prolyloxy arylamines have not been synthesized and the decomposition products of the esters generated in vitro have not yet been studied.

Although aminoacyl-tRNA synthetases are necessary for protein synthesis in all tissues, their importance in chemical carcino-genesis is difficult to assess. Mutation induction by this pathway has been studied extensively (123), yet metabolic activation in a carcinogen-target tissue has not been demonstrated. The only excep-tion is hepatic prolyl-tRNA synthetase activation of N-hydroxy-Trp-P-2; however, hepatic O-acetylation of this substrate also occurs to an appreciable extent (12). Further investigations involving the use of specific enzyme inhibitors would be helpful in addressing this problem.

Protonation of N-Hydroxy Arylamines (VIII)

The formation of O-protonated N-hydroxy arylamines (Figure 6) under acidic conditions has been well documented as an intermediate step in the Bamberger rearrangement to form aminophenols and other ortho- or para-substituted products (124-128). From absorption spectral data involving protonation equilibria (128), the exchange experi-ments of [^{18}O]H$_2$O into products or starting material (126,127), and from studies of reaction kinetics (125,128), the protonated hy-droxylamines appear to be relatively stable species whose rearrange-ment proceeds by an S_N1 mechanism with elimination of water as the rate-determining step. The electrophilic nature of this intermedi-ate was initially considered by Heller et al. (125); while Kriek (15), who proposed that elimination of water resulted in a highly electrophilic arylnitrenium ion, first demonstrated reactions with biological nucleophiles. Since that time, the reaction of car-cinogenic N-hydroxy arylamines with nucleic acids under mildly acidic conditions has been shown to be an effective procedure for preparation and identification of arylamine-nucleoside adducts and both ortho- and N-substituted products have been obtained (reviewed in 110). These include o-(guan-N^2-yl), o-(guan-O^6-yl), o-(aden-N^6-yl), N-(guan-8-yl), and N-(aden-8-yl) adducts. Of these, the N-(guan-8-yl) derivatives have usually been the major reaction product.

In contrast to the reactivity of N-sulfonyloxy and N-acetoxy esters of arylamides and arylamines, the relative reactivity of pro-tonated N-hydroxy arylamines with nucleophiles generally decreases in the order: DNA > denatured DNA > rRNA ≅ protein > tRNA >> nucleo-tides ≅ nucleosides ≅ methionine ≅ GSH (2,13-17,30,36,40,127,129, 130). Furthermore, the rate of reaction with DNA was found to be not only first order with respect to N-hydroxy arylamine concen-tration, but also first order with respect to DNA concentration (127,129,131). These data suggested that the reaction mechanism was either S_N2 or S_N1 with the involvement of an intermediate in the

rate-determining step (132). In view of the relatively slow exchange (40%/hr) of [^{18}O]H_2O into N-hydroxy-1-naphthylamine at pH 5,
the slower conversion of N-hydroxy-1-naphthylamine to aminonaphthols
(1%/hr), its rapid reaction with DNA (25%/hr), and the established
S_N1 reaction mechanism for the Bamberger reaction, a partially
delocalized hydrated nitrenium/carbenium ion intermediate (Figure 6)
was proposed (127,132,133). This intermediate is analogous to an
intimate ion pair formed under neutral conditions as described for
the electrophilic O-esters of N-hydroxy arylamides and arylamines
(vide supra).

    Thus, the overall reactivity and selectivity (N- vs. ring-
substitution) of protonated N-hydroxy arylamines should be determined by ability of the nucleophile to desolvate the hydrated ion,
to delocalize further the positive charge, and to result in product
formation (132,134). Such a mechanism is consistent with the preferential formation of N-substituted products from reaction with
nucleic acids (110) and from solvolysis of N-hydroxy arylamines in
benzene/trifluoroacetic acid (135); while aryl ring-substituted
products are preferentially obtained on solvolysis of N-hydroxy
arylamines (135) or 1-naphthylazide (136) in benzene/trifluoro-
methanesulfonic acid. Alternatively, upon desolvation, a true ion
pair could be formed between a negatively charged nucleophile or
catalyst and the nitrenium/carbenium cation, which could collapse to
the product or undergo internal rearrangement. However, the latter
mechanism seems improbable since, unlike the electrophilic O-esters,
the reactivity of protonated N-hydroxy arylamines with DNA is unaffected by reducing agents and their reaction with strong, low
molecular-weight nucleophiles such as 4-(p-nitrobenzyl)pyridine
cannot be detected (127,129,131).

    The exceptional reactivity of DNA for protonated N-hydroxy
arylamines can be rationalized by at least two mechanisms. First,
intercalation of the electrophilic intermediate between DNA bases
could sterically assist in desolvation and in directing the electrophilic center of the carcinogen over the nucleophilic region of
the DNA base. This seems unlikely, however, as pretreatment of DNA
with cis-Pt, which decreased the DNA contour length by 50%, failed
to reduce the reactivity of N-hydroxy-1-naphthylamine for the DNA
(137). A second possibility involves an electrostatic attraction
between the electrophile and the phosphate backbone of the DNA (77).
This seems more probable since either high ionic strength or stoichiometric (to DNA-P) amounts of Mg^{++} strongly inhibit DNA adduct
formation (77,137). In addition, evidence has been presented that
N-hydroxy arylamine-DNA/RNA phosphotriesters may be formed which
induce strand breaks (137,138) and could serve as a catalyst for
desolvation and subsequent adduct formation.

    The importance of protonated N-hydroxy arylamines as ultimate
carcinogens has been suggested for some time (28,40,139). From
studies on their reactivity with nucleic acids at different pH's
(2,15,16,63,130,131), the pK$_a$ for protonation of the N-hydroxy group
appears to be between pH 5 and 6; thus, a significant proportion
(1-10%) of the N-hydroxy derivative exists as the protonated form
even under neutral conditions. This would account for the significant levels of covalent modification of DNA observed in vitro by
reaction with N-hydroxy arylamines at neutral pH. Consequently, it
has been proposed that in vivo formation of non-acetylated aryl

amine-DNA adducts may arise, at least in part, by the direct reaction with protonated N-hydroxy arylamines (2,28,139). This hypothesis is further supported by the observation that synthetic or metabolically-generated N-OH-AF reacts appreciably with DNA in isolated liver nuclei to yield detectable levels of N-(deoxyguanosin-8-yl)-AF (2). This is consistent with the high concentration of DNA within cell nuclei (ca. 50 mg/ml) and with the first order relation between reaction rates and DNA concentrations.

Protonated N-hydroxy arylamines have also been proposed to be ultimate carcinogens for the urinary bladder (16,17,140,141) since urine pH is slightly acidic in a number of species (14,142). Furthermore, pharmacokinetic studies have shown that increased urine acidity and decreased frequency of urination are predictive of relative species susceptibility to urinary bladder carcinogenesis (142); and neoplastic transformation of cultured human fibroblasts by N-hydroxy arylamines is greatly enhanced by incubation at pH 5 as compared to pH 7 (143).

Nitrosoarenes (IX)

Nitrosoarenes are readily formed by the oxidation of primary N-hydroxy arylamines and several mechanisms appear to be involved. These include: 1) the metal-catalyzed oxidation/reduction to nitrosoarenes, azoxyarenes and arylamines (144); 2) the O_2-dependent, metal-catalyzed oxidation to nitrosoarenes (145); 3) the O_2-dependent, hemoglobin-mediated co-oxidation to nitrosoarenes and methemoglobin (146); and 4) the O_2-dependent conversion of N-hydroxy arylamines to nitrosoarenes, nitrosophenols and nitroarenes (147,148). Each of these processes can involve intermediate nitroxide radicals, superoxide anion radicals, hydrogen peroxide and hydroxyl radicals, all of which have been observed in model systems (149,151). Although these radicals are electrophilic and have been suggested to result in DNA damage (151,152), a causal relationship has not yet been established. Nitrosoarenes, on the other hand, are readily formed in in vitro metabolic incubations (2,153) and have been shown to react covalently with lipids (154), proteins (28,155) and GSH (17,156-159). Nitrosoarenes are also readily reduced to N-hydroxy arylamines by ascorbic acid (17,160) and by reduced pyridine nucleotides (9,161).

The mechanism of reaction of nitrosoarenes with GSH has been studied extensively and is known to involve an addition reaction with the thiol group to form an N-hydroxy-N-(glutathion-S-yl)-arylamine adduct. This intermediate can rearrange to an N-(glutathion-S-yl)-arylamine S-oxide or can be reduced to an N-hydroxy arylamine or an N-(glutathion-S-yl)-arylamine (Figure 7). It is interesting to note that 4-aminobiphenyl has recently been reported to form high levels of a hemoglobin adduct (5% of the dose) that appears to arise by addition of 4-nitrosobiphenyl to a cysteinyl sulfhydryl group in the protein, forming an N-S linkage (162). Binding of nitrosoarenes to nucleic acids has been suggested (43, 163), but negative results were obtained in subsequent studies (40,159). Thus, the role of nitrosoarenes as ultimate carcinogens per se seems unlikely, although modification of a critical cellular protein cannot be excluded.

A role for nitrosoarenes in arylamine carcinogenesis has been

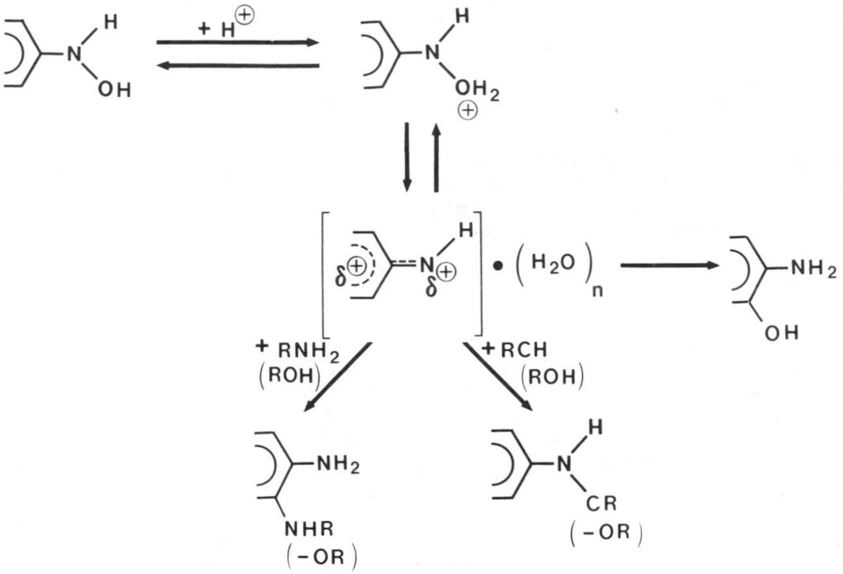

Figure 6.   Reaction Mechanism and Formation of Protonated N–Hydroxy
            Arylamines (VIII). RNH_2, N^2-guanine- and N^6-adenine-
            nucleic acids; ROH, O^6-guanine-nucleic acids; RCH,
            C8-guanine- and C8-adenine-nucleic acids.

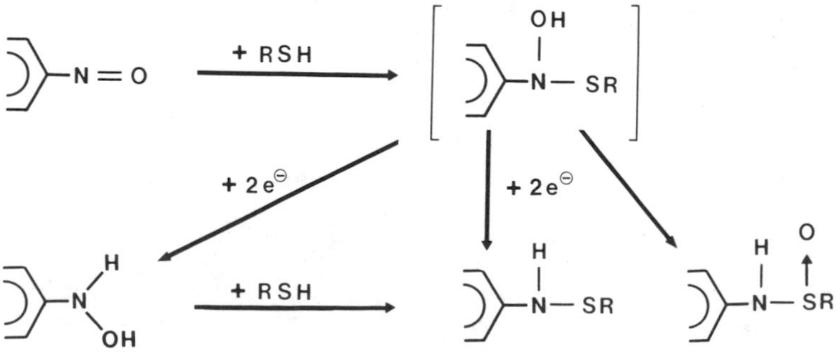

Figure 7.   Reaction Mechanism for Nitrosoarenes (IX).
            RSH, glutathione or cysteine.

suggested to be due to their facile interconversion with N-hydroxy arylamines by oxidation and reduction and their rapid detoxification by reaction with GSH (159). Consequently, addition of ascorbic acid significantly increased 2-nitrosofluorene mutagenicity (160); whereas, addition of GSH strongly inhibited mutagenic activity (164) and GSH depletion has resulted in increased DNA damage in hepatocytes by N-OH-AF (165).

Iminoquinones (X) and Diimines (XI)

The formation of iminoquinones (166,167) and diimines (20,168) as intermediates in the oxidation of aminophenols and aryldiamines has been well established. These intermediates readily undergo addition reactions with nucleophiles to yield N-, ortho-, or meta-substituted products (Figure 8). For example, 2-amino-1-naphthol, which has long been suggested as a proximate carcinogenic metabolite of 2-naphthylamine (169), is readily oxidized in air or by cytochrome c to 2-imino-1-naphthoquinone. This iminoquinone is electrophilic and can bind covalently to protein and DNA, can undergo reaction with aryl-NH_2 groups to give meta-substituted products, or can hydrolyze to form 2-amino-1,4-naphthoquinone (19,166,167,170-172). In this regard, the major reaction product of 2-imino-1-naphthoquinone with DNA has been recently identified as 4-(deoxyguanosin-N^2-yl)-2-amino-1,4-naphthoquinoneimine (19).

Diimines are formed directly by peroxidative metabolism of aryldiamines. For example, 4,4'-diiminobiphenyl (or benzidine-diimine), a product of benzidine peroxidation whose formation involves a cation radical intermediate (20,168), readily binds to protein and nucleic acid (173,174). This diimine also reacts with itself to form an azo dimer (20) or reacts with GSH to give an ortho-substituted glutathion-S-yl conjugate (175), with phenols to give an N-substituted indo dye (176), and with DNA to give N-(deoxyguanosin-8-yl)-benzidine (174). Other similarly reactive imines and iminoquinones have been shown to be formed in biological systems, notably N-acetyl-p-benzoquinone imine, which has been identified as the major hepatotoxic metabolite of acetaminophen and phenacetin (reviewed in 91).

Over the last few years, the significance of these intermediates as ultimate carcinogens has received new impetus since prostaglandin H synthase, a mammalian peroxidase which is widely distributed in extrahepatic tissues (177), can mediate the cooxidation of several carcinogenic arylamines to intermediates that bind covalently to protein and nucleic acid (20,168,178,179). For the urinary bladder carcinogen, 2-naphthylamine, the formation of 2-amino-1-naphthol and its subsequent oxidation to 2-imino-1-naphthoquinone have been shown to be primarily responsible for DNA binding in the in vitro peroxidase system (180). Furthermore, about 20-30% of the 2-naphthylamine-DNA adducts formed in the dog urinary bladder, which contains high levels of prostaglandin H synthase (84), appears to be derived from the addition reaction of 2-imino-1-naphthoquinone (19). For benzidine, another urinary bladder carcinogen, the major benzidine-DNA adduct formed in the prostaglandin H synthase-mediated reaction and in the urinary bladder of dogs given benzidine was shown to be identical to the N-(guan-8-yl) derivative that was prepared by reaction with synthetic 4,4'-diiminobiphenyl (174). Recently, the

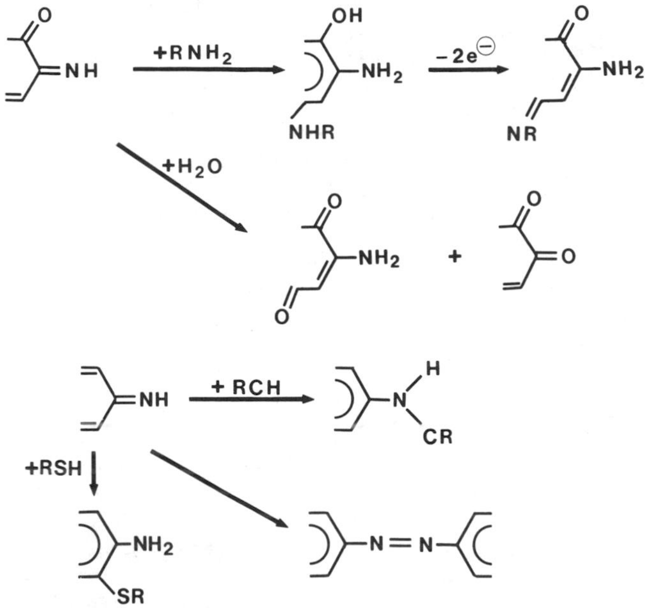

Figure 8.   Reaction Mechanisms for Iminoquinones (X) and Imines (XI).
RNH$_2$, N^z-guanine-nucleic acids or arylamines; RCH, C8-
guanine-nucleic acids or p-substituted phenols; RSH,
glutathione or cysteine.

peroxidative metabolism of AF has been carefully studied and found to result in the formation of a "head-to-tail" dimer, 2-amino-difluorenylamine, whose further oxidation to a reactive diimine may be responsible for macromolecular binding (181-183). However, for each of these carcinogens, there is also good evidence that electro-philic radical cations (20,150,168,182) can be produced and that these may yield covalent adducts with protein and nucleic acids. Further studies on the identification of these adducts should provide useful information on the role of radical intermediates in arylamine carcinogenesis.

N-Sulfonyloxy (XII) and N-Arylnitrone (XIII) Derivatives of N-Methyl Arylamines

Although several N-methyl-substituted arylamines have been shown to be carcinogenic (184-186), metabolic activation pathways have been investigated primarily for the hepatocarcinogenic aminoazo dyes, N-methyl-4-aminoazobenzene (MAB) and its 3'-methyl derivative (9,21, 22,187,188). N-Hydroxy-N-methyl arylamines are generally regarded as proximate carcinogenic metabolites (22,187,189) and have been shown to be converted to electrophilic N-sulfonyloxy derivatives by hepatic sulfotransferases (9,187) or to reactive N-arylnitrones by air oxidation (21).

Metabolically-formed N-sulfonyloxy-MAB was found to react with methionine, guanosine, and GSH to give ortho-methylmercapto, guan-8-yl, and ortho-glutathion-S-yl products (9,190); and these were the same major adducts found in vivo in rat hepatic protein, nucleic acid, and bile, respectively, after MAB administration (191-193). These studies were aided by the availability of synthetic N-benzoyloxy esters which show similar reactivity toward nucleophiles and are potent, direct-acting carcinogens and mutagens (57,194,195). Additional experiments have shown that substituted guan-N²-yl and aden-N⁶-yl derivatives are also formed in DNA after reaction in vitro with N-benzoyloxy-MAB and after dosing with MAB in vivo (196-198), which suggests a similar reactivity for metabolically-formed N-sulfonyloxy esters.

Thus, like the N-sulfonyloxy esters of arylamides and of primary arylamines (Figures 2 and 5), a reaction mechanism for N-sulfonyl-oxy-N-methyl arylamines is expected to involve formation of a nitrenium/carbenium cation-sulfate anion pair which reacts to give both N- or ring-substituted products, depending on the softness or hardness of the nucleophile (Figure 9). Recently, N-sulfonyloxy-MAB was prepared synthetically and its solvolysis and reaction with GSH was examined (199). In addition to the expected ring-substituted glutathion-S-yl adducts, a glutathion-S-methylene conjugate was obtained. This suggests that internal decomposition of the intimate ion pair involves loss of sulfuric acid and formation of a methimine (Figure 9), which can hydrolyze to formaldehyde and the primary arylamine or can react with GSH via a Mannich condensation to yield the glutathion-S-methylene product (200).

In in vitro N-hydroxy-MAB sulfotransferase-activating systems, N-sulfonyloxy-MAB also appears to undergo rapid reduction to MAB (Figure 9) with the concomitant oxidation of N-hydroxy-MAB to the N-arylnitrone (9). The oxidizing properties of the N-sulfonyloxy-MAB ion pair is consistent with results obtained for the primary

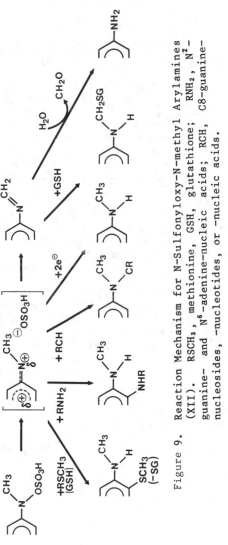

Figure 9. Reaction Mechanism for N-Sulfonyloxy-N-methyl Arylamines
(XII).   RSCH₃, methionine, GSH, glutathione;   RNH₂, N²-
guanine- and N⁶-adenine-nucleic acids;   RCH,   C8-guanine-
nucleosides, -nucleotides, or -nucleic acids.

N-sulfonyloxy arylamines (vide supra) and is supported by mechanistic studies using the analogous N-sulfonyloxy esters of purine N-oxides (201,202).

N-Arylnitrones (XIII) formed by oxidation of N-hydroxy-N-methyl arylamines, show high reactivity toward carbon–carbon and carbon–nitrogen double bonds in non-aqueous media (21,203) (Figure 10). Under physiological conditions, however, it appears that N-arylnitrones exist as protonated salts that readily hydrolyze to formaldehyde and a primary N-hydroxy arylamine; and efforts to detect N-arylnitrone addition products in cellular lipid, protein or nucleic acids have not been successful (204). Nitroxide radicals derived from N-hydroxy-MAB have also been suggested as reactive intermediates (150), but their direct covalent reaction with nucleic acids has been excluded (21).

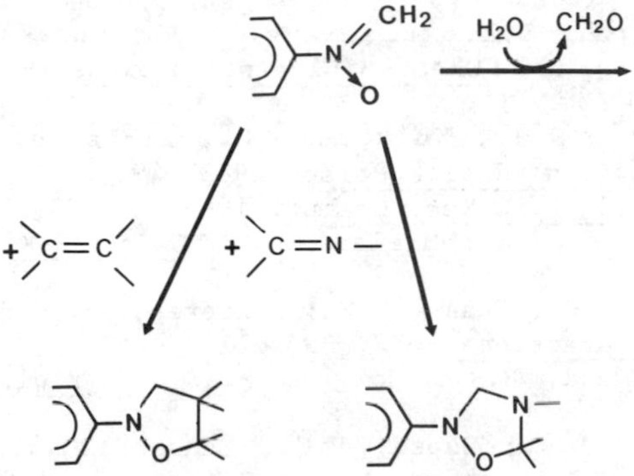

Figure 10. Reaction Mechanism for N-Arylnitrones (XIII).   Addition to C=C and C=N bonds yields isozaxolidines and oxadiazolidines, respectively.

For N-methyl arylamines, electrophilic N-sulfonyloxy esters appear to be strong candidates for the ultimate carcinogenic metabolites. However, additional studies are required as these conclusions are circumstantially based on their comparative reactivity with nucleophiles and on the failure of other metabolic conjugation systems to esterify N-hydroxy-N-methyl arylamines (9,187).

Acknowledgment

We thank Linda Amspaugh and Cindy Hartwick for helping prepare this review.

Literature Cited

1. King, C.M.; Glowinski, I.B.   Environ. Health Persp. 1983, 49, 43.

2. Frederick, C.B.; Mays, J.B.; Ziegler, D.M.; Guengerich, F.P.; Kadlubar, F.F. Cancer Res. 1982, 42, 2671.

3. Miller, E.C.; Miller, J.A. Cancer 1981, 47, 1055.

4. Mulder, G.J.; Meerman, J.H.N. Environ. Health Persp. 1983, 49, 27.

5. Malejka-Giganti, D.; Ritter, C.L.; Ryzewski, C.N. Environ. Health Persp. 1983, 49, 175.
6. Lotlikar, P.D.; Luha, L. Mol. Pharmacol. 1971, 7, 381.
7. Flammang, T.J.; Kadlubar, F.F. In "Microsomes and Drug Oxidation"; Boobis, A.R.; Caldwell, J.; deMattheis, F.; Elcombe, C.R., Eds; Taylor and Francis: London, 1985; in press.
8. Saito, K.; Yamazoe, Y.; Kamataki, T.; Kato, R. Biochem. Biophys. Res. Commun. 1983, 116, 141.
9. Kadlubar, F.F.; Miller, J.A.; Miller, E.C. Cancer Res. 1976, 36, 2350.
10. Lai, C.-C.; Miller, E.C.; Miller, J.A.; Liem, A. Proc. Amer. Assoc. Cancer Res. 1984, 25, 85.
11. Hashimoto, Y.; Degawa, M.; Watanabe, H.K.; Tada, M. Gann 1981, 72, 937.
12. Yamazoe, Y.; Shimada, M.; Kamataki, T.; Kato, R. Biochem. Biophys. Res. Commun. 1982, 107, 165.
13. Kadlubar, F.F.; Unruh, L.E.; Flammang, T.J.; Sparks, D.; Mitchum, R.K.; Mulder, G.J. Chem.-Biol. Interactions 1981, 33, 129.
14. Nussbaum, M.; Fiala, E.S.; Kulkarni, B.; El-Bayoumy, K.; Weisburger, J.H. Environ. Health Persp. 1983, 49, 223.
15. Kriek, E. Biochem. Biophys. Res. Commun. 1965, 20, 793.
16. Kadlubar, F.F.; Miller, J.A.; Miller, E.C. Cancer Res. 1977, 37, 805.
17. Mulder, G.J.; Unruh, L.E.; Evans, F.E.; Ketterer, B.; Kadlubar, F.F. Chem.-Biol. Interactions 1982, 39, 111.
18. Nagasawa, H.T.; Gutmann, H.R.; Morgan, M.A. J. Biol. Chem. 1959, 234, 1600.
19. Yamazoe, Y.; Miller, D.W.; Gupta, R.C.; Zenser, T.V.; Weis, C.C.; Kadlubar, F.F. Proc. Amer. Assoc. Cancer Res. 1984, 25, 91.
20. Josephy, P.D.; Eling, T.E.; Mason, R.P. J. Biol. Chem. 1983, 258, 5561.
21. Kadlubar, F.F.; Miller, J.A.; Miller, E.C. Cancer Res. 1976, 36, 1196.
22. Kimura, T.; Kodama, M.; Nagata, C. Carcinogenesis 1982, 3, 1393.
23. King, C.M.; Phillips, B. Science 1968, 159, 1351.
24. DeBaun, J.R.; Rowley, J.Y.; Miller, E.C.; Miller, J.A. Proc. Soc. Exp. Biol. Med. 1968, 129, 268.
25. Mulder, G.J. In "Sulfation of Drugs and Related Compounds"; Mulder, G.J., Ed; CRC Press: Boca Raton, 1981; p. 213.
26. Meerman, J.H.N.; Beland, F.A.; Ketterer, B.; Srai, S.K.S.; Bruins, A.P.; Mulder, G.J. Chem.-Biol. Interactions 1982, 39, 149.
27. Meerman, J.H.N.; Tijdens, R.B. Cancer Res. 1985, in press.
28. Barry, E.J.; Malejka-Giganti, D.; Gutmann, H.R. Chem.-Biol. Interactions 1969/1970, 1, 139.
29. Irving, C.C.; Veazey, R.A. Cancer Res. 1971, 31, 19.
30. Kriek, E. Biochim. Biophys. Acta 1974, 355, 177.
31. Kriek, E.; Hengeveld, G.M. Chem.-Biol. Interactions 1978, 21, 179.
32. Kennelly, J.C.; Beland, F.A.; Kadlubar, F.F.; Martin, C.N. Carcinogenesis 1984, 5, 407.

33. Gaugler, B.J.M.; Neumann, H.-G. Chem.-Biol. Interactions 1979, 24, 355.
34. Scribner, J.D.; Scribner, N.K.; Koponen, G. Chem.-Biol. Interactions 1982, 40, 27.
35. Scribner, J.D.; Koponen, G. Chem.-Biol. Interactions 1979, 28, 201.
36. Westra, J.G.; Flammang, T.J.; Fullerton, N.F.; Beland, F.A.; Weis, C.C.; Kadlubar, F.F. Carcinogenesis 1985, in press.
37. DeBaun, J.R.; Miller, E.C.; Miller, J.A. Cancer Res. 1970, 30, 577.
38. Beland, F.A.; Dooley, K.L.; Jackson, C.D. Cancer Res. 1982, 42, 1348.
39. Beland, F.A.; Miller, D.W.; Mitchum, R.K. J. Chem. Soc. Chem. Commun. 1983, 30.
40. King, C.M.; Phillips, B. J. Biol. Chem. 1969, 244, 6209.
41. Westra, J.G.; Kriek, E.; Hittenhausen, H. Chem.-Biol. Interactions 1976, 15, 149.
42. Kriek, E.; Westra, J.G. In "Chemical Carcinogens and DNA"; Vol. II; Grover, P.L., Ed.; CRC Press: Boca Raton, 1979; p. 1.
43. Vaught, J.B.; McGarvey, P.B.; Lee, M.-S.; Garner, C.D.; Wang, C.Y.; Linsmaier-Bednar, E.M.; King, C.M. Cancer Res. 1981, 41, 3424.
44. Gassman, P.G.; Granrud, J.E. J. Amer. Chem. Soc. 1984, 106, 1498.
45. Gassman, P.G.; Granrud, J.E. J. Amer. Chem. Soc. 1984, 106, 2448.
46. Novak, M.; Pelecanou, M.; Roy, A.K.; Adronico, A.F.; Plourde, F.M.; Olefirowicz, T.M.; Curtin, T.J. J. Amer. Chem. Soc. 1984, 106, 5623.
47. Scribner, J.D.; Miller, J.A.; Miller, E.C. Cancer Res. 1970, 30, 1570.
48. Ford, G.P.; Scribner, J.D. J. Amer. Chem. Soc. 1981, 103, 4281.
49. Mulder, G.J.; Hinson, J.A.; Gillette, J.R. Biochem. Pharm. 1978, 27, 1641.
50. Marques, M.M.; Beland, F.A. unpublished studies.
51. Andrews, L.S.; Pohl, L.R.; Hinson, J.A.; Fisk, C.L.; Gillette, J.R. Drug Metab. Disp. 1979, 7, 296.
52. Andrews, L.S.; Hinson, J.A.; Gillette, J.R. Biochem. Pharm. 1978, 27, 2399.
53. Ho, T.-L. Chem. Rev. 1975, 75, 1.
54. Moschel, R.C.; Hudgins, W.R.; Dipple, A. J. Org. Chem. 1979, 44, 3324.
55. Kriek, E. Cancer Letters 1979, 7, 141.
56. Bartsch, H.; Malaveille, C.; Stich, H.F.; Miller, E.C.; Miller, J.A. Cancer Res. 1977, 37, 1461.
57. Maher, V.M.; Miller, E.C.; Miller, J.A.; Szybalski, W. Mol. Pharmacol. 1968, 4, 411.
58. DeBaun, J.R.; Smith, J.Y.R.; Miller, E.C.; Miller, J.A. Science 1970, 167, 184.
59. Mulder, G.J.; Hinson, J.A.; Nelson, W.L.; Thorgeirsson, S.S. Biochem. Pharm. 1977, 26, 1356.
60. Wirth, P.J.; Thorgeirsson, S.S. Mol. Pharmacol. 1981, 19, 337.
61. Jackson, C.D.; Irving, C.C. Cancer Res. 1972, 32, 1590.

62. Poirier, M.C.; Hunt, J.M.; True, B.; Laishes, B.A., Young, J.F.; Beland, F.A. Carcinogenesis 1985, 5, 1591.

63. Beland, F.A.; Beranek, D.T.; Dooley, K.L.; Heflich, R.H.; Kadlubar, F.F. Environ. Health Persp. 1983, 49, 125.

64. Allaben, W.T.; Weis, C.C.; Fullerton, N.F.; Beland, F.A. Carcinogenesis 1983, 4, 1067.

65. Miller, J.A.; Miller, E.C. Prog. Exptl. Tumor Res. 1969, 11, 273.

66. Yost, Y.; Gutmann, H.R.; Rydell, R.E. Cancer Res. 1975, 35, 447.

67. Scribner, J.D.; Naimy, N.K. Cancer Res. 1973, 33, 1159.

68. Smith, B.A.; Gutmann, H.R.; Springfield, J.R. Carcinogenesis 1985, in press.

69. Calder, I.C.; Williams, P.J. Chem.-Biol. Interactions 1975, 11, 27.

70. Kriek, E. Chem.-Biol. Interactions 1971, 3, 19.

71. Scribner, J.D.; Naimy, N.K. Cancer Res. 1975, 35, 1416.

72. Scribner, J.D.; Scribner, N.K.; Smith, D.L.; Jenkins, E.; McCloskey, J.A. J. Org. Chem. 1982, 47, 3143.

73. Scribner, J.D.; Naimy, N.K. Experientia 1975, 31, 470.

74. Calder, I.C.; Creek, M.J. Austr. J. Chem. 1976, 29, 1801.

75. Scribner, J.D. J. Amer. Chem. Soc. 1977, 99, 7383.

76. Barry, E.J.; Gutmann, H.R. J. Biol. Chem. 1973, 248, 2730.

77. Lang, M.C.E.; Fuchs, R.P.P.; Daune, M.P. Cancer Res. 1977, 37, 3887.

78. Parham, J.C.; Templeton, M.A. Tetrahedron 1980, 36, 709.

79. Zady, M.F.; Wong, J.L. J. Org. Chem. 1980, 45, 2373.

80. Rayshell, M.; Ross, J.; Werbin, H. Carcinogenesis 1983, 4, 501.

81. Novak, M.; Brodeur, B.A. J. Org. Chem. 1984, 49, 1142.

82. Scott, C.M.; Underwood, G.R.; Kirsch, R.B. Tetrahedron Letters 1984, 25, 499.

83. Reigh, D.L.; Stuart, M.; Floyd, R.A. Experientia 1978, 34, 107.

84. Wise, R.W.; Zenser, T.V.; Kadlubar, F.F.; Davis, B.B. Cancer Res. 1984, 44, 1893.

85. Poirier, M.C.; Williams, G.M.; Yuspa, S.H. Mol. Pharmacol. 1980, 18, 581.

86. Maher, V.M.; Hazard, R.M.; Beland, F.A.; Corner, R.; Mendrala, A.L.; Levinson, J.W.; Heflich, R.H.; McCormick, J.J. Proc. Amer. Assoc. Cancer Res. 1980, 21, 71.

87. Irving, C.C. Natl. Cancer Inst. Monogr. 1981, 58, 109.

88. Camus, A.-M.; Friesen, M.; Croisy, A.; Bartsch. H. Cancer Res. 1982, 42, 3201.

89. Miller, E.C.; Lotlikar, P.D.; Miller, J.A.; Butler, B.W.; Irving, C.C.; Hill, J.T. Mol. Pharmacol. 1968, 4, 147.

90. Irving, C.C. Cancer Res. 1977, 37, 524.

91. Hinson, J.A. Environ. Health Persp. 1983, 49, 71.

92. Hinson, J.A.; Andrews, L.S.; Gillette, J.R. Pharmacology 1979, 19, 237.

93. Hill, J.T.; Irving, C.C. Biochemistry 1967, 6, 3816.

94. Cardona, R.A.; King, C.M. Biochem. Pharmacol. 1976, 25, 1051.

95. Irving, C.C.; Russell, L.T. Biochemistry 1970, 9, 2471.

96. Maher, V.M.; Reuter. M.A. Mutation Res. 1973, 21, 63.

97. Bartsch, H.; Dworkin, M.; Miller, J.A.; Miller, E.C. Biochim. Biophys. Acta 1972, 286, 272.
98. King, C.M. Cancer Res. 1974, 34, 1503.
99. Mangold, B.L.K.; Hanna, P.E. J. Med. Chem. 1982, 25, 630.
100. Allaben, W.T.; King, C.M. J. Biol. Chem. 1984, 259, 12128.
101. Flammang, T.J.; Westra, J.G.; Kadlubar, F.F.; Beland, F.A. Carcinogenesis 1985, in press.
102. Kawazoe, Y.; Ogawa, O.; Huang, G.-F. Tetrahedron 1980, 36, 2933.
103. Bailleul, B.; Galiegue, S.; Loucheux-Lefebvre, M.-H. Cancer Res. 1981, 41, 4559.
104. Delclos, K.B.; Tarpley, W.G.; Miller, E.C.; Miller, J.A. Cancer Res. 1984, 44, 2540.
105. Hashimoto, Y.; Shudo, K.; Okamoto, T. J. Amer. Chem. Soc. 1982, 104, 7636.
106. Hashimoto, Y.; Shudo, K.; Okamoto, T. Biochem. Biophys. Res. Commun. 1980, 96, 355.
107. Kadlubar, F.F.; Weis, C.C. unpublished data.
108. Lee, M.-S.; King, C.M. Chem.-Biol. Interactions 1981, 34, 239.
109. Juneja, T.R.; Garg, D.K.; Schafer, W. Tetrahedron 1982, 38, 551.
110. Beland, F.A.; Kadlubar, F.F. Environ. Health Persp. 1985, in press.
111. Enomoto, M.; Miller, E.C.; Miller, J.A. Proc. Soc. Exp. Biol. Med. 1971, 136, 1206.
112. Sakai, S.; Reinhold, C.E.; Wirth, P.J.; Thorgeirsson, S.S. Cancer Res. 1978, 38, 2058.
113. McCoy, E.C.; McCoy, G.D.; Rosenkranz, H.S. Biochem. Biophys. Res. Commun. 1982, 108, 1362.
114. Frederick, C.B.; Weis, C.C.; Flammang, T.J.; Martin, C.N.; Kadlubar, F.F. Carcinogenesis 1985, in press.
115. Delclos, K.B. personal communication.
116. Kawazoe, Y.; Huang, G.-F. Chem. Pharm. Bull. 1972, 20, 2073.
117. Boyland, E.; Nery, R. J. Chem. Soc. 1962, 5217.
118. Manson, D. J. Chem. Soc. 1971, 1508.
119. Tada, M.; Tada, M. Nature 1975, 255, 510.
120. Tada, M.; Tada, M. Biochim. Biophys. Acta 1976, 454, 558.
121. Yamazoe, Y.; Tada, M.; Kamataki, T.; Kato, R. Biochem. Biophys. Res. Comm. 1981, 102, 432.
122. Yamazoe, Y.; Kadlubar, F.F. unpublished data.
123. Ikenaga, M.; Ichikawa-Ryo, H.; Kondo, S. J. Mol. Biol. 1975, 92, 341.
124. Yukawa, T. Jap. J. Chem. (Nippon Kagaku Zasshi) 1950, 71, 603.
125. Heller, H.E.; Hughes, E.D.; Ingold, C.K. Nature 1951, 168, 909.
126. Kukhtenko, I.I. Zh. Organicheskoi Khim. 1971, 7, 330
127. Kadlubar, F.F.; Miller, J.A.; Miller, E.C. Cancer Res. 1978, 38, 3628.
128. Sone, T.; Tokuda, Y.; Sakai, T.; Shinkai, S.; Manabe, O. J. Chem. Soc. (Perkin Trans. II) 1981, 298.
129. Kadlubar, F.F.; Unruh, L.E.; Beland, F.A.; Straub, K.M.; Evans, F.E. Carcinogenesis 1980, 1, 139.
130. Martin, C.N.; Beland, F.A.; Roth, R.W.; Kadlubar, F.F. Cancer Res. 1982, 42, 2678.

131. Kadlubar, F.F.   unpublished studies with N-hydroxy-4-amino-biphenyl and N-hydroxy-N'-acetylbenzidine.
132. Bentley, T.W.; Schleyer, P.v.R.   J. Amer. Chem. Soc. 1976, 98, 7658.
133. Schulman, S.G.; Sturgeon, R.J.   J. Amer. Chem. Soc. 1977, 99, 7209.
134. Pross, A.   J. Amer. Chem. Soc. 1976, 98, 776.
135. Shudo, K.; Ohta, T.; Okamoto, T.   J. Amer. Chem. Soc. 1981, 103, 645.
136. Takeuchi, H.; Takano, K.   J. Chem. Soc. Chem. Commun. 1983, 447.
137. Kadlubar, F.F.; Melchior, W.B., Jr.; Flammang, T.J.; Springgate, C.; Moss, A.J., Jr.; Nagle, W.A.   J. Supramol. Struct. Cell. Biochem. 1981, 171.
138. Vaught, J.B.; Lee, M.-S.; Shayman, M.A.; Thissen, M.R.; King, C.M.   Chem.-Biol. Interactions 1981, 34,109.
139. Kriek, E.   Chem.-Biol. Interactions 1969/1970, 1,3.
140. Radomski, J.L.   Ann. Rev. Pharmacol. Toxicol. 1979, 19, 129.
141. Poirier, L.A.; Miller, J.A.; Miller, E.C.   Cancer Res. 1963, 23, 790.
142. Young, J.F.; Kadlubar, F.F.   Drug Metab. Disp. 1982, 10, 641.
143. Oldham, J.W.; Kadlubar, F.F.; Milo, G.E.   Carcinogenesis 1981, 2, 937.
144. Mulvey, D.; Waters, W.A.   J. Chem. Soc. (Perkin Trans. II) 1977, 1868.
145. Lindeke, B.   Drug Metab. Rev. 1982, 13, 71.
146. Kiese, M.   Pharmacol. Rev. 1966, 18, 1091.
147. Manson, D.   J. Chem. Soc. (Perkin Trans. I) 1974, 192.
148. Becker, A.R.; Sternson, L.A.   Proc. Natl. Acad. Sci. 1981, 78, 2003.
149. Stier, A.; Clauss, R.; Lucke, A.; Reitz, I.   Xenobiotica 1980, 10, 661.
150. Nagata, C.; Kodama, M.; Ioki, Y.; Kimura, T.   In "Free Radicals and Cancer"; Floyd, R.A., Ed.;   Marcel Dekker: New York, 1983; p. 1.
151. Nakayama, T.; Kimura, T.; Kodama, M.; Nagata, C.   Carcinogenesis 1983, 4, 765.
152. Kaneko, M.; Nakayama, T.; Kodama, M.; Nagata, C.   Gann 1984, 75, 349.
153. Lenk, W.; Scharmer, U.   Xenobiotica 1980, 10, 573.
154. Floyd, R.A.; Soong, L.M.; Stuart, M.A.; Reigh, D.L.   Arch. Biochem. Biophys. 1978, 185, 450.
155. Grantham, P.H.; Weisburger, E.K.; Weisburger, J.H.   Biochim. Biophys. Acta 1965, 107, 414.
156. Dolle, B.; Topner, W.; Neumann, H.-G.   Xenobiotica 1980, 10, 527.
157. Eyer, P.   Chem.-Biol. Interactions 1979, 24, 227.
158. Diepold, C.; Eyer, P.; Kampffmeyer, H.; Reinhardt, K.   Adv. Exp. Med. Biol. 1982, 136B, 1173.
159. Mulder, G.J.; Kadlubar, F.F.; Mays, J.B.; Hinson, J.A.   Mol. Pharmacol. 1984, 26, 342.
160. Wirth, P.J.; Dybing, E.; von Bahr, C.; Thorgeirsson, S.S.   Mol. Pharmacol. 1980, 18, 117.
161. Becker, A.R.; Sternson, L.A.   Bioorg. Chem. 1980, 9, 305.

162. Green, L.C.; Skipper, P.L.; Turesky, R.J.; Bryant, M.S.; Tannenbaum, S.R. Cancer Res. 1984, 44, 4254.
163. Kriek, E. In "Carcinogenesis, A Broad Critique"'; M.D. Anderson Hospital and Tumor Institute; Williams and Wilkins: Baltimore, 1967; p. 441.
164. Hongslo, J.; Haug, L.T.; Wirth, P.J.; Moller, M.; Dybing, E.; Thorgeirsson, S.S. Mutation Res. 1983, 107, 239.
165. Moller, M.E.; Glowinski, I.B.; Thorgeirsson, S.S. Carcinogenesis 1984, 5, 797.
166. Nagasawa, H.T.; Gutmann, H.R. J. Biol. Chem. 1959, 234, 1593.
167. Belman, S.; Troll, W. J. Biol. Chem. 1962, 237, 746.
168. Wise, R.W.; Zenser, T.V.; Davis, B.B. Carcinogenesis 1983, 4, 285.
169. Bonser, G.M.; Clayson, D.B.; Jull, J.W.; Pyrah, L.N. Br. J. Cancer 1952, 6, 412.
170. Troll, W.; Belman, S.; Levine, E. Cancer Res. 1963, 23, 841.
171. King, C.M.; Kriek, E. Biochim. Biophys. Acta 1965, 111, 147.
172. Hammons, G.J.; Guengerich, F.P.; Weis, C.C.; Beland, F.A.; Kadlubar, F.F. Cancer Res. 1985, in press.
173. Wise, R.W.; Zenser, T.V.; Davis, B.B. Carcinogenesis 1984, 5, 1499.
174. Yamazoe, Y.; Kadlubar, F.F. Environ. Health Persp. 1985, in press.
175. Rice, J.R.; Kissinger, P.T. Biochem. Biophys. Res. Commun. 1982, 104, 1312.
176. Josephy, P.D.; Mason, R.P.; Eling, T. Carcinogenesis 1982, 3, 1227.
177. Marnett, L.J. Life Sciences 1981, 29, 531.
178. Kadlubar, F.F.; Frederick, C.B.; Weis, C.C.; Zenser, T.V. Biochem. Biophys. Res. Commun. 1982, 108, 253.
179. Morton, K.C.; King, C.M.; Vaught, J.B.; Wang, C.Y.; Lee, M.-S.; Marnett, L.J. Biochem. Biophys. Res. Commun. 1983, 111, 96.
180. Frederick, C.B.; Hammons, G.J.; Beland, F.A.; Yamazoe, Y.; Guengerich, F.P.; Zenser, T.V.; Ziegler, D.M.; Kadlubar, F.F. In "Biological Oxidation of Nitrogen in Organic Molecules"; Gorrod, J.W.; Damani, L.A., Eds.; Taylor and Francis: London; 1985, in press.
181. Boyd, J.A.; Harvan, D.J.; Eling, T.E. J. Biol. Chem. 1983, 258, 8246.
182. Boyd, J.A.; Eling, T.E. J. Biol. Chem. 1984, 259, 13885.
183. Krauss, R.S.; Reed, G.A.; Eling, T.E. Proc. Amer. Assoc. Cancer Res. 1984, 25, 84.
184. Clayson, D.B.; Garner, R.C. In "Chemical Carcinogens"; Searle, C.E., Ed.; ACS Monograph No. 173, American Chemical Society: Washington, D.C., 1976; pp. 366-461.
185. Kawazoe, Y.; Ogawa, O.; Takahashi, K.; Sawanishi, H.; Ito, N. Gann 1978, 69, 835.
186. Miller, J.A.; Miller, E.C. Advances Cancer Res. 1953, 1, 339.
187. Labuc, G.E.; Blunck, J.M. Biochem. Pharmacol. 1979, 28, 2367.
188. Kimura, T.; Kodama, M.; Nagata, C. Gann 1984, 75, 895.
189. Miller, E.C.; Kadlubar, F.F.; Miller, J.A.; Pitot, H.C.; Drinkwater, N.R. Cancer Res. 1979, 39, 3411.
190. Kadlubar, F.F.; Ketterer, B.; Flammang, T.J.; Christodoulides, L. Chem.-Biol. Interactions 1980, 31, 265.

191. Scribner, J.D.; Miller, J.A.; Miller, E.C. <u>Biochem. Biophys.</u>
     <u>Res. Commun.</u> 1965, 20, 560.
192. Lin, J.-K.; Miller, J.A.; Miller, E.C.; <u>Cancer Res</u>. 1975, 35,
     844.
193. Ketterer, B.; Kadlubar, F.F.; Flammang, T.; Carne, T.; Enderby,
     G. <u>Chem.-Biol. Interactions</u> 1979, 25, 7.
194. Poirier, L.A.; Miller, J.A.; Miller, E.C.; Sato, K. <u>Cancer</u>
     <u>Res</u>. 1967, 27, 1600.
195. Wislocki, P.G.; Miller, J.A.; Miller, E.C. <u>Cancer Res</u>. 1975,
     35, 880.
196. Beland, F.A.; Tullis, D.L.; Kadlubar, F.F.; Straub, K.M.;
     Evans, F.E. <u>Chem.-Biol. Interactions</u> 1980, 31, 1.
197. Tullis, D.L.; Straub, K.M.; Kadlubar, F.F. <u>Chem.-Biol.</u>
     <u>Interactions</u> 1981, 38, 15.
198. Tarpley, W.G.; Miller, J.A.; Miller, E.C. <u>Cancer Res</u>. 1980, 40,
     2493.
199. Coles, B.; Ketterer, B.; Beland, F.A.; Kadlubar, F.F.
     <u>Carcinogenesis</u> 1984, 5, 917.
200. Ketterer, B.; Srai, S.K.S.; Waynforth, B.; Tullis, D.L.; Evans,
     F.E.; Kadlubar, F.F. <u>Chem.-Biol. Interactions</u> 1982, 38, 287.
201. Stohrer, G.; Salemnick, G. <u>Cancer Res</u>. 1975, 35, 122.
202. Parham, J.C.; Templeton, M.A. <u>Cancer Res</u>. 1980, 40, 1475.
203. Hamer, J.; Macaluso, A. <u>Chem. Rev</u>. 1964, 64, 473.
204. Kadlubar, F.F., unpublished data.

RECEIVED May 13, 1985

The In Vitro Metabolic Activation of Nitro Polycyclic Aromatic Hydrocarbons

FREDERICK A. BELAND[1], ROBERT H. HEFLICH[1], PAUL C. HOWARD[2], and PETER P. FU[1]

[1] National Center for Toxicological Research, Jefferson, AR 72079
[2] Center for Environmental Health Sciences, Case Western Reserve University, Cleveland, OH 44106

Nitro polycyclic aromatic hydrocarbons are environmental contaminants which have been detected in
airborne particulates, coal fly ash, diesel emission
and carbon black photocopier toners. These compounds
are metabolized in vitro to genotoxic agents through
ring oxidation and/or nitroreduction. The details of
these metabolic pathways are considered using
4-nitrobiphenyl, 1- and 2-nitronaphthalene, 5-nitro-
acenaphthene, 7-nitrobenz[a]anthracene, 6-nitro-
chrysene, 1-nitropyrene, 1,3-, 1,6- and 1,8-dinitro-
pyrene, and 1-, 3- and 6-nitrobenzo[a]pyrene as
examples.

It was over a century ago when the synthesis of 1-nitropyrene (1)
and 6-nitrochrysene (2) was described and since then a wide variety
of nitro polycyclic aromatic hydrocarbons (PAHs) have been prepared. In general, nitro substitution has been reported to inhibit
the carcinogenicity of PAHs (3), and it was not until 1950 that a
nitro PAH was found which would induce tumor formation (4).
Twenty-five years later, McCann et al. (5) observed that a number
of nitro PAHs were mutagenic in the Salmonella typhimurium assay
and, interestingly, that they did not require a mammalian post-
mitochondrial supernatant (S9) in order to be genotoxic. These
findings were of academic interest until 1978 when Pitts and
coworkers (6) reported that mutagenic nitro PAHs could be formed in
model atmospheres containing trace quantities of PAHs, nitrogen
oxide and nitric acid. At the same time, Wang et al. (7) found
that urban air particulates contained direct-acting bacterial
mutagens which they suggested might be nitro PAHs, and Jager (8)
detected 6-nitrobenzo[a]pyrene as an air pollutant. Since these
initial reports, numerous papers have appeared which show that
nitro PAHs are widespread environmental contaminants which may pose
a significant human health hazard. In this review we discuss the
synthesis, environmental occurrence, biological effects, and
biotransformation of these compounds with emphasis on their in
vitro metabolic activation pathways. A recent review by Rosenkranz

0097–6156/85/0283–0371$07.25/0
© 1985 American Chemical Society

and Mermelstein (9) considers the mutagenicity of nitro PAHs in greater detail.

Synthesis

Nitration is one of the most common reactions of PAHs. Under mild conditions, substitution will occur at the most reactive carbon to give the kinetically-controlled product as the predominant isomer (10). Thus, nitration of anthracene, pyrene, chrysene, perylene, benz[a]anthracene, benzo[a]pyrene (BaP) and dibenz[a,h]anthracene yields 9-nitroanthracene, 1-nitropyrene, 6-nitrochrysene, 3-nitroperylene, 7-nitrobenz[a]anthracene, 6-nitro-BaP and 7-nitrodibenz-[a,h]anthracene, respectively (1-2,10-15). In addition to these major products, other isomers plus more extensively nitrated derivatives are nearly always produced. Removal of the undesired compounds can be quite tedious and in most instances classical procedures such as recrystallization and column chromatography do not give material sufficiently pure for use in biological studies. The use of high pressure liquid chromatography is strongly encouraged for the purification of nitro PAHs because even trace quantities of the undesired isomers can lead to erroneous conclusions concerning biological activity. For example, the nitration of BaP gives 6-nitro-BaP accompanied by small amounts of 1- and 3-nitro-BaP (13). Pure 6-nitro-BaP is not a direct-acting bacterial mutagen whereas both 1- and 3-nitro-BaP are quite active (15-18). Therefore, reports of 6-nitro-BaP being a direct-acting bacterial mutagen (6,19) are probably a result of contamination by the latter two isomers. 1-Nitropyrene is another case; although it is a direct-acting bacterial mutagen, its activity is at least 100-fold less than that observed with 1,3-, 1,6- or 1,8-dinitropyrene (9). The wide variation in bacterial mutagenicity that has been reported for 1-nitropyrene (9) is, therefore, probably due to the presence of these dinitropyrenes as trace impurities.

A large number of reagents are available for the preparation of nitro PAHs. These include fuming nitric acid in acetic acid (20) or acetic anhydride (13), sodium nitrate in trifluoroacetic acid (21) or trifluoroacetic acid and acetic anhydride (17), dinitrogen tetroxide in carbon tetrachloride (22), sodium nitrate in trimethyl phosphate and phosphorus pentoxide (23), and nitronium tetrafluoroborate in anhydrous acetonitrile (24). Alternative approaches must be used to synthesize nitro PAHs substituted at positions other than the most reactive carbon. For instance, 4-nitropyrene has been prepared by nitration of 4,5,9,10-tetrahydropyrene followed by dehydrogenation (25-26).

Environmental Occurrence

A wide variety of nitro PAHs have been isolated from different environmental sources including airborne particulates (27-34), coal fly ash (35-37), diesel emission particulates (38-41) and carbon black photocopier toners (42-43). Their presence has also been suggested in the smoke from nitrate-fortified cigarettes (44). The structures of the most commonly detected nitro PAHs are shown in Figure 1 and in each instance it is the kinetically-favored isomer that is found.

There are at least two routes which could result in the forma-

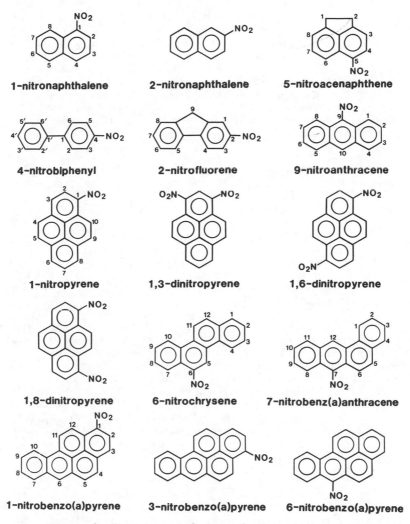

Figure 1. Structures of commonly detected nitro PAHs.

tion of nitro PAHs in the environment. Since PAHs are products of
the incomplete combustion of organic material, if nitrogen is
present, the nitro PAHs may be formed during the combustion process
itself (45). Nitro PAHs may also result from atmospheric reactions
of PAHs with nitrogen oxides (6,46-49), as evidenced by the observation that the predominant nitro PAHs detected in air samples are
derivatives of the most abundant PAHs found in the environment
(50-51). The quantity of nitro PAHs found in air samples may not
reflect their initial concentration because these compounds appear
to undergo photolytic decomposition. For example, when 9-nitro-
anthracene, 1-nitro-BaP, 6-nitro-BaP and 1-nitropyrene were exposed
to sunlight, each decomposed to yield quinones but the rates varied
markedly between compounds (52). It should also be noted that
nitro PAHs have been reported to be formed as artifacts during the
collection of air samples (53).

Biological Effects

Nitro PAHs have been shown to exhibit a large variety of biological
activities. Included in these are: the induction of mutations in
bacterial (Table I) and eukaryotic cells (9,17,54-57), the neoplastic transformation of cultured mammalian cells (58-59), and the
induction of DNA strand breaks (60), DNA repair (61-62), sister
chromatid exchanges (63-64), and chromosomal aberrations (65-66).
Nitro PAHs have also been demonstrated to bind cellular DNA in
bacteria (67-73) and mammalian cells (74-77), to inhibit preferentially the growth of repair-deficient bacteria (78), to have
recombinogenic activity in yeast (66,79-80) and to induce tumors in
experimental animals (Table II).

Mutagenicity. The mutagenicity of nitro PAHs has been studied most
extensively in the Ames' Salmonella typhimurium reversion assay
(5). The mutagenicities of representative nitro PAHs in this assay
are shown in Table I. Some of the more important features regarding their mutagenicity can be summarized as follows:
   i.   Nitro PAHs generally exhibit their highest mutagenicity
        in strain TA98 (a frameshift detector) in the absence of
        an S9 activating system. This direct mutagenic activity
        contrasts with the response observed with PAHs and amino
        PAHs which normally require S9 to induce mutations.
   ii. Some of the nitro PAHs (e.g., 1,3-, 1,6-, and 1,8-dini-
        tropyrene) are among the most mutagenic compounds ever
        tested in the S. typhimurium reversion assay.
   iii. Nitro PAHs which show greater activity in the presence
        of S9 (e.g., 6-nitro-BaP) may have a fundamentally
        different reactive intermediate than the direct-acting
        nitro PAHs.
   iv. There can be dramatic variations in mutagenic potential
        between nitro PAH isomers. For instance, 6-nitro-BaP is
        not a direct-acting mutagen whereas 1- and 3-nitro-BaP
        are potent direct-acting mutagens.
   v.   Dinitro PAHs appear to be more mutagenic than their
        mononitrated analogues.

Tumorigenicity. A number of nitro PAHs have been shown to be

Table I. Mutagenicity of Representative Nitro PAHs in Salmonella typhimurium strains TA98 and TA100 (revertants per nmole)

Compound	TA98		TA100		References
	-S9	+S9	-S9	+S9	
1-nitronaphthalene	0.01-1	0.2-0.3	0.8-10	0.7-1	19,81-83
2-nitronaphthalene	0.2-1	0.5	1-5	1	27,81-83
5-nitroacenaphthene	2-6	12-17	4-40	16-18	19,84-87
4-nitrobiphenyl	0.6-1	0.6	2-4	2	27,81-82
2-nitrofluorene	14-88	26-433	6-205	25	16,19,27,82,88
9-nitroanthracene	0.01-3	0.01-2	0.9-4	0.9-3	16,19,89-90
1-nitropyrene	470-4360	40-1506	150-908	25-45	16,18,19,27,91
1,3-dinitropyrene	28600-163800	-	8350	-	43,92-94
1,6-dinitropyrene	36000-192000	380	28000	60	19,43,92,94
1,8-dinitropyrene	72900-275000	120	21500	42	19,43,92-94
6-nitrochrysene	27-270	22-27	97-100	38-190	19,95
7-nitrobenz[a]-anthracene	<0.3	1	<0.6	2-11	90,96
1-nitrobenzo[a]-pyrene	80-1574	150-1000	60-2400	60	15-18
3-nitrobenzo[a]-pyrene	100-1900	250-1010	60-3100	60	15-18
6-nitrobenzo[a]-pyrene	0-31	50-450	0-5	30-450	15-19

Table II. Tumorigenicity of Representative Nitro PAHs

Compound	Species	Tumor or Site	Reference
2-nitronaphthalene	monkey	bladder	97
5-nitroacenaphthene	rat	intestine, mammary gland	98
	hamster	cholangiomas	98
4-nitrobiphenyl	dog	bladder	99
2-nitrofluorene	rat	mammary gland	4
	rat	liver, mammary gland	100
	rat	intestine, forestomach	100
1-nitropyrene	rat	injection site	101
	rat	injection site, mammary gland	102
	mouse	lung	103
1,3-dinitropyrene	rat	injection site	104
1,6-dinitropyrene	mouse	injection site	105
1,8-dinitropyrene	rat	injection site	104
6-nitrochrysene	mouse	skin	106

tumorigenic in experimental animals (Table II). In general, nitro PAHs demonstrate the same target tissue specificity as their amino PAH analogues. Thus, 2-nitronaphthalene (97) and 2-naphthylamine (107) induce bladder tumors in monkeys, 4-nitrobiphenyl (99) and 4-aminobiphenyl (108-109) are bladder carcinogens in dogs, and 2-nitrofluorene (4,100) and 2-aminofluorene (100) are carcinogenic for the liver, mammary gland and small intestine in rats. Although 6-nitrochrysene (106) and 6-aminochrysene (110) are tumorigenic on mouse skin, this tissue is not normally a target for aromatic amine carcinogens. Since mouse skin is a target tissue for PAHs, including chrysene itself (106), this suggests that 6-nitrochrysene and 6-aminochrysene behave like PAH carcinogens. The situation regarding nitropyrenes is more complex. When tested on mouse skin, neither pyrene nor 1-nitropyrene was active (106,111), whereas a mixture of 1,3-, 1,6- and 1,8-dinitropyrene was tumorigenic (111). In addition, these nitrated pyrenes induce tumors at the injection site in rats (101-102,104-105).

Microbial Metabolism

Salmonella typhimurium. Although most nitro PAHs are direct-acting mutagens in Salmonella typhimurium, these compounds must be metabolized to bind covalently to DNA (71,92,112). S. typhimurium contains a family of nitroreductases which are capable of reducing nitro PAHs, and strains which are deficient in these enzymes generally show decreased sensitivity toward nitro PAH-induced mutations (27,92,113-114). These observations suggest that reduced metabolic intermediates may be the critical reactive electrophiles.

Compared to the extensive data that have been obtained on the mutagenicity of nitro PAHs in S. typhimurium, relatively little is known about the metabolism of these compounds in this organism. Messier et al. (67) reported that incubation of 1-nitropyrene with S. typhimurium TA98 yielded 1-aminopyrene and 1-acetylaminopyrene as major and minor metabolites, respectively. The reduction of 1-nitropyrene was slow and was accompanied by a slow formation of DNA adducts. When incubations were conducted with the nitroreductase-deficient strain, TA100 F50, both the extent of 1-aminopyrene formation and DNA binding decreased. Howard et al. (71,115) also found reduction of 1-nitropyrene to 1-aminopyrene in strains TA98, TA1538 and ATCC 14028.

Although 1-aminopyrene is a reduced metabolite of 1-nitropyrene, this arylamine will not covalently bind to DNA in vitro (72). In contrast, when incubations were conducted with the intermediate reduction product, N-hydroxy-1-aminopyrene, extensive covalent binding to DNA was detected (72). This observation is consistent with the previous report that several N-hydroxy arylamines formed DNA adducts and induced mutations in S. typhimurium (116), and suggests that, at least for 1-nitropyrene, reduction to N-hydroxy-1-aminopyrene is a critical step in mutation induction.

Recently, Bryant et al. (70) examined the metabolism of 1,8-dinitropyrene in several S. typhimurium strains and found reduction to 1-amino-8-nitropyrene and 1,8-diaminopyrene. In addition, other unidentified metabolites were detected in strains which were sensitive to 1,8-dinitropyrene-induced mutations (TA98 and TA98NR) but not in the resistant strains, TA98/1,8-DNP$_6$ and TA98NR/1,8-DNP$_6$.

Certain nitro PAHs appear to require metabolism of their
N-hydroxy arylamine derivatives in order to induce mutations. For
example, while 2-nitrofluorene showed decreased mutagenicity in the
nitroreductase-deficient   mutant,   TA98NR,   and   in   strain
TA98/1,8-DNP$_6$, its presumed ultimate mutagenic derivative,
N-hydroxy-2-aminofluorene was inactive in only strain TA98/1,8-DNP$_6$
(117).   Observations such as these led McCoy et al. (117) to
propose that certain N-hydroxy arylamines require esterification to
N-acetoxy arylamines to induce mutations and that this capability
was lacking in strain TA98/1,8-DNP$_6$. Since N-hydroxy-2-acetyl-
aminofluorene was less mutagenic than N-hydroxy-2-aminofluorene in
all three strains, this suggests that direct O-acetylation occurs
to yield N-acetoxy-2-aminofluorene rather than N-acetylation
followed by N,O-acyltransfer. Although N-acetoxy arylamines are
generally too reactive to be detected, indirect evidence for their
formation was obtained by noting the formation of N-acetylated
derivatives of 2-aminofluorene in sensitive, but not resistant, S.
typhimurium strains (118). Thus, by comparing the differential
sensitivity in the three strains, it appears that the N-hydroxy
arylamine derived from 1-nitropyrene reacts directly with the
bacterial genome to induce mutations, whereas the analogous
derivatives of 2-nitrofluorene and 1,8-dinitropyrene are further
activated by O-acetylation (118). This suggests that the unidentified 1,8-dinitropyrene metabolites detected by Bryant et al. (70)
in mutation-sensitive (TA98 and TA98NR), but not resistant
(TA98/1,8-DNP$_6$), S. typhimurium strains may be acetylated products
(73).

Other bacteria. Intestinal bacteria may play a critical role in
the metabolic activation of certain nitroaromatic compounds in
animals (119) and several reports have appeared on the metabolism
of nitro PAHs by rat and human intestinal contents and microflora
(120-123). Kinouchi et al. (120) found that 1-nitropyrene was
reduced to 1-aminopyrene when incubated with human feces or
anaerobic bacteria. More recently, Kinouchi and Ohnishi (121)
isolated four nitroreductases from one of these anaerobic bacteria
(Bacteroides fragilis). Each nitroreductase was capable of con-
verting 1-nitropyrene into 1-aminopyrene, and one form catalyzed
the formation of a reactive intermediate capable of binding DNA.
Howard et al. (116) confirmed the reduction of 1-nitropyrene to
1-aminopyrene by both mixed and purified cultures of intestinal
bacteria. Two additional metabolites were also detected, one of
which appeared to be 1-hydroxypyrene. Recently, similar experiments have demonstrated the rapid reduction of 6-nitro-BaP to
6-amino-BaP (123).

Mammalian Metabolism
General considerations. Nitroaromatic compounds, such as nitro-
furans and nitrobenzenes are commercially important chemicals used
as drugs, food additives or synthetic intermediates. Since there
is widespread human exposure to these chemicals, their metabolism
has been studied extensively. Nitroaromatic compounds are reduced
by both hepatic cytosol and microsomes. The microsomal activity

appears to be associated with the cytochrome P-450 and its reductase, while cytosolic reductions can be catalyzed by a number of enzymes, including xanthine oxidase, DT-diaphorase, and aldehyde oxidase.

The in vitro metabolic reduction of nitro PAHs was first reported in 1967 (124). Since then a number of similar investigations have been conducted; however, most of these studies have been performed under only anaerobic conditions. The oxidative metabolism of nitro PAHs has been examined only recently, but as considered in the following sections, both reductive and oxidative metabolic pathways may be important in the metabolic activation of nitro PAHs.

4-Nitrobiphenyl, 1-nitronaphthalene, 2-nitronaphthalene and 5-nitroacenaphthene. Poirier and Weisburger (125) examined the reductive metabolism of 4-nitrobiphenyl, 1-nitronaphthalene and 2-nitronaphthalene by subcellular fractions of rat and mouse liver. Under anaerobic conditions, hepatic S9 and cytosol catalyzed the stoichiometric conversion of each of these nitro PAHs to their corresponding arylamines without any detectable formation of N-hydroxy arylamine intermediates. The rate of reduction was stimulated by NADPH and was further accelerated by the addition of both NADPH and FMN. These cofactors also greatly increased microsome-catalyzed amine formation.

El-Bayoumy and Hecht (84-85) have compared the metabolism of the carcinogen 5-nitroacenaphthene with its noncarcinogenic structural analogue, 1-nitronaphthalene. With rat liver S9, oxidation of the 2-carbon bridge of 5-nitroacenaphthene yielded a series of keto, hydroxy and dihydroxy metabolites which retained their nitro functions (Figure 2). Interestingly, 1-hydroxy-5-nitroacenaphthene and 1-oxo-5-nitroacenaphthene were better direct-acting mutagens in the S. typhimurium reversion assay than 5-nitroacenaphthene. Hepatic S9-catalyzed nitroreduction did not occur with the less mutagenic 1-nitronaphthalene, and only phenolic and dihydrodiol metabolites were detected.

These observations are significant for several reasons. First, it appears that 5-nitroacenaphthene must be oxidized before it can undergo nitroreduction because 5-aminoacenaphthene was not detected as a metabolite. Second, since 1-hydroxy- and 1-oxo-5-nitroacenaphthene are direct-acting mutagens, these metabolites may be proximate mutagenic forms of 5-nitroacenaphthene. Third, since 1-hydroxy- and 1-oxo-5-aminoacenaphthene were not direct-acting mutagens, intermediate reduction products, presumably the N-hydroxy arylamine derivatives, may be the ultimate mutagens.

McCoy et al. (86) have provided additional evidence that oxidation must precede nitroreduction in the metabolic activation of 5-nitroacenaphthene. In addition, they noted that this compound was less mutagenic in S. typhimurium TA98/1,8-DNP$_6$ which led them to suggest that the ring-oxidized N-hydroxy arylamine intermediates (Figure 2) are further activated by an "esterificase", which is apparently lacking in this particular strain of Salmonella.

Although El-Bayoumy and Hecht (84) did not detect the reduction of 1-nitronaphthalene to 1-naphthylamine in their S9 incubations, Poirier and Weisburger (125) did find reduction under

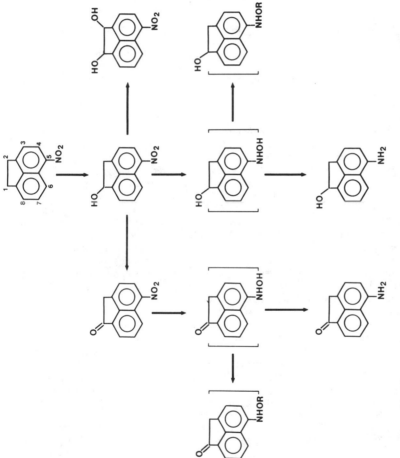

Figure 2. S9-catalyzed metabolites of 5-nitroacenaphthene. Similar oxidation occurs at carbon 2. Metabolites in brackets have not been detected, but may be ultimate mutagens.

anaerobic conditions in the presence of FMN. This suggests that
1-nitronaphthalene is less mutagenic than 5-nitroacenaphthene
because it is more resistant to nitroreduction. It is also
significant that when nitroreduction did occur, N-hydroxy-1-
naphthylamine was not detected as a metabolite (125) since this
compound is both mutagenic (5,126-127) and carcinogenic (127-129).
It is possible that during the S9-catalyzed reduction of 1-nitro-
naphthalene only the terminally-reduced species, 1-naphthylamine,
is released from the enzyme complex. Recent studies have indicated
that 1-naphthylamine is not oxidized by mixed function oxidases to
its N-hydroxy arylamine derivative (130). Taken together, these
data help explain why N-hydroxy-1-naphthylamine can be a potent
carcinogen while 1-nitronaphthalene and 1-naphthylamine are
apparently noncarcinogenic.

9-Nitroanthracene, 7-nitrobenz[a]anthracene and 6-nitrochrysene.
Rat liver microsomes oxidized 9-nitroanthracene primarily to
9-nitroanthracene-trans-3,4-dihydrodiol, with 9-nitroanthracene-
trans-1,2-dihydrodiol and 9,10-anthraquinone being detected as
minor metabolites (Figure 3; 89). The preferential oxidation of
the 3- and 4-carbons, as opposed to carbons 1 and 2, suggests that
metabolism is inhibited in regions peri to the nitro substituent.
9-Nitroanthracene and its two dihydrodiol metabolites were not
direct-acting mutagens in S. typhimurium and were only weakly
mutagenic in the presence of an S9-activating system. When
microsomal incubations were conducted under anaerobic conditions,
reduction to 9-aminoanthracene was not observed.

The primary product formed during the microsomal metabolism of
7-nitrobenz[a]anthracene was 7-nitrobenz[a]anthracene-trans-3,4-
dihydrodiol (Figure 4; 131). A small amount of 7-nitrobenz[a]an-
thracene-trans-8,9-dihydrodiol was also detected which is consistent with the previous observation (89) that nitro substitution
inhibits peri-region oxidation. Both dihydrodiol metabolites were
formed in a stereoselective manner with the R,R enantiomers
predominating. Since the same trans-3,4- and 8,9-enantiomers are
formed during the microsomal metabolism of benz[a]anthracene, it
appears that nitro substitution affects the regioselectivity but
not the stereoselectivity of metabolism. Nitro substitution also
affects the conformation of the resultant dihydrodiol metabolites.
Thus, while the hydroxyl groups of 7-nitrobenz[a]anthracene-trans-
3,4-dihydrodiol preferentially adopt a quasi-diequatorial conformation (131), the 8,9-isomer has a significant population with a
quasi-diaxial conformation. The effect of these conformations upon
the further metabolism of 7-nitrobenz[a]anthracene is presently not
known, although with other PAH derivatives, quasi-diaxial conformations tend to inhibit metabolism to diol epoxides (132). As was
observed with 9-nitroanthracene (89), 7-nitrobenz[a]anthracene and
its two dihydrodiol metabolites were not direct-acting mutagens in
S. typhimurium and were only weakly mutagenic in the presence of S9
(96).

El-Bayoumy and Hecht (95) have recently examined the metabolism of 6-nitrochrysene by rat liver S9. 6-Nitrochrysene-trans-
1,2-dihydrodiol was detected as the major metabolite and this was
further metabolized to a product tentatively identified as

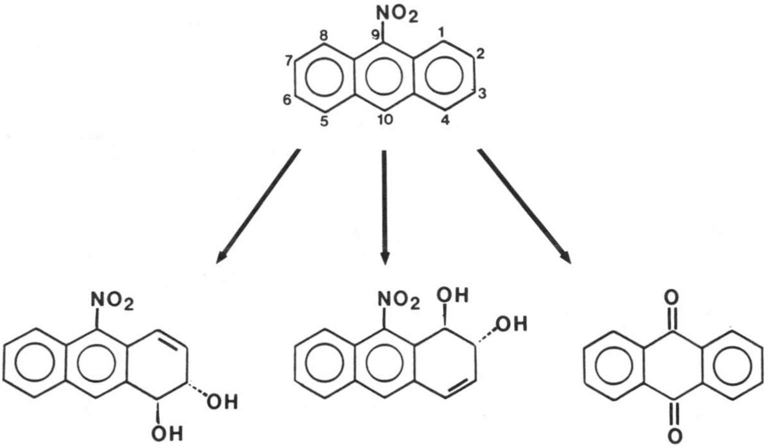

Figure 3. Microsomal metabolites of 9-nitroanthracene.

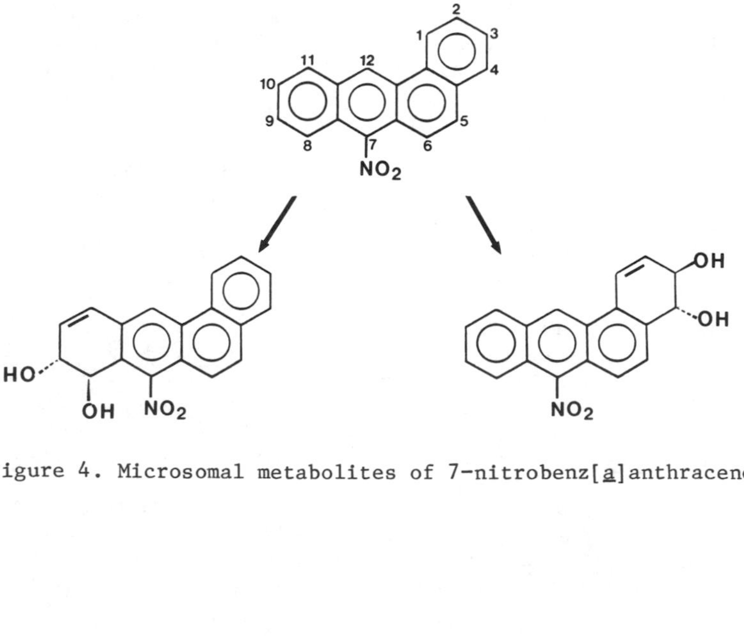

Figure 4. Microsomal metabolites of 7-nitrobenz[a]anthracene.

1,2-dihydroxy-6-nitrochrysene (Figure 5). Oxidation in the region
peri to the nitro function (i.e., carbons 7 and 8) was not observed
and reduction to 6-aminochrysene only occurred when the O_2 concentration was decreased. In S. typhimurium TA100, 6-nitrochrysene-
trans-1,2-dihydrodiol was a better direct-acting mutagen than
6-nitrochrysene or the other two metabolites. This latter
observation suggests that the dihydrodiol may be a proximate
mutagenic form of 6-nitrochrysene and that it is further activated
by bacterial nitroreduction. In the presence of S9, all of the
metabolites were mutagenic in strain TA100, with 6-aminochrysene
being the most active.

1-Nitropyrene. 1-Nitropyrene is the principal nitro PAH found in
diesel exhaust (40) and, therefore, has been the subject of intense
study. Nachtman and Wei (133) found that under anaerobic conditions, 1-nitropyrene was reduced by hepatic S9, cytosol or
microsomes to principally 1-aminopyrene. Only limited reduction
occurred in the absence of cofactors, while maximum metabolism was
observed in the presence of both FMN and NADPH.   Although the
microsomal fraction had the greatest specific activity toward
1-nitropyrene metabolism, the cytosol had 30 times the total
activity.
        Saito et al. (134) found that the cytosolic nitroreductase
activity was due to DT-diaphorase, aldehyde oxidase, xanthine
oxidase plus other unidentified nitroreductases.   As anticipated,
the microsomal reduction of 1-nitropyrene was inhibited by O_2 and
stimulated by FMN which was attributed to this cofactor acting as
an electron shuttle between NADPH-cytochrome P-450 reductase and
cytochrome P-450.   Carbon monoxide and type II cytochrome P-450
inhibitors decreased the rate of nitroreduction which was
consistent with the involvement of cytochrome P-450.   Induction of
cytochromes P-450 increased rates of 1-aminopyrene formation and
nitroreduction was demonstrated in a reconstituted cytochrome P-450
system, with isozyme P-448-IId catalyzing the reduction most
efficiently.
        El-Bayoumy and Hecht (91) were the first to report the oxida-
tive metabolism of 1-nitropyrene. Using rat liver S9, they found
3-, 6-, and 8-hydroxy-1-nitropyrene and 1-nitropyrene-trans-4,5-di-
hydrodiol (Figure 6). A small amount of 1-aminopyrene was also
formed and this increased in concentration as the O_2 concentration
was decreased. Similar metabolites have recently been detected
with mouse liver and lung S9 (103). When assayed in S. typhimurium
TA98 and TA100, 3- and 6-hydroxy-1-nitropyrene were found to be
better direct-acting mutagens than 1-nitropyrene. Thus, as was
found with 6-nitrochrysene (95) and 5-nitroacenaphthene (84-86),
ring hydroxylation does not necessarily represent a detoxification
process. Interestingly, when the mutagenic assays were conducted
in the presence of S9, 1-nitropyrene was more mutagenic than its
phenolic metabolites. The reason for this apparent dichotomy is
not known.
        Bond (135) found that S9 preparations from rat nasal tissue
have twice the specific activity for the oxidative metabolism of
1-nitropyrene as liver S9 and 10 times the activity of lung S9.
Each S9 preparation gave similar metabolic profiles with the

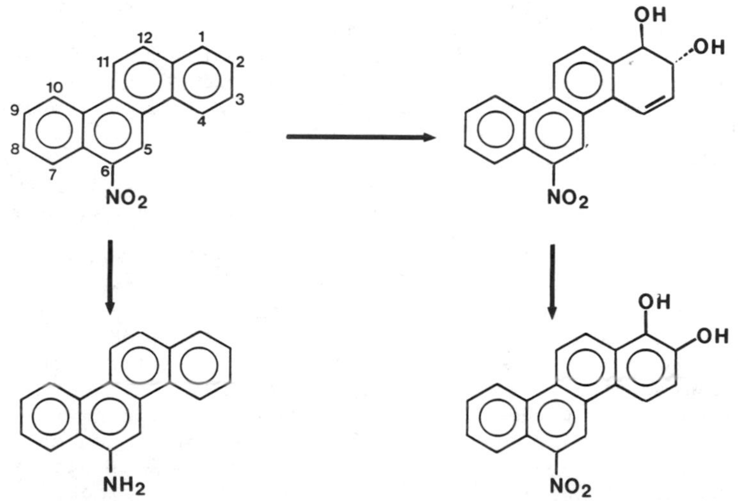

Figure 5. S9-catalyzed metabolites of 6-nitrochrysene.

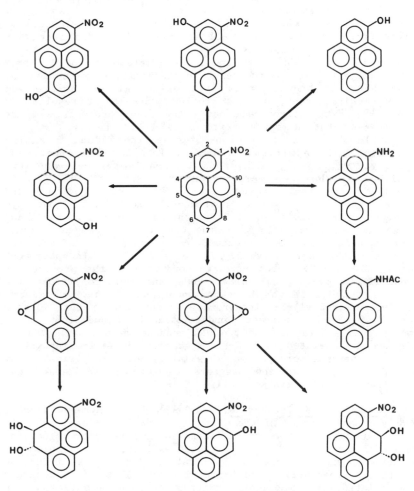

Figure 6. 1-Nitropyrene metabolites detected in in vitro incubations.

predominant metabolites being 3-, 6-, and 8-hydroxy-1-nitropyrene
plus at least three other unidentified products. King et al. (136)
conducted similar studies using S9 from rabbit liver. At least 12
metabolites were detected and, based upon cochromatography with
known standards, the major compounds appeared to be K-region (i.e.,
4,5- and 9,10-) dihydrodiols, 10-hydroxy-1-nitropyrene and other
phenols. Bond and Mauderly (137) also reported the presence of
10-hydroxy-1-nitropyrene as well as N-acetyl-1-aminopyrene in
perfused rat lung preparations.

     Howard et al. (138) have studied the metabolism of 1-nitro-
pyrene using rat liver microsomes. In microsomes from control and
phenobarbital-pretreated rats, the principal metabolite was
3-hydroxy-1-nitropyrene, whereas Aroclor pretreatment resulted in
6- and 8-hydroxy-1-nitropyrene being the major products. These
data suggest that different cytochrome P-450 isozymes may be
responsible for the formation of the individual phenolic
metabolites. In addition to these hydroxylated derivatives,
1-nitropyrene-trans-4,5-dihydrodiol, 1-aminopyrene, and two
additional metabolites were detected. More recently, this
microsomal metabolism has been examined in greater detail and
1-nitropyrene-4,5-oxide, 1-nitropyrene-9,10-oxide and 1-hydroxy-
pyrene were identified as metabolites through spectral analysis and
by comparison to synthetic standards (139). 1-Nitropyrene-trans-
9,10-dihydrodiol was also detected, which confirmed the original
isolation of this derivative by King et al. (136). In subsequent
studies, microsomal incubations were conducted in the presence of
Chinese hamster ovary (CHO) cells. Under anaerobic conditions, the
principal metabolite was 1-aminopyrene and one major adduct,
N-(deoxyguanosin-8-yl)-1-aminopyrene, was detected in the CHO cell
genome (140).   In contrast, under oxidative conditions amine
formation was suppressed and two DNA adducts were found. One of
these adducts coeluted with N-(deoxyguanosin-8-yl)-1-aminopyrene,
while the other was more polar. This latter adduct may result from
1-nitropyrene-4,5-oxide and suggests that genotoxic damage can
result from the oxidative metabolism of 1-nitropyrene.

1,3-, 1,6- and 1,8-Dinitropyrene. Although dinitropyrenes account
for only a small amount of the nitro PAHs found in diesel exhaust,
they make a significant contribution to the mutagenicity associated
with diesel particulates (40). As noted earlier, in S. typhimurium
these dinitropyrenes appear to be metabolically activated through
sequential nitroreduction and O-acetylation. This is in contrast
to the related, less mutagenic 1-nitropyrene, which only requires
nitroreduction for its mutagenicity (70,73,118). A similar pathway
has recently been found using mammalian nitroreductases and
acetylases (141). Incubation of rat liver cytosol with 1-nitro-
pyrene or 1,3-, 1,6- or 1,8-dinitropyrene resulted in the formation
of 1-aminopyrene and the respective aminonitropyrenes. When DNA
was included in the incubations, only a low level of DNA binding
was detected. However, addition of acetyl coenzyme A (AcCoA)
increased the binding of the dinitropyrenes to DNA 20- to 40-fold
while the binding of 1-nitropyrene was only slightly affected.
This increase in binding of dinitropyrenes to DNA in the presence
of AcCoA was not detected when using dog liver cytosol which is

known to be deficient in acetylases. These data indicate that mammalian cytosolic nitroreductases catalyze the formation of N-hydroxy arylamine intermediates which in the case of dinitro-pyrenes are converted to reactive N-acetoxy arylamines by cytosolic AcCoA-dependent acetylases.

6-Nitrobenzo[a]pyrene. 6-Nitro-BaP is also a component of diesel exhaust (16,40); however, in contrast to the nitrated pyrenes, it is only mutagenic in the presence of S9 (15-18). This suggests that oxidized metabolites of 6-nitro-BaP may be responsible for mutation induction, and therefore, these products have been characterized. When incubated with liver microsomes from 3-methyl-cholanthrene-pretreated rats, 6-nitro-BaP was converted into 1- and 3-hydroxy-6-nitro-BaP, 6-nitro-BaP-1,9- and -3,9-hydroquinone and BaP-3,6-quinone (Figure 7, 142). The same metabolites were found when control or phenobarbital-induced microsomes were used, with 1- and 3-hydroxy-6-nitro-BaP being the predominant products in each instance (143). The latter two phenols were more mutagenic than 6-nitro-BaP, but as was observed with 6-nitro-BaP, they required S9 to be active (142). In contrast, 6-nitro-BaP-1,9- and -3,9-hydro-quinones appear to be direct-acting bacterial mutagens (96).

Several conclusions can be drawn from these results. First, as has been noted with 5-nitroacenaphthene and 6-nitrochrysene, there appears to be no peri-region (i.e., 4,5- or 7,8-) oxidation. Second, although ring oxidation appears to be an activation step, S9 is still required for 1- and 3-hydroxy-6-nitro-BaP to be mutagenic. In addition, essentially the same mutagenic activity was found in TA98 and TA100 and their respective nitroreductase-deficient derivatives. Thus, in contrast to the nitro PAHs considered previously, the metabolic activation of 6-nitro-BaP may not involve nitroreduction, but only ring oxidation. Third, since 6-nitro-BaP-1,9- and -3,9-hydroquinones are direct-acting mutagens, these metabolites may be the ultimate mutagenic forms of 6-nitro-BaP.

Recently, the mechanism of 6-nitro-BaP ring hydroxylation has been elucidated by using 3-deutero-6-nitro-BaP (144). When incu-bated with 3-methylcholanthrene-induced rat liver microsomes, this deuterated analogue yielded the same metabolite profile previously observed with 6-nitro-BaP. Spectroscopic analysis of 3-hydroxy-6-nitro-BaP and 6-nitro-BaP-3,9-hydroquinone indicated that 30% of the deuterium label had migrated to carbon 2, presumably via an NIH shift. Therefore, it appears that 6-nitro-BaP-2,3-oxide is a common intermediate for these two metabolites.

1- and 3-Nitrobenzo[a]pyrene. 1- and 3-Nitro-BaP also appear to be components of diesel exhaust, but unlike 6-nitro-BaP, these nitro PAHs are direct-acting bacterial mutagens (15-18). When 1- and 3-nitro-BaP were incubated with liver microsomes from 3-methyl-cholanthrene-pretreated rats, 7,8-trans-dihydrodiols, 9,10-trans-dihydrodiols, and 7,8,9,10-tetrahydrotetrols were formed as major metabolites (Figure 8; 145-146), while amine formation was detected only under anaerobic incubations (147). These results are in contrast to those found with 6-nitro-BaP; with 1- and 3-nitro-BaP, all of the oxidation occurred in the terminal benzo ring (i.e.,

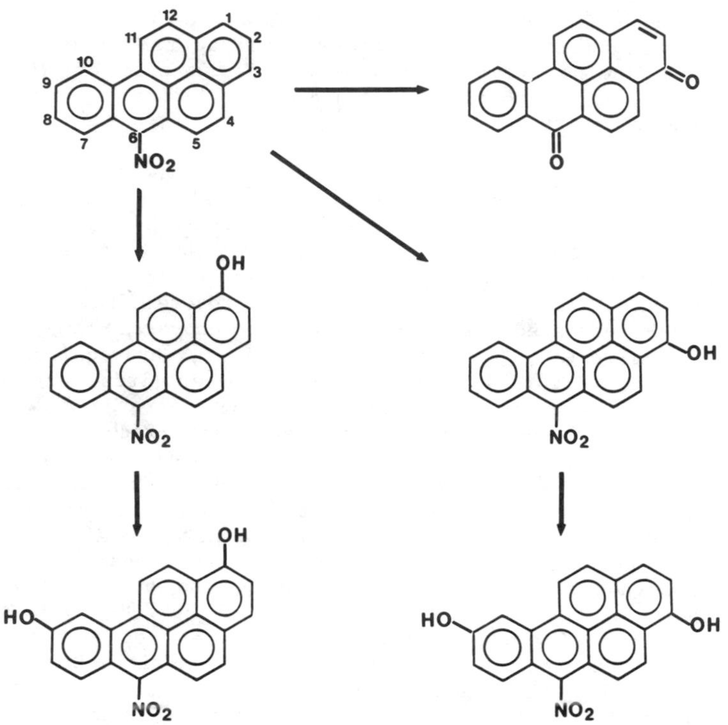

Figure 7. Microsomal metabolites of 6-nitrobenzo[a]pyrene.

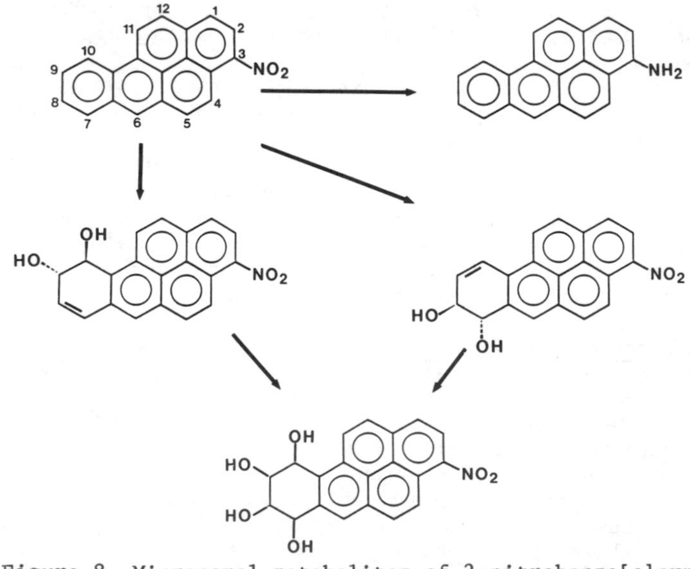

Figure 8. Microsomal metabolites of 3-nitrobenzo[a]pyrene.

carbons 7, 8, 9, and 10), while 6-nitro-BaP metabolism initially occurs at carbons 1 and 3. Thus, the nitro group has a significant effect upon the regioselectivity of metabolic oxidation. The absence of hydroxylation at carbon 3 in 1-nitro-BaP is noteworthy because, in addition to being a major site of oxidation in 6-nitro-BaP (142), 3-hydroxy-BaP is a predominant microsomal metabolite of BaP (148). Furthermore, the analogous carbon in 1-nitropyrene is a site for extensive metabolism (91,138). The reason for this marked regioselectivity is not known.

The conformations of the 1- and 3-nitro-BaP metabolites were determined through analysis of their NMR spectra (145-146). Both 1- and 3-nitro-BaP-trans-7,8-dihydrodiols existed predominantly in quasi-diequatorial conformations, which corresponds to the preferred conformation of the proximate carcinogen BaP-trans-7,8-dihydrodiol (149). This suggests that these dihydrodiol metabolites may be converted into electrophilic diol epoxides and in support of this contention, the stereochemistries of 1- and 3-nitro-BaP-7,8,9,10-tetrols were indicative of trans-7,8-dihydrodiol-anti-9,10-epoxide intermediates. It is possible, however, that a proportion of the tetrol metabolites were formed from trans-9,10-dihydrodiol-anti-7,8-epoxides.

The mutagenicities of 3-nitro-BaP, 3-amino-BaP, and 3-nitro-BaP-7,8- and -9,10-dihydrodiols have recently been compared in S. typhimurium strains TA98, TA98NR, and TA98/1,8-DNP$_6$ (147). In the absence of S9, 3-nitro-BaP showed decreased activity in TA98NR and slightly higher mutagenicity in TA98/1,8-DNP$_6$ compared to TA98, which is consistent with nitroreduction, but not esterification, being essential for mutation induction. This conclusion is supported by the observation that the direct-acting mutagenicity of 3-amino-BaP was the same in all three strains, probably as a result of it being spontaneously oxidized to N-hydroxy-3-amino-BaP. The dihydrodiols behaved similarly to one another: their direct-acting mutagenicity was similar to that of 3-nitro-BaP in TA98, decreased by about 50% in TA98NR, and was substantially decreased in TA98/1,8-DNP$_6$. These results suggest that both nitroreduction and esterification are required for the direct-acting mutagenicity of these dihydrodiol metabolites.

Conclusions

The data presented in this review show that both oxidative and reductive pathways are involved in the metabolic activation of nitro PAHs to genotoxic agents in vitro. The precise pathways depend upon the particular compound, but in each instance the nitro group appears to play a critical role in the activation process. With regard to oxidative metabolism, the location of the nitro function influences the regioselectivity of ring oxidation since metabolism is generally inhibited in peri regions. When peri-region oxidation does occur, it is quite limited in extent and a significant proportion of the dihydrodiol metabolite exists in a quasi-diaxial conformation. Thus, in addition to influencing the regioselectivity of oxidation, the nitro group can have a marked effect upon the conformation of the resultant metabolite which, in turn, may influence subsequent metabolism. The nitro function may also be directly involved in the activation sequence via reduction to reactive N-hydroxy arylamines. N-Hydroxy arylamines can react directly with DNA, or in some instances they may be further activated by O-esterification. Although the factors that allow an N-hydroxy arylamine to undergo O-esterification are not known, ester formation generally results in the production of a compound which is more genotoxic than the N-hydroxy arylamine precursor. Finally, some nitro PAHs are metabolized to ultimate mutagens by a combination of oxidative and reductive pathways. With these compounds, ring oxidation generally precedes nitroreduction and the net effect appears to be the formation of an N-hydroxy arylamine metabolite which can serve as a substrate for O-esterification.

The data from these in vitro studies can provide insight into the tumorigenicity of nitro PAHs. Thus, nitro derivatives of noncarcinogenic PAHs, such as 2-nitronaphthalene or 4-nitrobiphenyl, are extensively reduced in vitro which is consistent with their showing the same target specificity as their aromatic amine analogues. This view is strengthened by the observation that identical DNA adducts have been found in the bladder epithelium of dogs administered either 4-nitrobiphenyl or 4-aminobiphenyl (116). Nitro derivatives of carcinogenic PAHs demonstrate varied tumorigenic responses which may be associated with differences in metabolic pathways. For example, 6-nitrochrysene is tumorigenic on mouse skin and undergoes extensive dihydrodiol formation in vitro which suggests that it is metabolized to a reactive diol epoxide. In contrast, dihydrodiol metabolites are apparently not formed from 6-nitrobenzo[a]pyrene and this compound gives a negative tumorigenic response on mouse skin. In addition, 6-nitrobenzo[a]pyrene has two protons peri to its nitro function, which appears to restrict its ability to be enzymatically converted to a reactive N-hydroxy arylamine (150). These in vitro metabolic studies also indicate that a combination of oxidative and reductive pathways may be involved in the tumorigenicity of certain nitro PAHs. This may be particularly important with the ring-oxidized metabolites of 1-nitropyrene, 5-nitroacenaphthene and 3-nitrobenzo[a]pyrene, which are at least as genotoxic as their parent nitro PAHs. These observations suggest that in order to assess the human health risk from nitro PAHs, tumorigenicity assays should be conducted not only with the parent compounds and their nitroreduction products, but also with the ring-oxidized metabolites that have been detected in in vitro incubations.

Acknowledgment
We thank Ruth York for helping prepare this review.

Literature Cited

1. Graebe, C. Liebigs Ann. 1871, 158, 292.
2. Schmidt, J. J. prakt. Chem. 1874, 9, 250.
3. Hartwell, J.L. "Survey of Compounds Which Have Been Tested for Carcinogenic Activity"; U.S. Public Health Service Publication No. 149: Washington, D.C., 1951.
4. Morris, H.P.; Dubnik, C.S.; Johnson, J.M. J. Natl. Cancer Inst. 1950, 10, 1201.
5. McCann, J.; Choi, E.; Yamasaki, E.; Ames, B.N. Proc. Natl. Acad. Sci. USA 1975, 72, 5135.

6. Pitts, J.N., Jr.; van Cauwenberghe, K.A.; Grosjean, D.; Schmid, J.P.; Fitz, D.R.; Belser, W.J., Jr.; Knudson, G.B.; Hynds, P.M. Science 1978, 202, 515.
7. Wang, Y.Y.; Rappaport, S.M.; Sawyer, R.F.; Talcott, R.E.; Wei, E.T. Cancer Lett. 1978, 5, 39.
8. Jager, J. J. Chromatogr. 1978, 152, 575.
9. Rosenkranz, H.S.; Mermelstein, R. Mutation Res. 1983, 114, 217.
10. Clar, E. In "Polycyclic Hydrocarbons"; Academic: London, 1964; Vols. 1 and 2.
11. Meisenheimer, J. Liebigs Ann. 1904, 330, 133.
12. Bavin, P.M.G.; Dewar, M.J.S. J. Chem. Soc. 1955, 4477.
13. Dewar, M.J.S.; Mole, T.; Urch, D.S.; Warford, E.W.T. J. Chem. Soc. 1956, 3572.
14. Cook, J.W. J. Chem. Soc. 1931, 3273.
15. Pitts, J.N., Jr.; Zielinska, B.; Harger, W.P. Mutation Res. 1984, 140, 81.
16. Pitts, J.N., Jr.; Lokensgard, D.M.; Harger, W.; Fisher, T.S.; Mejia, V.; Schuler, J.J.; Scorziell, G.M.; Katzenstein, Y.A. Mutation Res. 1982, 103, 241.
17. Chou, M.W.; Heflich, R.H.; Casciano, D.A.; Miller, D.W.; Freeman, J.P.; Evans, F.E.; Fu, P.P. J. Med. Chem. 1984, 27, 1156.
18. Lofroth, G.; Toftgard, R.; Nilsson, L.; Agurell, E.; Gustafsson, J.-A. Carcinogenesis 1984, 5, 925.
19. Tokiwa, H.; Nakagawa, R.; Ohnishi, Y. Mutation Res. 1981, 91, 321.
20. Rodenburg, L.; Brandsma, R.; Tintel,; Cornelisse, J.; Lugtenburg, J. J. Chem. Soc. Chem. Commun. 1983, 1039.
21. Spitzer, U.A.; Stewart, R. J. Org. Chem. 1974, 39, 3936.
22. Radner, F. Acta Chem. Scand. 1983, 37, 65.
23. Pearson, D.E.; Frazer, J.G.; Frazer, V.S.; Washburn, L.C. Synthesis 1976, 621.
24. Vance, W.A.; Chan, R. Environ. Mutagenesis 1983, 5, 859.
25. Bavin, P.M.G. Can. J. Chem. 1959, 37, 1614.
26. Bodine, R.S.; Ruehle, P.H.; Roth, R.W.; Bosch, G.; Bosch, L.; Opperman, G.; Saugier, J.H. In "Polynuclear Aromatic Hydrocarbons: Formation, Metabolism and Measurement"; Cooke, M.; Dennis, A.J., Eds.; Battelle: Columbus, OH, 1983; p. 135.
27. Wang, C.Y.; Lee, M.-S.; King, C.M.; Warner, P.O. Chemosphere 1980, 9, 83.
28. Talcott, R.E.; Harger, W. Mutation Res. 1981, 91, 433.
29. Gibson, T.L. Atmospheric Environ. 1982, 16, 2037.
30. Fukino, H.; Mimura, S.; Inoue, K.; Yamane, Y. Mutation Res. 1982, 102, 237.
31. Pitts, J.N., Jr.; Harger, W.; Lokensgard, D.M.; Fitz, D.R.; Scorziell, G.M.; Mejia, V. Mutation Res. 1982, 104, 35.
32. Nielsen, T. Anal. Chem. 1983, 55, 286.
33. Ramdahl, T.; Becher, G.; Bjorseth, A. Environ. Sci. Technol. 1982, 16, 861.
34. Tokiwa, H.; Kitamori, S.; Nakagawa, R.; Horikawa, K.; Matamala, L. Mutation Res. 1983, 121, 107.
35. Li, A.P.; Clark, C.R.; Hanson, R.L.; Henderson, T.R.; Hobbs, C.H. Environ. Mutagenesis 1982, 4, 407.

36. Wei, C.-I.; Raabe, O.G.; Rosenblatt, L.S. Environ. Mutagenesis 1982, 4, 249.
37. Rappaport, S.M.; Wang, Y.Y.; Wei, E.T.; Sawyer, R.; Watkins, B.E.; Rapoport, H. Environ. Sci. Tech. 1980, 14, 1505.
38. Clark, D.R.; Brooks, A.L.; Li, A.P.; Hadley, W.M.; Hanson, R.L.; McClellan, R.O. Environ. Mutagenesis 1982, 4, 333.
39. Xu, X.B.; Nachtman, J.P.; Jin, Z.L.; Wei, E.T.; Rappaport, S.M.; Burlingame, A.L. Anal. Chim. Acta 1982, 136, 163.
40. Schuetzle, D. Environ. Health Perspect. 1983, 17, 65.
41. Paputa-Peck, M.C.; Marano, R.S.; Schuetzle, D.; Riley, T.L.; Hampton, C.V.; Prater, T.J.; Skewes, L.M.; Jensen, T.E.; Ruehle, P.H.; Bosch, L.C.; Duncan, W.P. Anal. Chem. 1983, 55, 1946.
42. Lofroth, G.; Hefner, E.; Alfheim, I.; Molle, M. Science 1980, 209, 1037.
43. Rosenkranz, H.S.; McCoy, E.C.; Sanders, D.R.; Butler, M.; Kiriazides, D.K.; Mermelstein, R. Science 1980, 209, 1039.
44. McCoy, E.C.; Rosenkranz, H.S. Cancer Lett. 1982, 15, 9.
45. Yergey, J.A.; Risby, T.H.; Lestz, S.S. Anal. Chem. 1982, 54, 354.
46. Fukui, S.; Hirayama, T.; Shindo, H.; Nohara, M. Chemosphere 1980, 9, 771.
47. Hirayama, T.; Nohara, M.; Shindo, H.; Fukui, S. Chemosphere 1981, 10, 223.
48. Tokiwa, H.; Nakagawa, R.; Morita, K.; Ohnishi, Y. Mutation Res. 1981, 85, 195.
49. Hirayama, T.; Nohara, M.; Ando, T.; Tanaka, M.; Nagano, K.; Fukui, S. Mutation Res. 1983, 122, 273.
50. Guerin, M.R. In "Polycyclic Hydrocarbons and Cancer"; Gelboin, H.V.; Ts'o, P.O.P., Eds.; Academic: New York, 1978; Vol. 1, p. 3.
51. Baum, E.J. In "Polycyclic Hydrocarbons and Cancer"; Gelboin, H.V.; Ts'o, P.O.P., Eds.; Academic: New York, 1978; Vol. 1, p. 45.
52. Chou, M.W.; Fu, P.P. unpublished data.
53. Risby, T.H.; Lestz, S.S. Environ. Sci. Technol. 1983, 17, 621.
54. Li, A.P.; Dutcher, J.S. Mutation Res. Lett. 1983, 119, 387.
55. Li, A.P.; Clark, C.R.; Hanson, R.L.; Henderson, T.R.; Hobbs, C.H. Environ. Mutagenesis 1983, 5, 263.
56. Cole, J.; Arlett, C.F.; Lowe, J.; Bridges, B.A. Mutation Res. 1982, 93, 213.
57. Takayama, S.; Tanaka, M.; Katoh, Y.; Terada, M.; Sugimura, T. Gann 1983, 74, 338.
58. Howard, P.C.; Gerrard, J.A.; Milo, G.E.; Fu, P.P.; Beland, F.A.; Kadlubar, F.F. Carcinogenesis 1983, 4, 353.
59. DiPaolo, J.A.; DeMarinis, A.J.; Chow, F.L.; Garner, R.C.; Martin, C.N.; Doniger, J. Carcinogenesis 1983, 4, 357.
60. Moller, M.; Thorgeirsson, S.S. Proc. Amer. Assoc. Cancer Res. 1984, 25, 110.
61. Campbell, J.; Crumplin, G.C.; Garner, J.V.; Garner, R.C.; Martin, C.N.; Rutter, A. Carcinogenesis 1981, 2, 559.
62. Butterworth, B.E.; Earle, L.L.; Strom, S.; Jirtle, R.; Michalopoulos, G. Mutation Res. 1983, 122, 73.

63. Nachtman, J.P.; Wolff, S. Environ. Mutagenesis 1982, 4, 1.
64. Marshall, T.C.; Royer, R.E.; Li, A.P.; Kusewitt, D.F.; Brooks, A.L. J. Toxicol. Environ. Health 1982, 10, 373.
65. Danford, N.; Wilcox, O.; Parry, J.M. Mutation Res. 1982, 105, 349.
66. Wilcox, P.; Danford, N.; Parry, J.M. In "Mutagens in Our Environment"; Sorso, M.; Vaino, H., Eds.; Liss: New York, 1982, p. 249.
67. Messier, F.; Lu, C.; Andrews, P.; McCarry, B.E.; Quilliam, M.A.; McCalla, D.R. Carcinogenesis 1981, 2, 1007.
68. Quilliam, M.A.; Messier, F.; Lu, C.; Andrews, P.A.; McCarry, B.E.; McCalla, D.R. In "Polynuclear Aromatic Hydrocarbons: Physical and Biological Chemistry"; Cooke, M.; Dennis, A.J.; Fisher, G.L., Eds.; Battelle: Columbus, OH, 1982; p. 667.
69. Andrews, P.A.; Bryant, D.; Vitakunas, S.; Gouin, M.; Anderson, G.; McCarry, B.E.; Quilliam, M.A.; McCalla, D.R. In "Polynuclear Aromatic Hydrocarbons: Formation, Metabolism and Measurement"; Cooke, M.; Dennis, A.J., Eds.; Battelle: Columbus, OH, 1983, p. 89.
70. Bryant, D.W.; McCalla, D.R.; Lultschik, P.; Quilliam, M.A.; McCarry, B.E. Chem.-Biol. Interact. 1984, 49, 351.
71. Howard, P.C.; Heflich, R.H.; Evans, F.E.; Beland, F.A. Cancer Res. 1983, 43, 2052.
72. Heflich, R.H.; Howard, P.C.; Beland, F.A. Mutation Res. in press.
73. Heflich, R.H.; Fifer, E.K.; Djuric, Z.; Beland, F.A. Environ. Health Perspect. in press.
74. Jackson, M.A.; King, L.C.; Ball, L.M. Drug Chem. Toxicol. 1983, 6, 549.
75. King, L.C.; Jackson, J.; Ball, L.M.; Lewtas, J. Cancer Lett. 1983, 19, 241.
76. Beland, F.A.; Heflich, R.H.; Howard, P.C.; Kurian, P.; Milo, G.E. Proc. Eur. Assoc. Cancer Res. 1983.
77. Heflich, R.H.; Fullerton, N.F.; Beland, F.A. in preparation.
78. Rosenkranz, H.S.; Poirier, L.A. J. Natl. Cancer Inst. 1979, 62, 873.
79. Wilcox, P.; Parry, J.M. Carcinogenesis 1981, 2, 1201.
80. McCoy, E.C.; Anders, M.; McCartney, M.; Howard, P.C.; Beland, F.A.; Rosenkranz, H.S. Mutation Res. 1984, 139, 115.
81. El-Bayoumy, K.; LaVoie, E.J.; Hecht, S.S.; Fow, E.A.; Hoffman, D. Mutation Res. 1981, 81, 143.
82. Scribner, J.D.; Fisk, S.R.; Scribner, N.K. Chem.-Biol. Interact. 1979, 26, 11.
83. McCoy, E.C.; Rosenkranz, E.J.; Petrullo, L.A.; Rosenkranz, H.S.; Mermelstein, R. Environ. Mutagenesis 1981, 3, 499.
84. El-Bayoumy, K.; Hecht, S.S. In "Polynuclear Aromatic Hydrocarbons: Physical and Biological Chemistry"; Cooke, M.; Dennis, A.J.; Fisher, G.L., Eds.; Battelle: Columbus, OH, 1982; p. 263.
85. El-Bayoumy, K.; Hecht, S.S. Cancer Res. 1982, 42, 1243.
86. McCoy, E.C.; De Marco, G.; Rosenkranz, E.J.; Anders, M.; Rosenkranz, H.S.; Mermelstein, R. Environ. Mutagenesis 1983, 5, 17.

87. McCoy, E.C.; Rosenkranz, E.J.; Mermelstein, R.; Rosenkranz, H.S. Mutation Res. 1983, 111, 61.

88. McCoy, E.C.; Rosenkranz, E.J.; Rosenkranz, H.S.; Mermelstein, R. Mutation Res. 1981, 90, 11.

89. Fu, P.P.; Von Tungeln, L.S.; Chou, M.W. submitted.

90. Greibrokk, T.; Lofroth, G.; Nilsson, L.; Toftgard, R.; Carlstedt-Duke, J.; Gustafsson, J.-A. In "Toxicity of Nitroaromatic Compounds"; Rickert, D.E.; Dent, J.G.; Gibson, J.E.; Popp, J.A.; Rosenkranz, H.S., Eds.; Hemisphere: Washington, D.C. in press.

91. El-Bayoumy, K.; Hecht, S.S. Cancer Res. 1983, 43, 3132.

92. Mermelstein, R.; Kiriazides, D.K.; Butler, M.; McCoy, E.C.; Rosenkranz, H.S. Mutation Res. 1981, 89, 187.

93. Pederson, T.C.; Siak, J.-S. J. Appl. Toxicol. 1981, 1, 54.

94. Nakayasu, M.; Sakamoto, H.; Wakabayashi, K.; Terada, M.; Sugimura, T.; Rosenkranz, H.S. Carcinogenesis 1982, 3, 917.

95. El-Bayoumy, K.; Hecht, S.S. Cancer Res. 1984, 44, 3408.

96. Fu, P.P.; Chou, M.W. unpublished results.

97. Conzelman, G.M., Jr.; Moulton, J.E.; Flanders, L.E. Gann 1970, 61, 79.

98. Takemura, N.; Hashida, C.; Terasawa, M. Brit. J. Cancer 1974, 30, 481.

99. Deichmann, W.B.; MacDonald, W.M.; Coplan, M.M.; Woods, F.M.; Anderson, W.A.D. Indust. Med. Surg. 1958, 27, 634.

100. Miller, J.A.; Sandin, R.B.; Miller, E.C.; Rusch, H.P. Cancer Res. 1955, 15, 188.

101. Ohgaki, H.; Matsukura, N.; Morino, K.; Kawachi, T.; Sugimura, T.; Morita, K.; Tokiwa, H.; Hirota, T. Cancer Lett. 1982, 15, 1.

102. Hirose, M.; Lee, M.-S.; Wang, C.Y.; King, C.M. Cancer Res. 1984, 44, 1158.

103. El-Bayoumy, K.; Hecht, S.S.; Sackl, T.; Stoner, G.D. Carcinogenesis 1984, 5, 1449.

104. Ohgaki, H.; Negishi, C.; Wakabayashi, K.; Kusama, K.; Sato, S.; Sugimura, T. Carcinogenesis 1984, 5, 583.

105. Ohnishi, Y.; Kinouchi, T.; Manabe, Y.; Tsutsui, H.; Otsuka, H.; Tokiwa, H.; Otofuji, T. In "Short-Term Genetic Bioassays in the Evaluation of Complex Environmental Mixtures"; Water, M.D., Ed.; Plenum: New York, in press.

106. El-Bayoumy, K.; Hecht, S.S.; Hoffman, D. Cancer Lett. 1982, 16, 333.

107. Conzelman, G.M., Jr.; Moulton, J.E.; Flanders, L.E.; Springer, K.; Crout, D. J. Natl. Cancer Inst. 1969, 42, 825.

108. Walpole, A.L.; Williams, M.H.C.; Roberts, D.C. Brit. J. Indust. Med. 1954, 11, 105.

109. Deichmann, W.B.; Radomski, J.L.; Anderson, W.A.D.; Coplan, M.M.; Woods, F.M. Indust. Med. Surg. 1958, 27, 25.

110. Lambelin, G.; Roba, J.; Roncucci, R.; Parmentier, R. Eur. J. Cancer 1975, 11, 327.

111. Nesnow, S.; Triplett, L.L.; Slaga, T.J. Cancer Lett. 1984, 23, 1.

112. Howard, P.C.; Beland, F.A. Biochem. Biophys. Res. Commun. 1982, 104, 727.

113. McCoy, E.C.; Rosenkranz, H.S.; Mermelstein, R. Environ. Mutagenesis 1981, 3, 421.
114. Rosenkranz, H.S.; McCoy, E.C.; Mermelstein, R.; Speck, W.T. Mutation Res. 1981, 91, 103.
115. Howard, P.C.; Beland, F.A.; Cerniglia, C.E. Carcinogenesis 1983, 4, 985.
116. Beland, F.A.; Beranek, D.T.; Dooley, K.L.; Heflich, R.H.; Kadlubar, F.F. Environ. Health Perspect. 1983, 49, 125.
117. McCoy, E.C.; McCoy, G.D.; Rosenkranz, H.S. Biochem. Biophys. Res. Commun. 1982, 108, 1362.
118. McCoy, E.C.; Anders, M.; Rosenkranz, H.S. Mutation Res. 1983, 121, 17.
119. Goldman, P. Ann. Rev. Pharmacol. Toxicol. 1978, 18, 523.
120. Kinouchi, T.; Manabe, Y.; Wakisaka, K.; Ohnishi, Y. Microbiol. Immunol. 1982, 26, 993.
121. Kinouchi, T.; Ohnishi, Y. Appl. Environ. Microbiol. 1983, 46, 596.
122. El-Bayoumy, K.; Sharma, C.; Louis, Y.M.; Reddy, B.; Hecht, S.S. Cancer Lett. 1983, 19, 311.
123. Cerniglia, C.E.; Howard, P.C.; Fu, P.P.; Franklin, W. Biochem. Biophys. Res. Commun. 1984, 123, 262.
124. Uehleke, H.; Nestel, K. Naunyn-Schmiedebergs Arch. Pharmakol. Exp. Pathol. 1967, 257, 151.
125. Poirier, L.A.; Weisburger, J.H. Biochem. Pharmacol. 1974, 23, 661.
126. Perez, G.; Radomski, J.L. Indust. Med. Surg. 1965, 34, 714.
127. Belman, S.; Troll, W.; Teebor, G.; Mukai, F. Cancer Res. 1968, 28, 535.
128. Radomski, J.L.; Brill, E.; Deichmann, W.B.; Glass, E.M. Cancer Res. 1971, 31, 1461.
129. Dooley, K.L.; Beland, F.A.; Bucci, T.J.; Kadlubar, F.F. Cancer Res. 1984, 44, 1172.
130. Hammons, G.J.; Guengerich, F.P.; Weis, C.C.; Beland, F.A.; Kadlubar, F.F. submitted.
131. Fu, P.P.; Yang, S.K. Biochem. Biophys. Res. Commun. 1983, 115, 123.
132. Yang, S.K.; Chou, M.W.; Fu, P.P. In "Carcinogenesis: Fundamental Mechanisms and Environmental Effects"; Pullman, B.; Ts'o, P.O.P.; Gelboin, H.V., Eds.; Reidel: Boston, 1980; p. 143.
133. Nachtman, J.P.; Wei, E.T. Experentia 1982, 38, 837.
134. Saito, K.; Kamataki, T.; Kato, R. Cancer Res. 1984, 44, 3169.
135. Bond, J.A. Mutation Res. 1983, 124, 315.
136. King, L.C.; Kohan, M.J.; Ball, L.M.; Lewtas, J. Cancer Lett. 1984, 22, 255.
137. Bond, J.A.; Mauderly, J.L. Cancer Res. 1984, 44, 3924.
138. Howard, P.C.; Flammang, T.J.; Beland, F.A. Carcinogenesis in press.
139. Djuric, Z.; Fifer, E.K.; Heflich, R.H.; Howard, P.C.; Beland, F.A. submitted.
140. Heflich, R.H.; Beland, F.A. unpublished observation.
141. Djuric, Z.; Beland, F.A. submitted.

142. Fu, P.P.; Chou, M.W.; Yang, S.K.; Beland, F.A.; Kadlubar, F.F.; Casciano, D.A.; Heflich, R.H.; Evans, F.E.   Biochem. Biophys. Res. Commun. 1982, 105, 1037.
143. Fu, P.P.; Chou, M.W.   In "Cytochrome P-450, Biochemistry, Biophysics and Environmental Implications"; Hietanen, E.; Laitinen, M.; Hanninen, O., Eds.; Elsevier: Amsterdam, 1982, p. 71.
144. Chou, M.W.; Evans, F.E.; Yang, S.K.; Fu, P.P.   Carcinogenesis 1983, 4, 699.
145. Chou, M.W.; Fu, P.P.   Biochem. Biophys. Res. Commun. 1983, 117, 541.
146. Chou, M.W.; Von Tungeln, L.S.; Unruh, L.E.; Fu, P.P. In "Eighth International Symposium on Polynuclear Aromatic Hydrocarbons"; Cooke, W.M.; Dennis, A.J., Eds.; Battelle: Columbus, OH, in press.
147. Chou, M.W.; Heflich, R.H.; Fu, P.P. submitted.
148. Holder, G.; Yagi, H.; Dansette, P.; Jerina, D.M.; Levin, W.; Lu, A.Y.H.; Conney, A.H.   Proc. Natl. Acad. Sci. USA 1974, 71, 4356.
149. Jerina, D.M.; Selander, H.; Yagi, H.; Wells, M.C.; Davey, J.F.; Mahadevan, V.; Gibson, D.T.   J. Amer. Chem. Soc. 1976, 98, 5988.
150. Fu, P.P.; Chou, M.W.; Miller, D.W.; White, G.L.; Heflich, R.H.; Beland, F.A. submitted.

RECEIVED May 2, 1985

Author Index

Amin, Shantu, 85
Beland, Frederick A., 341,371
Cavalieri, Ercole L., 289
Chang, Richard L., 63
Chiu, Pei-Lu, 19
Conney, Allan H., 63
Dipple, Anthony, 1
Dommen, Josef, 239
Eisenstadt, Eric, 327
Fu, Peter P., 371
Geacintov, Nicholas E., 107
Glusker, Jenny P., 125
Harvey, Ronald G., 35
Hecht, Stephen S., 85
Heflich, Robert H., 371
Hoffmann, Dietrich, 85
Howard, Paul C., 371
Jeffrey, Alan M., 185

Jerina, Donald M., 63
Kadlubar, Fred F., 341
Kumar, Subodh, 63
LaVoie, Edmond J., 85
LeBreton, P. R., 209
Lehr, Roland E., 63
Levin, Wayne, 63
Marnett, Lawrence J., 307
Melikian, Assieh A., 85
Miller, Kenneth J., 239
Mushtaq, Mohammad, 19
Rogan, Eleanor G., 289
Sayer, Jane M., 63
Taylor, Eric R., 239
Wood, Alexander W., 63
Yagi, Haruhiko, 63
Yang, Shen K., 19

Subject Index

A

N-Acetoxy arylamides
  carcinogenicity, 347
  electrophilic reactivity, 347
  properties, 347
  reaction mechanism, 347-48
  role as ultimate carcinogens, 347
N-Acetoxy arylamines
  isolation, 350
  metabolic formation, 350,352
  reaction mechanism, 350-52
  role as ultimate carcinogens, 352
Activated carcinogen–DNA interactions,
    computer modeling, 170,176
Activated carcinogen–nucleotide
    interactions
  discussion, 159,161,169
  ribose O atom—methyl group
      interaction, 170,172f
  stacking of polycyclic aromatic
      hydrocarbons and
      bases, 169-70,171f-72f

Activated carcinogen–polypeptide
    interactions, 157
Aflatoxin B$_1$
  crystal structure, 131
  mode of action, 131
Alkylated nucleotides,
    structures, 161,165f-68f,169-70,173f
Alkylated tripeptide,
    structure, 157-59
Alkylation of amino groups of DNA
    bases, 139,140f
Ångstrom units, definition, 128
Anthracene, carcinogenic activity, 45
Anthracenedihydrodiols, synthesis, 45
Arylnitrones, reaction mechanism, 363f

B

Bay region
  definition, 8,128
  structures, 8,11f

397

Bay-region dihydrodiol epoxide
    generalization,
    discussion, 8,12,15
Bay-region hypothesis, carcinogenic
    mechanism, 145
Bay-region methyl group enhanced
    effect on tumorigenicity,
    mechanistic basis, 91,96,97
Bay-region theory
    diagram, 65
    discussion, 64,65
    theoretical basis, 65
Benz[a]acridine
    effect of nitrogen substitution on
        tumorigenicity, 78
    structure, 78
    tumorigenicity, 78
Benz[a]acridine tetraepoxide
    effect of nitrogen substitution on
        mutagenicity, 79
    mutagenicity, 79,80f
Benz[a]acridinediol epoxide
    effect of nitrogen substitution on
        mutagenicity, 79
    mutagenicity, 78,79,80f
Benz[c]acridine
    effect of nitrogen substitution on
        tumorigenicity, 78
    structure, 78
    tumorigenicity, 78
Benz[c]acridine tetraepoxide
    effect of nitrogen substitution on
        mutagenicity, 79
    mutagenicity, 79,80f
Benz[c]acridinediol epoxide
    effect of nitrogen substitution on
        mutagenicity, 79
    mutagenicity, 78,79,80f
Benz[a]anthracenes
    activity, 5,7,9t
    carcinogenic activity, 38
    DNA adducts, 200-201
    effect of nitrogen substitution on
        tumorigenicity, 78
    metabolic activation pathway, 25
    metabolite adducts, 200-201
    stereoselective metabolism at the
        K region, 27
    structures, 6,78,126,138,140f
Benz[a]anthracene tetraepoxide
    effect of nitrogen substitution
        on mutagenicity, 79
    mutagenicity, 79,80f
Benz[a]anthracenedihydrodiols
    enantiomeric
        compositions, 25,28t,29t
    major enantiomer, 27,28t
    metabolic pathways, 25
    optical purity, 25

Benz[a]anthracenedihydrodiols--
    Continued
    separation, 25
    stereoselective metabolism, 29
    synthesis via Methods II and
        III, 40,41f
    synthesis via quinone
        reduction, 40,42
Benz[a]anthracenediols,
    structures, 145,146f,147f
Benz[a]anthracenediol epoxide
    effect of nitrogen substitution on
        mutagenicity, 79
    mutagenicity, 78,79,80f
Benz[a,h]anthracene, spectrum, 2
Benzene
    crystal structure, 128
    X-ray diffraction studies, 128
Benzofluoranthenes, carcinogenic
    activity, 56
Benzo[c]phenanthrenes
    carcinogenic activity, 5,46
    structure, 6,46,49
    synthesis via Method I, 46,48,49f
Benzo[a]pyrenes
    bay region, 8,11f
    bay-region dihydrodiol epoxide
        metabolism route, 12,13f
    bond lengths, interbond angles, and
        torsion angles, 128,129f
    carcinogenic activity, 7,12,13t,14t
    conformations, 389
    DNA adducts, 198-200
    levels in air, 3
    levels in drinking water,
    major activation pathways, 19,20f
    metabolic activation, 387-89
    metabolism, 64
    microsomal metabolites, 387,388f
    mutagenicity, 387-89
    nonreactive metabolites, 215
    physical binding
        interactions, 215-16
    stereoselective
        metabolism, 19-21
    structure, 6,126
    structure of bay-region dihydrodiol
        epoxide, 8,11f
    structure of K-region dihydrodiol
        epoxide, 10,11f
    structures of metabolites and
        metabolic model
        compounds, 212,213f
Benzo[a]pyrene epoxides
    structure, 139,141f
    isolation, 23
    stereochemistry, 23,25
Benzo[a]pyrene metabolism, classes of
    products, 296

Benzo[a]pyrene metabolites
  absolute configuration determined by
      X-ray diffraction, 145,148f
  carcinogenic activities, 12,13t,14t
  comparison of physical binding
      properties, 227
  physical binding to secondary sites
      on DNA, 225
  structures, 12,13t,14t
Benzo[a]pyrene oxidation
  addition of RNA or DNA, 308
  addition of Salmonella typhimurium
      strains, 310
  PGH synthase, 308
  quinone products, 308,309f

Benzo[a]pyrenedihydrodiol metabolites
  absolute configurations, 21
  excitation chirality CD
      spectra, 21,22f
  optical purity, 21,23t
  stereochemistry, 23,25
Benzo[a]pyrenediols, structures, 143

Benzo[a]pyrenediol epoxides
  association constants, 223-25
  conformations, 152,153f-54f
  conformational energies and average
      benzo ring conformations, 253-56
  conformational stabilization by
      hydrogen bonding, 256
  diastereoisomers, 240,252,253f
  intercalation to DNA, 223
  premutational lesion, 333,335
  properties of aromatic
      chromophore, 109
  stereoisomers, 107,108f
  structures of
      analogues, 149,151f-53f
  structures of
      stereoisomers, 149,150f
  synthesis, 64
  torsion angles, 152,155t
  UV absorption binding
      constants, 225,226t
Benzo[a]pyrenediol epoxide—DNA adducts
  binding energies via trans
      addition, 267-73
  binding energy, 249-51
  binding process steps, 248
  energy change for loss of electron
      pair conjugation on NHR, 259-61
  externally and internally bound
      adducts, 275
  formation mechanism, 251
  intercalation optimum binding
      orientations, 263-65,266f
  intercalative covalent
      binding, 267,268f-69f,272f,274
  intermolecular energy dependence for
      trans addition, 266,267f

Benzo[a]pyrenediol epoxide—DNA adducts
    Continued
  mechanism for
      stereoselectivity, 246-48,281-83
  possible reaction
      schemes, 112-13,115f
  reaction coordinates, 251,252f
  reaction mechanism, 212,215
  reaction pathways, 110,112
  rehybridization of amino groups, 259
  relative intercalation
      energies, 264t,265
  stereoselectivity, 266-69,272-75,276t
  structure of the covalent

      adducts, 110
  theoretical modeling, 246
Benzo[a]pyrenediol epoxide—DNA adduct
    reaction pathways
  covalent binding mechanism, 110
  fraction of diol epoxide molecules
      binding to DNA, 112
  ionic strength-level of covalent
      binding relationship, 112
  mechanism, 110,112
  rate constant, 112
  sequence specificity, 112-13
Benzo[a]pyrenediol epoxide—DNA
    covalent adducts
  absorbance and linear dichroism
      spectra of enantiomers bound to
      DNA, 120,121f
  fluorescence heterogeneity, 116-18
  fluorescence properties, 116-18,119f
  fluorescence quenching, 117-18
  linear spectra dichroism, 114
  solvent accessibility, 117
  stereoselective covalent
      binding, 120
  structures of model reactants that
      give site I adducts, 114,115f
Benzo[a]pyrenediol epoxide
    mutagenicity
  molecular genetic studies, 335,337
  previous investigations, 331-32
  specificity, 332
  two-step transformation
      process, 337-38
Benzo[a]pyrenediol epoxide mutagenic
    specificity
  base substitution events, 332,333t
  distribution of lacI nonsense
      mutations, 332,334t
  induced mutations in the lacI
      gene, 332,335t,336f
Benzo[a]pyrenediol epoxide—N2
    (guanine) adducts
  benzo ring conformations and confor-
      mational energies, 257,259t
  preferred conformations, 257

Benzo[a]pyrenediol epoxide–N2 (guanine)
    adducts—Continued
  relative energies of various
    conformations, 257,258t
Benzo[e]pyrene
  adduct formation, 200
  carcinogenic activity, 43
Benzo[e]pyrenedihydrodiols
  epoxidation, 43
  synthesis, 43,44f

                        C

Carcinogenesis, process, 4
Carcinogenic activity
  dependence on benzo[a]pyrene, 3
  determination, 2
  mouse skin monitoring system, 4
  route of administration–tissue
    affected relationship, 4
  structure relationship, 2
Carcinogenic arylamines and
    arylamides, metabolic activation
    pathways, 341,342f
Carcinogenic distillates,
    fluorescence, 2
Carcinogenic polycyclic aromatic
    hydrocarbons
  bay-region geometry, 133,135f
  buckling of the molecule, 133,136
  crystal structures, 131
  planarity distortions, 131,133
  shapes and sites of
    activation, 131,132f
  site of action, 139
  torsion angles, 133
  views of distortions due to steric
    effects, 133,134f-35f
Carcinogens
  carcinogenicity for man, 3
  characterization, 1,2
  examples, 1
  occupational exposure, 3
Chemical carcinogenesis, multistage
    process, 185-86
Chrysenes
  carcinogenic activity, 45
  DNA adducts, 201
Chrysenedihydrodiols
  epoxidation, 46
  synthesis via Methods III and
    IV, 46,47f,49f
Cocarcinogens, definition, 241
Computer-modeling studies
  docking a base into
    Z-DNA, 170,174f-75f
  model of quasi-intercalated
    polycyclic aromatic
    hydrocarbons, 176,177f

Cooxygenation, importance in chemical
    carcinogenesis, 323
Crystal structure, definition, 127
Cytochrome P-450
  definition, 307
  properties, 307
Cytochrome P-450c substrate binding
    site model
  description, 29,31
  schematic, 29,30f

                        D

Deoxyribonucleic acid—See DNA
Dibenz[a,c]anthracene,
    structure, 88
Dibenz[a,h]anthracene
  carcinogenic activity, 42
  spectrum, 2
  structure, 127
Dibenz[a,h]anthracenediol epoxides
  epoxidation, 42
  structures, 42,44f
  synthesis via Method II, 42,
Dibenzo[a,e]fluoranthene
  carcinogenic activity, 58
  principal metabolites, 58
Dibenzo[a,h]pyrene
  carcinogenic activity, 48
  structure, 49
  synthesis via Method I, 48
Dibenzo[a,i]pyrene
  carcinogenic activity, 48
  structure, 49
  synthesis via Method I, 48
Dihydrodiol epoxides
  carcinogenic initiation, 10
  structures, 10,11f
Dihydroxy-7,8-dihydro-
    benzo[a]pyrene oxidation
  correlation of the rate of
    BP-7,8-dihydrodiol oxidation,
    anti-diol epoxide formation,
    and mutagen
    generation, 310,312f
  diol epoxide products, 310-15
  mutation induction in Salmonella
    typhimurium, 310,311f
  PGH synthase, 310
  yields of adducts from diol
    epoxide versus those from
    dihydrodiol, 313t
Diimines
  metabolic formation, 359
  reaction mechanism, 359,360f
  role as ultimate
    carcinogens, 359,361

7,12-Dimethylbenz[a]anthracene
  bond lengths, interbond angles, and
      torsion angles, 128,129f
  breast tumor induction, 85,86
  carcinogenic activity, 50
  physical binding
      interactions, 215-16
  stereoselective metabolism at the
      K region, 28
  structure, 126,143
  structures of metabolites and
      metabolitic model
      compounds, 212,214f
  view of 5,6-cis-diol, 143,144f
  view of K-region oxide, 143,144f
7,12-Dimethylbenz[a]anthracene epoxide,
    structure, 139,141f
7,12-Dimethylbenz[a]anthracene
    metabolites, comparison
    of physical binding
    properties, 227
7,12-Dimethylbenz[a]anthracenedihydro-
    diols
  enantiomeric
      compositions, 27,28t,29t
  enantiomeric ratios, 28
  metabolic activation
      pathways, 27,30f
  optical purity, 27
  properties, 27
  stereoselective metabolism, 29
Dinitropyrenes, metabolic
    activation, 386
Diol epoxides
  conformational
      preferences, 69,71t,72
  conformational-reactivity
      relationship, 72,73
  diastereomeric series, 69
  formation, 127
  log k_0 versus $\Delta E_{deloc}/\beta$, 73,74f
  log relative mutagenicity versus
      $\Delta E_{deloc}/\beta$, 75-77
  mutagenicity, 73,75,77
  reaction mechanism, 65
  reactivity, 73
  tumorigenicity, 77
  types, 69
Diol epoxide adducts
  conformational-biological activity
      relationship, 120,122
  fluorescence quenching, 116-17
  linear dichroism spectra, 110,111f
  solvent accessibility, 110
Diol epoxide-DNA adduct conformations,
    theoretical modeling, 118,120
DNA
  alignment, 245
  alkylation by a diol
      epoxide, 176,178f

DNA--Continued
  binding constants, 210,211t
  crystal structure, 136
  diagrams of major
      conformations, 159,162f-63f
  intercalation, 136
  major conformations, 159,160t
  sites accessible to alkylating
      agents, 161,164f
DNA adducts
  conformational structure of the
      receptor site, 242-43
  covalent bond formation, 242
  detection, 189
  extent of reaction, 242
  intercalation complexes, 244-45
  intercalative covalent binding, 245
  orientation of the pyrene moiety and
      the base pairs, 243
DNA adduct detection
  direct methods, 189-90
  fluorescence measurements, 194-95
  immunological methods, 192-94
  indirect methods, 189
  postlabeling techniques, 188-90
DNA atomic coordinates, mathematical
    approaches, 248-49
DNA binding
  DMBA metabolites, 27,30f
  energy for insertion, 250-51
  metabolic steps, 12
  metabolites, 10,12
  receptor sites, 249-50
DNA binding-carcinogenicity
    relationship, 188-89
DNA binding sites
  definitions, 261t
  energy required to alter DNA to
      receptor sites, 261t,262-63
  kink sites, 261t,262-63
  properties, 109
  types, 109
DNA repair strategies
  mutagenic repair process, 328
  SOS response, 328
DNA site adducts
  biological activity, 120,122
  characteristics, 114

                          E

Electron energy parameter ($\Delta E_{deloc}/\beta$),
    estimation of relative
    reactivity, 65,66,68
Electronic theories of
    carcinogenesis, 7

Epoxide ring, averaged
  geometry, 139,142f
Epoxides, contours of hydrogen-bond
  positions, 139,142f,143
Externally bound benzo[a]pyrenediol
  epoxide-DNA adducts
  adduct with the pyrene moiety in the
    major groove, 279-80,281f
  anti and syn reorientation of G and
    A, 279,280f
  distribution, 279-80
  intercalations, 277,279
  orientations, 275-76,277f
  orientation angles and
    energies, 276-79

                    F

Fatty acid hydroperoxide dependent
  oxidation, pathways, 308,309f
Fatty acid hydroperoxide polycyclic
  aromatic hydrocarbon oxidation,
  significance, 320
Fluoranthenes
  carcinogenic activity, 56
  DNA adducts, 201
Fluoranthenedihydrodiols
  epoxidation, 56
  structure, 57
  synthesis, 56,57f
N-2-Fluorenylacetamide, role of metabolites in carcinogenic action, 10
Fluorescence quenching
  association constant for

    intercalation, 216
  fluorescence decay
    profiles, 218,221f
  fluorescence lifetimes, 218
  mechanism, 216
  quenching constants, 223-24
  Stern-Volmer plots, 218,219f-20f
Fluorine-probe approach,
  description, 86

                    G

N-Glucuronyloxy arylamides
  reaction mechanism, 348
  role in
    hepatocarcinogenesis, 348,350
N-Glucuronyloxy arylamines, reaction
  mechanism, 348,349f

                    H

Hydroperoxide-dependent epoxidation,
  unsaturated versus saturated fatty
  acid hydroperoxides, 317,318t
Hydroperoxide-dependent oxidations
  catalytic cycle of horseradish
    peroxidase, 314,315f
  hydroperoxide specificity of BP
    oxidation, 314
  identification of peroxidase-
    reducing substrates, 314-15
  oxidizing agent, 314,317
N-Hydroxy arylamides, reactions, 341
N-Hydroxy arylamides and N-hydroxy
    arylamines, carcinogenicity, 341
N-Hydroxy arylamines
  protonation, 355
  reactions, 343
N-Hydroxy arylamine O-seryl (O-prolyl)
    esters
  metabolic formation, 342f,353
  reaction mechanism, 355
  role in carcinogenesis, 355
Hydroxybenzo[a]pyrene
  alternate synthesis, 38,41f
  conformational and stereochemical
    assignments, 38,39f
  half-lives, 38
  purification, 38
  reactivity, 38
  synthesis via Method I, 36,38,39f
Hydroxy-3-methylcholanthrenes,
  synthesis, 53,54f,56
Hydroxy-5-methylchrysene,
  synthesis, 53,55f,56

                    I

Iball index, definition, 128,131
Iminoquinones
  metabolic formation, 359
  reaction mechanism, 359,360f
  role as ultimate
    carcinogens, 359,361
Initiation
  interaction of carcinogens with
    DNA, 10
  irreversibility, 4
  mutagenic mechanism, 5
  reversibility, 4
Initiation-promotion system,
  characteristics, 4
Intercalation of an acridine in a
  nucleic acid, views, 136-38

K

K- and L-region hypothesis
    description, 8
    exceptions, 8
    regions of benz[a]anthracene, 8,9f
K-region, definition, 128
K-region epoxides
    biological activities, 10
    metabolic action, 10

L

lacI system for analyzing nonsense
    mutations in Escherichia coli,
    description, 331
Linear dichoism, description, 109

M

Metabolic activation, one-electron
    oxidation, 290
Method I synthetic approach,
    description, 36
Method II synthetic approach,
    description, 40
Method III synthetic approach,
    description, 40
Method IV synthetic approach,
    description, 45
11-Methyl-15,16-dihydrocyclopenta[a]-
    phenanthrene, structure, 133
Methylated polycyclic aromatic
    hydrocarbons
    examples, 85,86
    highly tumorigenic examples having a
        methyl adjacent to an angular
        ring, 86,87f
    metabolic activation, 88
    occurrence, 85
    structural requirements favoring
        mutagenicity, 86,88
7-Methylbenz[a]anthracene, carcinogenic
    activity, 48
7-Methylbenz[a]anthracenedihydrodiol
    epoxidation, 50,52
    peracid epoxidation, 50
    synthesis via Methods
        I-IV, 48,50-52,54f
3-Methylcholanthrene
    carcinogenic activity, 52
    DNA adducts, 201
    structure, 126

3-Methylcholanthrenedihydrodiols
    structure, 54
    synthesis via Method IV, 52
3-Methylcholanthrenediol epoxides,
    structures, 54,55
3-Methylcholanthrenetriol epoxide,
    structures, 54
5-Methylchrysene
    carcinogenic activity, 53
    structure, 136
5-Methylchrysene metabolites
    half-lives of dihydrodiol
        epoxides, 96t
    half-lives of dihydrodiol epoxides
        versus extents of DNA
        bindings, 96,98f
    relative extents of binding of
        dihydrodiol epoxides to DNA, 91
    structures of bay-region dihydrodiol
        epoxides and
        dihydrodiols, 91,92f
    structures of major adducts formed
        upon DNA-metabolite
        reaction, 91,95f,96
    tumorigenicity in rats, 91,93t
    tumor-initiating activity on mouse
        skin, 91,94t
5-Methylchrysenedihydrodiols
    epoxidation, 56
    synthesis via Method IV, 53,55f,56
11-Methyl-15,16-dihydrocyclopenta[a]-
    phenanthrene, structure, 133
Mouse skin system
    advantages, 4
    tumor induction, 4
Mutagenesis
    base mispairing, 328
    processes of DNA damage, 328-29
    roles played by DNA lesions, 330
    SOS processing, 329
Mutagenic mechanism, description, 5

N

Naphthacene
    activity, 7
    structure, 6
Nitro polycyclic aromatic hydrocarbons
    biological effects, 374
    environmental occurrence, 372,374
    mammalian metabolism, 378
    microbial metabolism, 377
    mutagenicity, 374,375t
    purification, 372
    structures, 372,373f
    synthesis, 372
    tumorigenicity, 374,376t,377

Nitro polycyclic aromatic hydrocarbon
    mammalian metabolism
  effects of nitro substitution, 381
  microsomal metabolites, 381,382f
  oxidation and
      nitroreduction, 379,381,383,
      386-87,389
  reasons for study, 378
  S-9-catalyzed
      metabolites, 379,380f,384f
Nitro polycyclic aromatic hydrocarbon
    microbial metabolism
  other bacteria, 378
  Salmonella typhimurium, 377-78
1-Nitropyrene
  metabolites detected in in vitro
      incubations, 383,385f,386
  mutagenicity, 383
  oxidative metabolism, 383
  reduction, 383
Nitrosoarene
  metabolic formation, 357
  reaction mechanism, 357,358f
  role in carcinogenesis, 357,359

                    O

Optical purity determination
  CSP-HPLC separation of
      enantiomers, 23,24f
  hydration mechanism of K-region
      epoxide enantiomers, 21,22f
  methods, 21
Oxidation
  activation of polycyclic aromatic
      hydrocarbons to carcinogens, 307
  description, 307

                    P

Peri methyl group inhibitory effect on
    tumorigenicity, mechanistic
    basis, 99
Peroxidases
  carcinogenic activators, 307
  properties, 307-8
Peroxy radicals
  epoxidation, 317,319-20
  reactions, 317
Peroxy radical generation,
    mechanism, 317,319f
Perturbational molecular orbital
    calculations, 65,66
Phenanthrene, carcinogenic
    activity, 43

Phenanthrene epoxide,
    structure, 139,141f
Phenanthrenedihydrodiols, synthesis
    via Method IV, 45,47f
Physical binding properties,
    spectrocopic probes, 216
Polycyclic aromatic hydrocarbons
  addition of an epoxide group, 143
  biological properties, 19
  carcinogenic identification, 2
  carcinogenicity, 126,292-96
  carcinogenicity in mouse skin versus
      that in rat mammary gland, 300-302
  charge localization in the radical
      cation, 292,296
  classifications of metabolites, 241
  covalent binding factors, 292
  environmental exposure, 3,4
  examples, 198
  intercalation in DNA, 138-39,140f
  ionization potential, 292,293t-95t
  nitration, 372
  numbering, abbreviations, and
      $\Delta E_{deloc}/\beta$ values, 66-68
  reactive metabolites, 196
  structural formulas, 36,37f
  structure-activity relationship, 5
  structures, 292,293t-95t
  synthesis of dihydrodiol and diol
      epoxide derivatives, 36
  view of carcinogenic molecules
      showing K and bay
      regions, 128,130f
  X-ray diffraction studies, 128

Polycyclic aromatic hydrocarbon diol
    epoxide-DNA interaction,
    perpendicularity, 152,156f,157
Polycyclic aromatic hydrocarbon-DNA
    adducts
  activation by cellular
      peroxidases, 300
  binding constants, 210,211t,227,229
  binding of hydrocarbon
      metabolites, 210,212
  effects of DNA on solubility, 210
  electronic influence on stacking
      interactions, 227,228f
  evidence for one-electron oxidation
      activation, 300
  metabolitic reactivity-
      carcinogenicity relationship, 212
  photoelectron spectrum, 229,230f
  physical binding, 210
  structure-activity studies, 210
  structures, 197-98
Polycyclic aromatic hydrocarbon-DNA
    adduct structures,
    determination, 195-98

Polycyclic aromatic hydrocarbon—DNA
    intercalation, influence of DNA
    structure and
    environment, 229,231f,232
Polycyclic aromatic hydrocarbon
    metabolic activation, effects of
    fluorine substitution, 101
Polycyclic aromatic hydrocarbon
    oxidation, hydroperoxide and
    mixed-function oxidase
    differences, 320-23
Polycyclic aromatic hydrocarbon
    tumorigenicity, effect of fluorine
    substitution, 99-101
Primary carcinogens, definition, 241
Procarcinogen--See Secondary
    carcinogen
Prostaglandin H synthase
    biosynthesis of PGH$_2$, 308,309f
    heme-requiring activities, 313-14
    oxidation of polycyclic aromatic
    hydrocarbons, 308
    properties, 308
Proteins, polypeptide chain
    folding, 157
Protonated N-hydroxy arylamines
    reaction mechanism and
    formation, 355-56,358f
    reactivity and selectivity, 355-56
    role as ultimate carcinogens, 356-57

                        Q

Quantum chemical calculations, 65
Quinones
    metabolic formation by an initial
    one-electron oxidation of
    benzo[a]pyrene, 296-99
    metabolic formation for polycyclic
    aromatic hydrocarbons of various
    ionization potentials, 297t

                        R

Radical cations
    chemical properties, 290
    nucleophilic trapping
    mechanism, 290-91
    one-electron oxidation and
    subsequent nucleophilic
    trapping, 290-91
    specific reactivity with
    nucleophiles, 290,292

Radical cation perchlorates, synthesis
    and subsequent nucleophilic
    coupling, 290,292

                        S

Secondary carcinogens, definition, 241
SOS mutagenesis, occurrence at sites
    of DNA damage, 330
SOS-processing system,
    description, 329-30
Spectroscopic probes
    emission spectra of DNA, 216,217f
    fluorescence quenching, 216
    UV absorption, 218
Structural requirements favoring
    mutagenicity
    bay-region methyl group, 86,88,91
    comparative tumorigenicity of
        methylchrysenes, 88,90f
    free peri position, 86,88
    inhibition of tumorigenicity by
        peri-methyl substitution, 88,89f
    nonplanarity, 97
    unsubstituted angular ring, 86,88
Structure-activity relationships
    effect of a methyl group, 7,9t,48
    substitution, 7
    unsubstitution, 5,7
Sugar conformations,
    diagrams, 159,161,163f
N-Sulfonyloxy arylamides
    decomposition, 345
    electrophilic reactivity, 343
    electrophilic substitution, 345
    metabolic formation, 343-45
    reaction mechanism, 345,346f
    role in arylamide
        tumorigenesis, 345,347
N-Sulfonyloxy arylamines
    metabolic formation, 352
    reaction mechanism, 352-53,354f
    role as ultimate carcinogen, 353
N-(Sulfonyloxy)-N-methyl arylamines
    reaction mechanism, 361-63
    reactivity, 361
    role as ultimate carcinogens, 361

                        T

Tetrahydroepoxides
    log $\underline{k}_0$ versus $\Delta E_{deloc}/\beta$ plot, 68
    log relative mutagenicity toward S.
    typhimurium TA 100, 69,70f

Tetrahydroepoxides--Continued
  models for diol epoxide reactive
    site, 68
  mutagenicity, 69
  reactivity, 68
  tumorigenicity, 69
Tetraols, fluorescence, 116
Tripeptides, structures, 157
Triphenylene
  activity, 7
  carcinogenic activity, 43
  structure, 6
Triphenylenedihydrodiols
  epoxidation, 43,44f
  synthesis, 43

U

Ultimate carcinogen--See Primary
  carcinogen

Unstructured polycyclic aromatic
  hydrocarbons, structures, 5,6
UV absorption
  absorption spectra, 218,222f
  association constants, 223,224t

W

Weigle reactivation and mutagenesis,
  description, 328

X

X-ray diffraction analyses of
  crystals, description, 127-28

Production by Meg Marshall
Indexing by Deborah H. Steiner
Jacket design by Pamela Lewis

Elements typeset by Hot Type Ltd., Washington, D.C.
Printed and bound by Maple Press Co., York, Pa.